System-on-Chip For Real-Time Applications

THE KLUWER INTERNATIONAL SERIES
IN ENGINEERING AND COMPUTER SCIENCE

System-on-Chip For Real-Time Applications

edited by

Wael Badawy
University of Calgary, Canada

Graham Jullien
University of Calgary, Canada

KLUWER ACADEMIC PUBLISHERS
Boston / Dordrecht / London

Distributors for North, Central and South America:
Kluwer Academic Publishers
101 Philip Drive
Assinippi Park
Norwell, Massachusetts 02061 USA
Telephone (781) 871-6600
Fax (781) 681-9045
E-Mail < kluwer@wkap.com>

Distributors for all other countries:
Kluwer Academic Publishers Group
Post Office Box 322
3300 AH Dordrecht, THE NETHERLANDS
Telephone 31 78 6576 000
Fax 31 78 6576 254
E-Mail < services@wkap.nl>

Electronic Services < http://www.wkap.nl>

Library of Congress Cataloging-in-Publication Data

A C.I.P. Catalogue record for this book is available
from the Library of Congress.

Title: *System-on-Chip for Real-Time Applications*
Edited by Wael Badawy and Graham Jullien
ISBN: 1-4020-7254-6

Printed on acid-free paper.

Printed in the United States of America.

Contents

Preface xi.

Acknowledgements xiii.

Contributors xv.

Chapter 1: INTRODUCTION **1**

System on Chip: The Challenge and Opportunities 3
WAEL BADAWY

Electronic Product Innovation Direct Mapped Signal Processing SoC Cores 17
JOHN V MCCANNY

System-On-Chip Implementation Of Signal Processors 26
EARL E. SWARTZLANDER, JR.

Chapter 2: DESIGN REUSE **29**

Methodologies and Strategies for Effective Design-Reuse 31
LUDOVIC LOISEAU, YVON SAVARIA

A VHDL/SystemC Comparison in Handling Design Reuse 41
LUC CHAREST , EL MOSTAPHA ABOULHAMID

Aspect Partitioning for Hardware Verification Reuse 51
SÉBASTIEN REGIMBAL, JEAN-FRANÇOISLEMIRE, YVON SAVARIA, GUY BOIS, EL MOSTAPHA ABOULHAMID , ANDRÉ BARON

Reconfigurable Combinatorial Accelerators for Real Time Processing 61
I. SKLIAROVA , A.B. FERRARI

Tuning Methodologies for Parameterized Systems Design 71
GIUSEPPE ASCIA, VINCENZO CATANIA, MAURIZIO PALESI

Chapter 3: TESTAND VERIFICATION 83

Formal Verifications of Systems on Chips: Current and Future Directions 85
KHALED M. ELLEITHY

A Practical Approach to the Formal Verification of SoC's with Symbolic Model-Checking 98
EMIL DUMITRESCU

High Performance Verification Solutions for SOC Designs 111
MAKARAND JOSHI , ASHOK KUMAR

Novel Test Methodologies for SoC/IP Design: Implementation and Comparison 125
L. HONG, M. NAHVI, R. FUNG, A. IVANOV, R. SALEH

Chapter 4: MODELING 137

SOC Modeling and Simulation Based on Java 139
HENRY FU , RICHARD F. HOBSON

RTOS Modeling Using SystemC 150
MOHAMED ABD EL-SALAM, ASHRAF SALEM, GAMAL ALY

Modeling, Synthesis and Implementation of Communicating Hierarchical FSM 160
VALERY SKLYAROV

A Modeling Method for Reconfigurable Architectures 170
LILIAN BOSSUET, GUY GOGNIAT, JEAN-PHILIPPE DIGUET, JEAN-LUC PHILIPPE

Chapter 5: DESIGN TECHNIQUES 181

The Syslib-Picasso Methodology for the Co-Design Specification Capture Phrase 183
LUC FILION, JÉROME CHEVALIER, GUY BOIS, EL MOSTAPHA ABOULHAMID

Automatic Porting Of Binary File Descriptor Library 193
MAGHSOUD ABBASPOUR, JIANWEN ZHU

Code Compression on Transport Triggered Architectures 203
JARI HEIKKINEN, JARMO TAKALA, AND JAAKKO SERTAMO

An Approach To Flexible Multi-Level Network Design 214
V. CHEUSHEV, J. KOLODZIEJCZYK, T. LUBA,C. MORAGA, P. SAPIECHA, V. SHMERKO S. YANUSHKEVICH

Chapter 6: MEMORY 225

Survey of Emerging Nonvolatile Embedded Memory Technologies 227
BRUCE F. COCKBURN

Configurable Parallel Memory Implementation For System-on-Chip Designs 237
JARNO VANNE, EERO AHO, KIMMO KUUSILINNA, TIMO HÄMÄLÄINEN

XOR-scheme Implementations In Configurable Parallel Memory 249
EERO AHO, JARNO VANNE, KIMMO KUUSILINNA, TIMO HÄMÄLÄINEN

An Novel Low Power Embedded Memory Architecture for MPEG-4 Applications with Mobile Devices 262
MOHAMMED SAYED , WAEL BADAWY

Assessment of MPEG-4 VTC and JPEG2000 Dynamic Memory Requirements 273
GAUTHIER LAFRUIT, JAN BORMANS

Chapter 7: CIRCUIT TECHNIQUES **285**

Modified Distributed Arithmetic Architecture for Adiabatic DSP Systems 287
DUSAN SUVAKOVIC, C. ANDRE T. SALAMA

Design of a CMOS Wide Range Logarithmic Amplifier with a Modified Parallel Architecture 296
SANGHOON JOO, MINKYU SONG, HYUNYEON CHO, SANGKI KIM

Digital Hardware Implementation of Continuous & Discrete Chaotic Generators 305
MOHAMED I. SOBHY, MOHAMMED A. ASEERI , ALAA E. R. SHEHATA.

Novel 1-Bit Full Adder Cells For Low-Power System-On-Chip Applications 314
MOHAMMED SAYED , WAEL BADAWY

Chapter 8: LOW POWER **325**

A New Logic Method for considering Low Power and High Testibility 327
YOON-SIK SON, JONG-WHA CHONG

System Synthesis for Optically-Connected, Multiprocessors On-chip 339
NEAL K. BAMBHA, S.S. BHATTACHARYYA

Low Power System On Chip Platform Architecture for High Performance Applications 349
W.C. LO, A.T. ERDOGAN, T.ARSLAN

Chapter 9: INTERCONNECT AND TECHNOLOGY **357**

SOC Interconnect in Deep Submicron 359
MAGDY BAYOUMI

Optimizing Inductive Interconnect for Low Power 380
MAGDY A. EL-MOURSY AND EBY G. FRIEDMAN

Skin Effects in System on a Chip Interconnects 392
SHIZHONG MEI AND YEHEA I. ISMAIL

Chapter 10: MICRO ELECTRO MECHANICAL SYSTEMS **403**

Road Map Towards Designing MEMS Devices with High-Reliability 405
WALIED A. MOUSSA

A Mems Socket Interface For Soc Connectivity 411
SAZZADUR CHOWDHURY, M. AHMADI, G. A. JULLIEN, W. C. MILLER

On the Application of Finite Element to Investigate the Reliability of Electrostatic Comb-Drive Actuators Utilized in Micro-Fluidic and Space Systems 422
HESHAM AHMED, WALIED MOUSSA, WAEL BADAWY

An HDL Model For A Vacuum-Sealed Micromachined Pressure Sensor 429
RAFIK S. GUINDI

Performance Analysis of MEMS-based Inertial Sensors for Positioning Applications 440
WALID ABDEL-HAMID, ABOELMAGD NOURELDIN, DR. NASER EL-SHEIMY, DR. GERARD LACHAPELLEE

INDEX **451**

PREFACE

System-on-Chip (SoC) describes an evolving paradigm for timely design of integrated circuits (ICs) that contain tens to hundreds of millions of transistors. As of the date of writing this preface, the majority of the effort is directed towards the design of large digital chips, but the promise of the technology is to allow the design of large integrated microsystems involving such disparate technologies as CMOS IC's, Micro-Electro-Mechanical Systems (MEMS), Microfluidic Devices, RF transceivers and optical devices. For industry, the evolution of a timely design paradigm is essential to maintain the shrinking time to market in face of the exponential increase in device density for modern IC technologies. For researchers there is both a challenge and an opportunity. The challenge is to incorporate this exacting design paradigm into research projects that involve SOC design and fabrication; the design paradigm includes the cataloguing of steps in approved design flows, providing appropriate "wrappers" for I/O compatibility, incorporating standard test procedures, and adopting approved bus structures for chip-level data transfer. The opportunity is the easier acceptance of leading edge research ideas by industry through the development of custom "IP (Intellectual Property) Blocks" that can insulate the user of the blocks from their internal operation. The opportunities extend far beyond "all digital" applications, and researchers are beginning to look at microsystem integration that will eventually lead to ubiquitous sensing, processing communication and control devices application areas such as analysis and therapeutic devices in the health sciences areas and personal assistants embedded in clothing and appliances.

With these goals and applications in mind, Dr. Wael Badawy organized a workshop on SoC for Real-Time applications, at the 5th World Multiconference on Systemics, Cybernetics and Informatics in Orlando, Florida, July 2001. The success of this first workshop led to a second International Workshop being organized at the Banff Centre in Banff, Alberta during July 6-7, 2002. The program chair was Wael Badawy and the General Chair Graham Jullien. This workshop was coordinated with the annual Canadian Microelectronics Corporation workshop on SoC. The International SoC Workshop theme of Real-Time applications was intended to highlight the special needs of the signal processing community, which will be responsible for many of the future advances in small information processing devices, where performance (throughput, power, noise) is of particular interest. The workshop recognized the coexistence of both technology and design concerns, and papers were received from a large cross-section of research workers. Two distinguished plenary speakers, Prof. J. McCanny, CBE, FRS, and Prof. E.

Swartzlander, FIEEE, were also invited to the workshop in order to provide a mature perspective on the implementation of signal processing using SoC technology.

A selected set of papers from the workshop were collected and used to form the basis of this book. The book reflects the diversity of the workshop sessions and has been organized into the following chapters:

- Introduction
- Design Reuse
- Modeling
- Architecture
- Design Techniques
- Memory
- Circuits
- Low Power
- Interconnect and Technology
- Micro-Electro-Mechanical Systems

The introductory chapter contains an SoC overview by the program chair, Dr. Wael Badawy, and contributions from each of the plenary speakers. The contents of the book represent state-of-the-art research contributions by internationally known authors in the area of System-on-Chip technology, design and use, and we expect that the contents will be of considerable interest to both industry and academia as this technology advances over the next few years.

Graham Jullien and Wael Badawy
University of Calgary
July, 2002

ACKNOWLEDGEMENTS

The editors of this book are indebted to three special peoples. Brinda Prasad, Moahammed Sayed of the University of Calgary, who put in many hours of work and skill in organizing and formatting this book, and Liu Xiao, also of the University of Calgary, who developed and maintained the web-site that allowed the 2nd International Workshop on SoC for Real-Time Applications to be run efficiently. The authors also thank these and other students at the University of Calgary, for all their efforts during the running of the workshop, July 6-7, 2002.

We would also like to acknowledge the workshop sponsors: IEEE CAS society technical committee on Communications and VLSI, IEEE Canada, Micronet R&D, the Canadian Microelectronics Corporation, the Informatics Circle of Research Excellence (iCORE) in Alberta and the University of Calgary.

Contributors

Abbaspour, Maghsoud
Aboulhamid, El Mostapha
Aho, Eero
Aly, Gamal
Ascia, G.
Badawy, Wael
Baron, ANDRE
Bois Guy
Bormans, Jan
Bossuet, Lilian
Catania, Vinceuzo
Charest, Luc
Chevalier, Jerome
Cheushev, V
Cockburn, Bruce F
Das, Prodip. K
Diguet, Jean-Phillippe.
Dumitrescu, Emil
Elleithy, Khaled. M
El-Moursy, Madgy. A
El-Salam, Mohamed. Abd
El-Sheimy, Naser
Erdogan, A. T
Ferrari, A. B
Filion, Luc
Friedman, Eby. G
Fu, Henry
Fung, R
Gogniat, Guy
Guindi, Rafik. S
Hamalainan, T
Heikkinen, Jari
Hobson, Richard. F
Hong, L
Ismail, Yehea. I
Ivanov, A
Joshi, Makarand
Jullien, G. A
Kolodziejczy, J
Kumar, Ashok
Lachapellee, G
Lafruit, Gauthier
Lo, W. C
Loiseau, Ludovic
Luba, T
Mei, Shizhong
Miller, W. C
Moraga, C
Moussa, Medhat
Moussa, Walied
Nahvi, M
Noureldin, A
Palesi, Maurizio
Philippe, Jean-Luc
Regimbal, S.
Russell, G.
Safavi-Naeini, S.
Salama, T.
Saleh, R.
Salem, A.
Sapiecha, P.
Savaria, Y.
Sayed, M.
Sertamo, J.
Shehata, A. E. R.
Shemrko, V. P.
Skliarova, I.
Sklyarov, V.
Sobhy, M. I.
Son, Y.
Sotiropoulos, A.
Stergiou, S.
Suvakovic, D.
Takala, J.
Thornton, M.
Turner, L. E,
Vanne, J.
Vilarino, D. L.
Wang, Q.
Weeks, M.
Yanushkevich, S. N.
Zhang, N.
Zhu, J.

1

INTRODUCTION

This book is concerned with building Systems-on-Chips for real-time systems, where processing typically takes place on streaming data. Many of the applications are in the area of signal processing systems, and so it is appropriate that we start with a chapter devoted to exploring the issues associated with building such systems using SOC concepts.

There are three sections in this chapter. The first section, "System on Chip. The Challenge" by Wael Badawy of the University of Calgary, provides an overview of the challenges and opportunities associated with the design of circuits for real-time applications where the number of logic gates is in excess of 10 million. In particular, this section deals with the paradigm shift that is occurring from the current cell library driven application specific integrated circuit (ASIC) design to SoC design techniques based on reusable intellectual property.

The second section, "Electronic Product Innovation Using Direct Mapped Signal Processing SoC Cores" by John McCanny of the Queen's University of Belfast and Amphion Semiconductor Inc., highlights the need to look beyond the use of programmable DSP processors to implement high performance signal processing applications at SoC densities. Current generations of many electronics products (e.g. 2G cell phones, set-top-boxes) incorporate increasing levels of signal processing functionality. In the main, this has been provided by and led to the development of increasingly sophisticated families of programmable DSP processors. Although this approach has been very successful in the past, user demand for new devices and enhanced functionality, flexibility and portability (e.g. wireless PDAs, set-top-boxes, 3G wireless handsets, Wireless LAN, digital VCR's etc.) is requiring system performance where the reliance on the programmable DSP solution will be increasingly inadequate. This section looks at alternative techniques, based on high performance signal processing intellectual property (IP) blocks, that can be tailored to a variety of performance requirements.

The final section, "System on Chip Implementation of Signal Processors" by Earl Swartzlander Jr., of the University of Texas at Austin, looks at the evolution of the building of special purpose signal processors and the advantages that SoC densities bring to the increase in performance of these systems. This section develops on the theme that signal processing systems are excellent candidates for implementation via System-on-Chip design. This is due to the high throughput and low power demands of signal processing systems and the well-structured nature of the algorithms. Realizing a complete system on a single chip eliminates many chip-to-chip connections that are required for conventional implementations. Replacing a chip-to-chip connection with a connection that stays on chip reduces the capacitive load by at least two orders of magnitude. Since the power consumption of a driver circuit varies in direct proportion with the capacitive load (at a fixed speed), the SoC design will have much lower power consumption than the conventional version, which provides the potential to achieve high throughput, small size, and low power for dedicated signal processing systems.

Section 1.1

System on Chip

The Challenge and Opportunities

Wael Badawy
Dept. of Electrical and Computer Engineering
University of Calgary
Calgary, Canada T2N 1N4

Abstract: The increase in the number of logic gates (10 million to 20 million gates) that can be implemented on a single integrated chip allows for design of circuits with great functional complexity. A whole system previously integrated into a board is now being integrated into a single chip leading to System on Chip (SoC) design. A paradigm shift from the current cell library -driven application specific integrated circuit (ASIC) design to a new System on Chip (SoC) technology is now occurring. This paper outlines the challenges in the development of SoC design for real time systems.

Keyword: System on Chip

1. INTRODUCTION

With the increase in the number of logic gates that can be implemented on a single chip, various functionalities, like memories, logic, analog, CPU, DSP, MPEG decoder/encoders, etc. can be integrated into a single silicon chip, realizing an entire system on chip (SoC). According to the virtual socket interface (VSI) alliance, SoC is defined as a highly integrated device. It is also known as a system on silicon, a system-on-chip, system-LSI, system-ASIC, and as a system-level integration (SLI) device [1]. An SLI device is defined as "greater than 100 thousand gates with at least one programmable core and on-chip memory"[2]. Another definition can be found in [2]: "A complex IC that integrates the major functional elements of a complete end-product into a single chip or chipset" In this paper the SoC is defined as a complex IC that consists of one or more programmable cores, connected using an on-chip bus with signal processor cores (digital or analog), and that may incorporate on-chip memory. The proposed definition describes the difference between a SoC microprocessor and as well it distinguishes SoCs and complex application specific integrated circuits (ASICs).

With the rapid achievement of mass production volumes in consumer products, SoC development faces a continuous decrease of the time to market window. It is expected that in the near future new consumer products will take less than one year to achieve a market penetration of ten million units. All the components that today are part of a system will have to be integrated together. Therefore one of the key factors for the success of ultra-deep sub-micron technologies will be the capability of integrating into a single piece of silicon different components, such as logic functions, analog functions, and memories of different types (static memories with either dynamic or non-volatile). Design methodologies will evolve and Concurrent Engineering, Intellectual Properties (IP) reuse, HW/SW co-design and co-verification, fast prototyping and new CAD tools will be common practice in the coming years. SoC as an emerging technology has its own challenges, issues and trends. This chapter outlines the issues, challenges and trends for the new SoC era in the key areas out lined in Figure1.

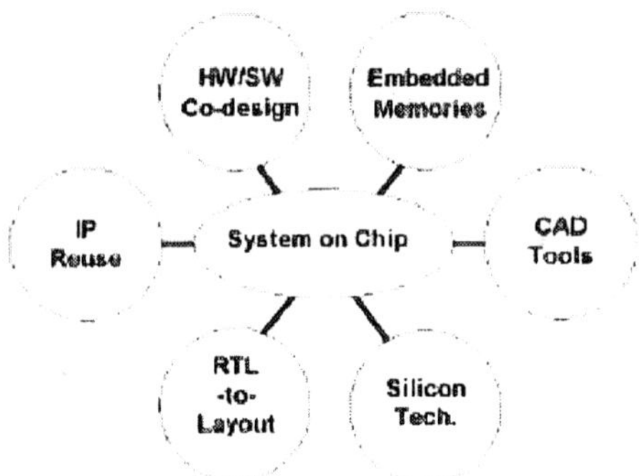

Figure 1. The main challenges and trends in SoC development

2. TECHNOLOGY CHALLENGES

SoC requires designing and integrating digital modules with analog blocks, special memories and RF designs for some applications. Due to the design complexity of the SoC, its architecture is based on the reuse of the Intellectual Property (IP) core to meet time-to-market requirements. Virtual Component (VC) or IP modules are often originated from different vendors with different formats.

Virtual Socket Interface Alliance (VSIA) [3], Virtual Component Exchange (VCX) [4] and Reusable Application Specific Intellectual Property Developers (RAPID) [5] are used to define the IP cores that allow exchange between vendors.

The Reuse Methodology Manual (RMM) and Open MORE (Measure Of Reuse Excellence) [6] developed by Mentor Graphics Corp. and Synopsys Inc. [7] provide guidelines for design, test-benches, model creation and physical integration [8]. These guidelines provide the underlying methodology to establish a best practice IP reuse process for SoC designs.

Modular processing for implementing SoC designs starts with a high-speed digital CMOS process followed by processing masks to implement technologies such as high density SRAM, DRAM, flash memory, and BiCMOS for RF applications. While digital design among different vendors is quit similar, adding analog capabilities to a baseline digital process has a huge overhead since the process is different among vendors.

Feature size reduction is the driving force for device scaling, which has led to increased integration. However this scaling raises short channel issues [9] in transistors. The process of integrating analog components like ADCs, DACs, and PLLs along with digital ones is not only very complex, but it will also be extremely difficult to obtain a good yield ratio. An alternative solution is to use a multi-die approach where different functions are confined to different dies, diffused in different processes and later combined in one package. Issues in the fabrication of SoCs are derived from the mixing of each individual technology developed separately for each application. For correct functionality, MOS scaling requires gate oxide thinning and reduction in supply voltage. As a result of the increasing gap between transistor and I/O voltage, and also because many different voltages have to be considered in the same design, innovations in process technology have become necessary [10]. The multiple gate oxide solutions, which was first used for non-volatile memories are now used in the CMOS process. Introduction of muli-gate oxide (dual and triple) solutions are the most recent process modifications for SoC architectures [11].

Among the technological challenges is the development of multiple threshold voltage transistors that require leakage management strategies for high performance SoCs. Moreover, a thick gate oxide must be developed that can sustain the high voltage required by different embedded memory components. Embedded memories consume the most power and the largest area of SoC

3. DESIGN METHODOLOGIES

Embedded telecommunications and multimedia define new design challenges such as heterogeneity, dynamic behavior and dynamic memory management, architectural and timing problems due to IP reuse, changing requirements and compatibility to a number of standards. Integrating both HW and SW components and the ability to perform complete system simulation rapidly and accurately is one of the main issues for the SoC development. SoC changes the design space from ASIC cores to platforms for running embedded software. Most embedded real time systems have some kind of Real Time Operating System (RTOS) or programmable kernel. An RTOS can be described as multi-threaded and preemptable, supporting thread priority and synchronization. Multithreaded processor cores are required to handle interrupts and maintain the required throughput [12]. The interrupt latency and context switching time depend on the application and should be accurately modeled

High level modelling languages

Due to the complexity of the SoC, new design process with higher level of abstraction to capture both hardware and software are required to model the hardware/software interaction. New languages with greater reliance on tools are required to express the interaction of hardware and software components.

New classes are added to C++ that allow concurrency and reactivity to describe hardware/software interactions [13], [14]. This library extension approach allows for the development of a high-level system model, which captures functionality, structure and timing. It allows refinement to proceed from a functional model to a state machine model and finally to a synthesizable RTL model. To describe hardware using programming languages requires Concurrency, Instantiation, Reusability, Reactivity, Hardware data types and operators, and HW/SW co-verification.

· **Concurrency** - Hardware is essentially concurrent i.e. all cores execute in parallel with respect to each other. Concurrency is implemented using parallel threads or an event queue.

· **Instantiation** – Architecture is modeled into simpler components with less complexity, which can be effectively managed. Instantiation creates a hierarchy of components (objects) where each component runs concurrently and emulates the parallelism in a circuit.

· **Reusability** - Creation of smaller, less complex kernel components allows for reusability.

· **Reactivity** - The ability of a functional block to get executed based on the occurrence of a stimulus. Reactivity allows the modeling of combinational elements that react to the clock edges, which allow modeling of sequential elements.

· Hardware data types support flexibility to reflect the size of data (similar to 'length in VHDL) and operations. Operators such as concatenation, bit selection and non-blocking assignments are required to describe hardware effectively.

· Using high-level languages allow HW/SW co-verification. Reusability is done by the constructor of a class that defines the interface of the class with other parts of the design. The definition of an IP can be encapsulated within a constructor. This means the users of the IP are provided with the constructor. This enables the users to create an instance of the IP within their design along with the interfaces.

SystemC [15] and SpecC are popular languages for system level design. SystemC is a modeling platform consisting of C++ class libraries for system level design. The main goal of using SystemC is to promote and accelerate the exchange of IP models and executable specifications using C (C++) based platform. Through an open source language and reference implementation, designers can create, validate and share models with other vendors. SpecC, originally developed by UC Irvine, is a language-specification description language based on C designed to support integration of product development, from determining specifications to implementing the design. SpecC methodology is an IP centric codesign methodology for specifying, modeling and designing embedded and SoC systems at system level. Moreover, System Level Design Languages (SLDLs) including Dataflow style languages (static and dynamic) which are which are used for signal processing applications can be used for embedded systems and control dominated languages such as StateCharts, CoCentric System Studio and Esterel [16].

Platform Based Design

Platform Based Design (PBD) is used for rapid SoC development. A stable core-based architecture can be rapidly extended and customized for a range of applications. SoCs have to be designed with flexibility in mind to meet changing standards and market trends. Changes required for the next generation SoCs must be made quickly and efficiently to meet several market requirements. PBD takes the best out of HW/SW co-design. Instead of designing SoC architectures from scratch, architectural platforms are developed for particular applications. Integration of IP cores is used to customize system components with both hardware and software to suit different applications.

CoWareN2C [17] is a plaform with many derivatives of a core that can be customized by adding more HW and/or SW. PBD enables designers to rapidly turn around platforms for many derivatives for a particular product line. In PBD, design reuse is planned for a specific family of products and is based on pre-verified blocks with standard interfaces. CoWare methodology separates function from communications, making the replacement of one piece of IP with another easier. It also allows the evaluation of the potential of one piece of IP versus another. CowareN2C can also be used to create virtual platforms for integrating HW and SW using high-level languages like SystemC.

The main issue with PBD is concerned with providing a platform that satisfies the needs of a broad base of designs while providing capability for customization and product differentiation.

Cost evaluation will be used to analyze performance benefits of different modifications. Creating new architectures may require changes in software.

Dynamic binary translation will be used in order to avoid expensive and time-consuming software re-design. Binary translation techniques aim at translating object code (binary code) from one architecture to another [18]. Dynamic binary translation uses a translator to execute the program, i.e. the input program is dynamically analyzed and translated to target code. Dynamic binary translation overcomes some of the disadvantages of static translation. Some form of hardware wrapper will probably be required for dynamic translation in SoCs.

Memory management

One of the main performance challenges in computer systems today is the speed mismatch between microprocessors and memory. Processors use a large fraction of the transistors and area of the chips for large SRAM caches. In SoC designs, memory is the largest part of the system and it may be about 70-90% of the chip. Memory management is one of the critical design processes due to the limited bandwidth. SoC solutions will narrow this processor-memory performance gap due to the absence of off-chip bus transfers. In traditional systems, the operating system is responsible for all background memory and storage related tasks. On the other hand, embedded systems cannot use software run-time memory management since it is expensive in terms of area, performance and power. Moreover, embedded systems use dedicated distributed memory systems. Therefore, dynamic memory management must be synthesized in the embedded system implementation itself. Matisse [19] is a design environment that optimizes memory management. It allows the exploration of different architectures to select optimal memory management design.

Chinook [20] and Polis [21] and data flow dominated methodologies such as COSSAP (Commercial DSP synthesis tool from Synopsys) and Ptolemy [22] are frameworks for HW/SW co-synthesis and include control dominated methodologies like Telecom applications and hardware that involve lots of protocol processing. These applications are modeled as a set of concurrent processes that access data. Concurrent object-oriented models play a key role in large scale HW/SW system design for these applications.

Intelligent RAM (IRAM) or Computational RAM (CRAM) [23] offers several advantages such as less latency, increased bandwidth, reduced power/energy consumption, and reduced space and weight for embedded portable systems. In SoC design, issues such as scheduling, resource allocation, memory hierarchy and organization, manipulation and modification of stored data structures need to be addressed in terms of explicit memory I/O operations.

Power consumption

With the advances in the mobile communication and personal communication services (PCSs), there is increasing demand to run video applications on mobile systems. High-speed data and computation-intensive processing will lead to much higher power consumption than traditional portable applications. Due to the limited power-supply capability of current battery technology, current research efforts face two conflicting design challenges. One is to explore high-performance design and implementation techniques that can meet the stringent speed constraints for real-time systems. The second is to consider low-power design approaches to prolong the operating time of portable devices.

Power consumption has been introduced as a major design parameter due to the rapidly advancing technology of mobile communications, which require portable power supplies. There are three main sources of power dissipation in a CMOS circuit: (1) switching power, which is consumed by the circuit capacitance during transistor switching; (2) short-circuit power, which is due to current flowing from power supply to ground during transistor switching; and (3) static power, which is due to leakage and static currents. The switching and short circuit power are known as dynamic power, which constitutes about 99% of the total power dissipation in CMOS VLSI circuits. Dynamic power, Pdynamic, is given by the following equation:

$$Pdynamic= f_{clk}.V_{dd}.((C_i.V_{swing\,i}.p_i)+I_{sc}.p_i)$$

where fclk is the system clock frequency, Vdd is the power supply voltage, Ci is the capacitance at node I, Vswing i is the voltage swing at node i, pi is the switching probability at node i, and Isc is the short circuit current at node i.

As system speed increases, the increase in power consumption becomes a major problem. Thc increase in the power dissipation will generate extra heat and reduce the reliability of a circuit. The development of techniques allowing for the development of low-power circuits is crucial to the practicality of future products. Due to its quadratic dependence, lowering Vdd contributes a significant power saving. However, the delay in processing time will increase as Vdd decreases. To meet the throughput requirements, the delay must be compensated for at the system and algorithmic/architecture design level. Simply, a new version of the design that allows more latency is required.

Low-power system-level approaches involve two main techniques. One is partitioning the system into modules and shutting down modules that are not working such as multi-chip modules (MCMs) [24]. The other approach is to use a monitor scheduler to distribute the workload over several functional units. The scheduler can be a hardware or software monitor.

Besides the system-level low-power approaches, the algorithmic/architectural approach is the most promising one [25]. Reformulating the algorithms and mapping them to efficient low-power VLSI architectures to compensate for the speed penalty caused by low supply voltage can achieve the algorithmic/architectural low-power design. The power savings of the algorithmic/architectural level approach are in the range of 70-90% [26]. The current goal is to reduce the total power dissipation of the electronic systems to two orders of magnitude less than what it would have been with the conventional technology [27]. Approximations, parallel and pipelined design approaches are used to compensate for the delay penalty [25], [28].

Low-power and high-performance architectures can be achieved using approximation approaches which trade high performance with the cost of the circuits, pipelining which trades the overhead of extra buffers with latency, and parallel architectures which trade area with latency.

In the approximation approaches, the constraints are relaxed in order to reduce the cost and yield acceptable performance. Approximation approaches generally lead to much simplified architectures, which then add to the savings in total implementation cost. For example, three-step motion estimation [29] is usually employed to reduce the computational complexity while maintaining a performance comparable with full-search motion estimation schemes [30].

Pipelining is a commonly used technique to achieve high-speed data processing The main idea behind pipelining a design is shortening the delay through the critical path by inserting latches between consecutive pipeline stages. As a result, the speed of the system is faster than that of the original system at the penalty of increasing the latency by the number of pipelining stages. On the other hand, pipelining can be used to compensate for the delay incurred in the low-power design when the supply voltage drops significantly. Some derivations of the pipelining techniques are asynchronous circuit pipelining [31], two-level (word/bit-level) pipelining [32], and bit-level pipelining approaches [33].

Parallel data processing decomposes the desired function into independent and parallel small tasks. These tasks are executed concurrently, and individual results are combined to obtain the desired computation result. It is a "divide-and-conquer" strategy in solving problems. The Fast Fourier Transform (FFT) [34] is a famous example where the algorithm first decomposes an N-point FFT into two N/2-point FFTs, and expands the computation recursively to form a parallel computing network.

Power minimization for real time SoCs is addressed at the level of several synthesis tasks including scheduling, instruction and data cache size determination (application specific) and processor core selection. Voltage is treated as a variable. Non-preemptive scheduling of a set of independent tasks with arbitrary arrival times and deadlines on a variable voltage processor is a NP-complete task [33].

Power consumption is a new set of design requirements due to the limitation of the current battery technology. New methodologies for low-power core-based real-time SoCs are based on dynamic variable voltage. Effective scheduling techniques treat voltage as a variable whose value is to be determined in addition to conventional task scheduling and allocation [35]. Conventional methods for reducing power consumption in a core in CMOS technology involve lowering the supply voltage level, which exploits the quadratic dependence of power on voltage. Reducing the supply voltage however increases the circuit delay and decreases clock speed. This technique is therefore ineffective when tight deadlines are present in systems.

Networked SoCs such as pocket computers, wireless sensors, pagers and cell phones rely on efficient use of power, reliability and thermal management. These devices have a much wider dynamic range of power demand. Power consumption is of two types - instantaneous and average power. Average power is affected by communication protocols and power management strategies. Instantaneous power is affected by transceiver architecture and RF circuit design. The signaling protocols, choice of modulation, transceiver architecture and RF (IF) circuit design affect power. Power management is done at different modes of operation such as transmit, receive, idle, sleep and off modes. Effective power management requires power aware protocol design.

Reconfiguration in SoCs

Reconfigurable computing achieves significant speed on a general class of applications [36]. For SoCs, different configuration bit streams will be stored on chip. It will introduce overhead in the amount of storage, which should be minimal. Moreover, dynamic reconfiguration at run time introduces the overhead of the reconfiguration time. Multiple reconfigurable bit streams are stored for different contexts, which consumes memory space. In SoC implementations, the entire application needs to be stored in on-chip memory. A solution that minimizes this configuration overhead using distributed caches is presented in [37]. The flexibility requirement of SoC requires programmable devices with low power that support multiple standards of communication, audio and video. Traditional DSPs and processors offer high flexibility but are power inefficient, as opposed to ASICs, which are not flexible but are efficient in terms of speed and power. An architecture framework that is a tradeoff between ASICs and programmable processors is the Pleiades, which is a low-power reconfigurable DSP architecture composed of heterogeneous components [38].

4. ANALOG COMPONENTS IN SOC

Analog and RF components along with digital ones will be the main components of SoC. The 3G mobile telephony is one of the mail technology drivers for mixed signal processing and EDA tools. High performance demands and complexity at reduced cost are driving the integration of RF and analog components into digital designs. RF, digital and analog circuits have different design methodologies and therefore integrating these on mixed signal ICs is quite difficult. Some problems in integrating analog and mixed signal IP onto dies with logic functions include noise issues, dissimilar power requirements and actual implementation between two separate engineering organizations. IP core and packaging address only digital logic while many of the new ICs have analog and mixed signal interfaces. The main challenge is that analog space cannot be fully covered by any finite library of blocks. Analog components are often customized cells. Simulation-based synthesis is a new method that is proposed to synthesize a range of custom analog cells large enough to be compatible with industrial designs although the cost of a full circuit-level simulation of a design candidate is not practical [39].

The analog components exchange data with digital ones using A/D converters. One of the challenges of low power is that transducers usually generate signals that are in the microvolt to milli-volt range that is well below the resolution of a 10-bit A/D converter. Therefore, direct digitization is not possible. In addition, many digital applications like hard-disk servos exhibit signal characteristics those require extensive analog processing prior to the digital operation. The latency and delays to convert the analog signal into a digital one introduce a delay overhead.

Moreover, the proximity of analog and digital devices on the substrate create increasingly difficult design challenges. In a high accuracy and precision system, the analog circuitry is placed on a separate chip. To protect analog blocks from digitally induced noise, a SoC process must isolate analog and digital blocks. This introduces extra space between the blocks that increases the area; placing an oxide barrier between them; or surrounding them with heavily doped isolation rings. Analog cores are more susceptible than digital ones to thermal problems. Isolation of analog blocks from heat producing digital blocks such as clock generators or output buffers is necessary.

Using copper instead of aluminum metallization results in lower total chip power dissipation and smaller on chip temperature rise reducing analog component drift [37]. New tools are being developed to design and verify mixed signal circuits. ADVanceMS [38] supports top-down and bottom-up design methodologies for mixed signal SoC. It utilizes IEEE standard 1076. ADVanceMS uses capabilities of ModelSim and Eldo (a Mentor Graphics product for analog transistor simulation) for digital and analog circuit simulation respectively. Moreover, ADVanceRFIC extends features of Eldo to perform frequency domain RF analysis for mixed signal designs [40].

5. ON-CHIP BUSES

Traditional design methodologies when applied to SoCs lead to a very long design verification time due to the impact of "glue logic." Glue logic is the custom-designed hardware linking all functional IP cores and setting up communication between functional cores and off-chip peripherals. The verification time becomes longer since the glue logic function has to be verified with all other functional units connected. The interface functions and internal functions related to interfaces in IPs must also be verified [41]. To reduce the verification time of glue logic, SoC bus cores have been proposed by VSIA, IBM and other vendors. New bus architectures based on circuit and packet switching (different from classical buses) needs to be investigated. A solution to reduce the verification time is to remove all glue logic by using a bus controller for all components in the SoC.

Current hardware solutions for protocol processing are of two types -ASIC and programmable CPU. An ASIC-based solution is not flexible, and the performance of the programmable CPU is low. Network applications have to be processed in real time and the protocols have to be adaptive. Current hardware for such applications is not very good. Protocol processors that process framing-deframing on the fly is required. The IBM Coreconnects [42] eases integration and reuse of processor, system and peripheral cores within standard product and custom SoC designs. Other popular bus architectures are the C bus and Lexra Bus Controller (LBC) [43]. Another architecture is the Integration Architecture - SonicsIA micronetwork [44]. A micro-network is an on-chip network that decouples, unifies, and manages all communications between heterogeneous cores such as the microprocessor, digital signal processor (DSP), and/or I/O functions. There is no single bus architecture that serves as a universal solution for a SoC design targeted at communication applications.

The IEEE 1394 standard originally developed by Apple is a very fast external bus standard that supports data transfer rates of up to 400Mbps. In addition to its high speed, 1394 also supports isochronously data-delivering data at a guaranteed rate. This makes it ideal for devices that need to transfer high levels of data in real-time, such as video devices. Like USB (Universal Serial Bus), IEEE 1394 supports both Plug-and-Play and hot plugging. The main difference between 1394 and USB is that 1394 supports faster data transfer rates and is more expensive. For this reason, it is expected to be used mostly for devices that require large throughputs, such as video cameras, whereas USB will be used to connect most other peripheral devices.

6. SOC VERIFICATION

Verification of today's SoCs occur at low levels of abstraction, typically RTL. As the complexity of SoCs grows, it is important to move verification to higher levels of abstraction. Hardware/software co-simulation is a step in this direction but is not sufficient due to deficient processor models and slow hardware simulation speeds. It is also difficult to model the real-world environment around the SoC. New verification languages such as Vera from Synopsys, and RAVE from Chronology [45] have enabled easier and precise functional verification along with testbench generation.

Support for system-level verification already exists for the software part of SoC, where a common technique is to use software monitors to instrument code. Instrumentation of software utilizes target resources such as memory and execution time. When the instrumentation code is removed the system may change behavior due to timing differences. Hardware based monitoring makes a SoC observable at different abstraction levels in both hardware and software. A small hardware component is integrated into the SoC [43]. It works like a probe either listening to logic or system level events in a passive manner or is activated by software that writes to a specific register. When using a hardware based RTOS kernel for management of SoC software, the process-level events can be monitored without software overhead and with no intrusion on the systems' timing behavior. Since it is nonintrusive, it gives the SoC developer the free choice of either keeping or removing the monitor component in the final product. A hardware RTOS kernel implements traditional (software) RTOS functions (scheduling, task management, interprocess communication, synchronization etc.) in hardware. The RTU (Real Time Unit) is such a component that has proven to be successful for increasing RTOS performance and operational predictability [46].

Traditional test methods are not suitable for core-based designs since the core user is responsible for manufacturing and testing the core. The IEEE P1500 standard [47] addresses testing embedded core based systems while the IEEE ISO Nexus 5001 [48] forum addresses real time debugging. The IEEE 1149.4 [49] mixed signal test bus is being developed to support mixed signal built-in structures for improving controllability and observability so that test development time is reduced.

BIST is an appropriate method for testing complex systems. Deterministic BIST methods are particularly important as they do not require any modifications of the core under test thus protecting intellectual property. The main difficulties in testing SoC designs are due to reduced accessibility of the cores, increased complexity of the chip and heterogeneity of the modules. Testing SoC designs involves individual core testing and system testing. Test-per-scan and Test-per-clock are the most popular BIST schemes [50]. The Test-per-scan method uses a complete or partial scan path that is serially filled by the ATPG (Automatic Test Pattern Generator). The scan chain contents are applied to the Unit Under Test (UUT) and the response is loaded back into the scan chain in parallel. Then a new bit stream is shifted in and the scan output is compressed. In the Test-per-clock method, special registers are used which work in four modes. In the system mode, the registers operate as ordinary flip-flops.

In the pattern-generation mode, they perform autonomous state transitions, and the states are the test patterns. In the response evaluation mode, the responses of the UUT are compressed and in the Shift mode the registers work as a scan path. In the pattern generation mode, the registers and logic are configured as a linear feedback shift register. This scheme leads to shorter test times as a new pattern is generated in each clock cycle at least for a part of the circuit [50].

REFERENCES

[1] VSI Alliance Glossary, VSI Alliance, March 98, www.vsi.org
[2] H. Chang et al., Surviving the SOC Revolution, Readings, Kluwer Academic Publisher, 1999.
[3] VSIA Alliance Architecture Document, Ver 1.0, www.vsi.org
[4] Virtual Component Exchange, www.vcx.org
[5] RAPID, www.rapid.org
[6] www.mentorg.com/openmore
[7] www.synopsys.com
[8] S. Dey, C. N. Taylor et al., "Using a Soft Core in a SoC Design: Experiences with picoJava", IEEE Design and Test, July -September 2000, pp. 60 -71.
[9] J. P. Bruines, "Process Outlook for Analog and RF applications", Microelectronic Engineering, 54 (2000), Elsevier, pp. 35-48.
[10] M. Haond, "Issues for CMOS Integration in Systems on a Chip", Journal of Microelectronic Engineering , 54 (2000), Elsevier, pp. 23 -34.
[11] P. Senn, "Telecommunication Applications: New Challenges for Microelectronics", Journal of Microelectronic Engineering, 54 (2000), Elsevier, pp. 5-14.
[12] C. Zilles, J. Elmer, G. Sohi, "The use o f Multithreading for Exception Handling", MICRO-32, Nov 1999, pp. 219-229.
[13] D. Verkest, J. Cockx et al., "On the use of C++ for System on Chip Design", Proc of IEEE Workshop on VLSI, April 1999.
[14] R. Roth, D. Ramanathan, "A High Level Hardware Design Methodology Using C++", Fourth High Level Design Validation and Test Workshop, 1999, pp. 73-80.
[15] www.systemc.org
[16] L. Blanc, S. Dissoubray, "Esterel Methodology for Complex System Design", Microelectronic Engineering, 54 (2000), Elsevier, pp. 163-170.
[17] CoWareN2C for Platform -based Design, www.coware.com
[18] K. Ebcioglu, E. R. Altman, "DAISY: Dynamic Compilation for 100% Architectural Compati-bility", IBM Research Reprt, www.research.ibm.com
[19] D. Verkest, J. da Silva et al., "Matisse: A System on Chip Design Methodology Emphasizing Dynamic Memory Management", Journal of VLSI Signal Processing, Vol 21, No. 3, July 1999, pp. 277-291.
[20] P. Chou, R. Ortega et al., "The Chinook Hardware/Software Co -Synthesis System", International Symposium on System Synthesis, September 1995, pp. 22-27.
[21] F. Balarin et al., "POLIS: A Design Environment for Control-dominated Embedded Systems", Technical Report, UC Berkeley, 1996.
[22] J. Buck et al., "Ptolemy: A Framework for Simulating and Prototyping Heterogenous Systems", International Journal of Computer Simulation, special issue on simulation software development, Vol 4, 1994.
[23] IRAM Project, http://iram.cs.berkeley.edu
[24] S. A. Khan and V. K. Madisetti, "System Partitioning of MCM for Low Power," IEEE Design Test Comput. Mug., vol. 12, pp.41-52, Spring 1995.
[25] A. P. Chandrakasan and R. W. Brodersen, "Minimizing Power Consumption in Digital CMOS Circuits," Proc. IEEE, vol. 83, pp. 498-523, Apr. 1995.
[26] K. J. Ray Liu et al. "Algorithm-Based Low power and High-Performance Multimedia Signal Processing", Proc. of IEEE, vol. 86, No. 6, June 1998, pp. 1155-1202.
[27] H. K. Thapar and J. Cioffi, "A Block Processing Method for Designing High Speed Viterbi Decoders," in Proc. IEEE Communications Conf, June 1989, pp. 1096-1100.

[28] A. P. Chandrakasan, S. Sheng, and R. W. Brodersen, "Low-Power CMOS Digital Design," IEEE J. Solid-State Circuits, vol. 27, PP. 473-484, Apr. 1992.
[29] T. Koga, K. Iinuma, A. Hirano, Y. Iijima, and T. Ishiguro. "Motion-compensated Interframe Coding for Video Conferencing", Proc. National Telecommun. Conf, New Orleans, LA, 1981, pp. G.5.3.l-G.5.3.5.
[30] J. R. Jam and A. K. Jam, "Displacement Measurement and its Application in Interframe Image Coding," IEEE Trans. Commun. vol. COM-29, pp. 1799—1808, Dec. 1981.
[31] T. H. Meng, R. Broderson, and D. Messerschmitt, "Asynchronous Design for Programmable Digital Signal Processors. IEEE Trans. Signal Processing. vol. 39. pp. 939—952, Apr. 1991.
[32] K. I. R. Liu, S.-F. Hsieh, and K. Yao, "Systolic Block Householder Transformation for RLS Algorithm with Two-level Pipelined Implementation," IEEE Trans. Acoust., Speech. Signal Processing, vol. 40, pp. 946-958. Apr. 1992.
[33] C. L. Wang, "Bit-serial VLSI Implementation of Delayed LMS Adaptive FIR filter," IEEE Trans. Signal Processing, vol. 42, pp. 2169—2175, Aug. 1994.
[34] A. V. Oppenheim and R. W. Schafer. Discrete-Time Signal Processing. Englewood Cliffs, NJ: Prentice-Hall. 1989
[35] I. Hong, D. Kirovski et al., "Power Optimization of Variable Voltage Core Based Systems", Proceedings of Design Automation Conference, 1998, pp. 176-181.
[36] Y. Li, T. Callahan et al., "Hardware-Software Co-Design of Embedded Reconfigurable Architectures", Proceedings, 37th Design Automation Conference, 2000.
[37] Z. Huang and S. Malik, "Managing Dynamic Reconfiguration Overhead in Systems on Chip Design using Reconfigurable Datapaths and Optimized Interconnection Networks", DATE 2001, March 2001.
[38] www.ee.princeton.edu/ESRG/pleiades
[39] R. Phelps, M. J. Krasnicki et al., "A Case Study of Synthesis for Industrial-Scale Analog IP: Re-design of the Equalizer/Filter Frontend for an ADSL CODEC", Proceedings of Design Automation Conference, 2000, pp. 1-6.
[40] Mixed signal design, www.mentor.com/ams
[41] Vijay K. Madiseti and Lan Shen, "Interface Design for Core based Systems", IEEE Design & Test of Computers, pp. 42-51, 1997.
[42] www.chips.ibm.com/products/coreconnect
[43] www.lexra.com/lx4180.html
[44] www.sonicsinc.com
[45] www.chronology.com
[46] M. E. Shobaki, L. Lindh"A Hardware and Software Monitor for High level System on Chip Verification", Proc IEEE Symposium on Quality Electronic Design, 2001.
[47] IEEE P1500, http://grouper.ieee .org/groups/1500
[48] http://www.ieee-isto.org/Nexus5001
[49] Mixed Signal Test Bus, www.manta.ieee.org/groups/1149/4
[50] H. J. Wunderlich, "BIST for systems on chip", Integration, the VLSI Journal, 26 (1998) Elsevier, pp. 55-78

Section 1.2

Electronic Product Innovation Using Direct Mapped Signal Processing SoC Cores

JOHN V McCANNY

CTO, Amphion Semiconductor Ltd ,50 Malone Road, Belfast, BT9 5BS, Northern Ireland, UK

1. INTRODUCTION

Current generations of many electronics products (e.g. 2G cell phones, set-top-boxes) incorporate increasing levels of signal processing functionality. In the main, this has been provided by and led to the development of increasingly sophisticated families of programmable DSP processors. This approach has been very successful. However, with user demand for new devices and enhanced functionality, flexibility and portability (e.g. wireless PDAs, set-top-boxes, 3G wireless handsets, Wireless LAN, digital VCR's etc.) this is now proving to be limited and, almost certainly, will increasingly be insufficient, as new product aspirations run way ahead of technology's apparent ability to deliver.

The main reason is that the exponential rise in computational performance and efficiency that is required cannot be delivered through advances in semiconductor technology i.e. through Moore's Law alone - see figure 1.

Combined Approach

The only practical way to bridge this gap is to employ a heterogeneous approach in which RISC and DSP programmable processors are combined with **highly innovative, special purpose architectures** whose function is to perform computationally intensive and power hungry operations.

Special Purpose Architectures

Recently, companies, such as Amphion (www.amphion.com) have produced comprehensive portfolios of signal processing SoC platforms and accelerators.

These are **close to optimal** in terms of silicon performance, power dissipation and gate count – **up to several orders of magnitude computationally more efficient** than equivalent functions implemented on a programmable processor. They also interface directly to most common RISC and DSP programmable processors and can be easily migrated through different generations of silicon technology, helping to preserve and **enhance current design investments**. Alternatively, these can be implemented as stand-alone systems.

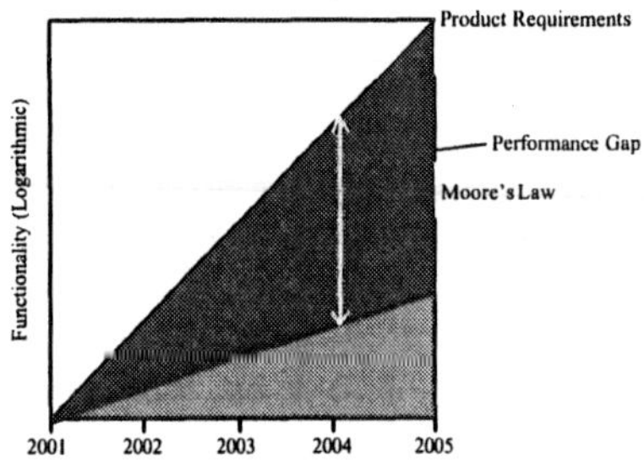

Figure 1. Product Functionality Performance Gap

The enhanced performance and reduced power consumption provided by such specialized platforms **liberates processor and software engineers** and allows **increased functionality and flexibility** to be introduced into new electronic products. This combination of this new technology with existing technology provides a **quantum leap in product innovation.** This is the key to the introduction of much more advanced products much earlier in effect "fast-forwarding Moore's Law!"

These off-the-shelf SoC solutions also avoid the need to code complex signal processing functions on DSP processors, providing higher software productivity.

2. SIGNAL PROCESSING – TECHNICAL CHALLENGES

Over the past two decades the computational requirements of many signal processing applications (e.g. speech, audio, video) have tended to outstrip those of general purpose processors. This led to the development and evolution of the programmable DSP processor.

Early 2G phones, for example, which were predominantly voice based, required 30-50 MOPS (Million of Operations per second) for functions such as voice compression, bit error correction etc. These were therefore typically implemented using a DSP processor along with a RISC processor with the latter being used to provide the user interface capability, the handling of communications protocols and so on.

As we move ahead, next generations of mobile device (e.g. 3G cell phones, wireless LAN devices, wireless PDAs) increasingly must incorporate many other functions in addition to voice processing. Examples include MP3 capability, the handling of streaming video, navigation, digital imaging. These features significantly enhance product functionality and thus product differentiation. They also significantly increase signal processing, and thus computational, requirements. In tandem, these new products require much more sophisticated communications capabilities - modulation/demodulation, streaming video bit error recovery, encryption and decryption etc. Processing rates of several GOPS can then become the signal-processing requirement. This is in addition to increasing operating system and applications finctionality, such as word-processing, spreadsheets, computer games and WEB browsing. For portable devices all this must to be achieved within extremely stringent power budgets.

Other products such as next generation set top boxes, Digital Video Recorders, integrated digital TV/HDTV receivers and WEB-TV also have extremely demanding signal processing requirements (GOPS+). Moreover, some of these new devices are employing combinations of both TV and computer technologies, for example, features such as Intemet access, video conferencing and games capability.

3. ARCHITECTURAL STYLES

As discussed, the approach that has been successfully applied in previous and current generation products is to use architectures that combine RISC and programmable DSP processors. This combination has the attraction of programmability and flexibility. However its main limitation is that programmable processors are inherently inefficient in dealing with computationally demanding tasks such as video processing and increasingly demanding channel communications requirements.

An alternative and much more attractive strategy is to assign computationally intensive and power demanding tasks to dedicated hardware (e.g. MPEG4, JPEG/JPEG2000 video/image compression). The technical benefits of using special purpose architectures are very clear from figure 2. This is based on internal research at Amphion [1] and concurs with parallel research at Phillips [2] and UC Berkley [3,4], for example. This shows that the computational efficiency (for example measured in MIPS/W) of a dedicated signal processing architecture can be up to several orders of magnitude greater compared with the same function implemented in software on a programmable processor, with this gap increasing significantly with adva nces in silicon fabrication technology.

This use of special purpose architectures [1] provides major improvements over more general IP block approaches. The improvements are accomplished by mapping complex signal processing computations directly into silicon – mainly without software or through clever hardware/ software combinations. Typical architectures take the form of richly interconnected arrays containing 100's/1000's of elementary processors operating parallel. The structure and data flow of information through these data paths is tuned exactly to the flow graph of the signal processing functions to be computed and choreographed, without the need for external memory, to produce new results, typically, on every clock cycle. These features lead to extre mely high performance and the avoidance of regular transfers to and from memory, which leads to major reductions in power dissipation. Through this, SoC solutions achieve close to the maximum theoretical compute efficiency of silicon and often up to severa l orders of magnitude greater than the same function coded in software on a programmable processor.

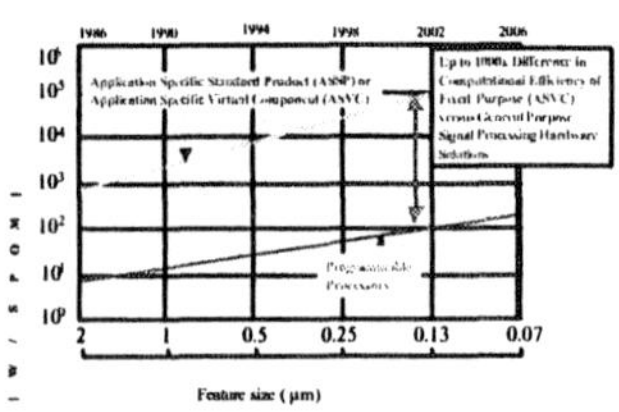

Figure 2. Special Purpose versus Programmable Processors

The structured nature of special purpose signal processing solutions also means that these are easily captured in hardware description languages and using parameterization can be extensively re -used. These can also be used hierarchically to create even more complex signal processing SoC solutions, usually in a fraction of the time previously thought possible. Indeed the time required to produce new SoC hardware is often less than that required to produce equivalent software implementations on DSP processors. These cores can also easily be retargeted to different silicon fabrication processes, including FPGA and PLD implementation.

Major enhancements to new product functionality

Of course, in general, there often needs to be a trade-off between ultimate performance and flexibility. Increasingly as we move to more complex "Platform SoC" solutions the major architectural style that is emerging is a heterogeneous one combining programmable RISC and DSP processors, with computationally intensive, repetitive and power sensitive operations being assigned to special purpose hardware – figure 3. Assigning such tasks (e.g. video processing) to special purpose hardware liberates the processors so that these can be used to provide enhanced functionality at a much earlier stage in development of new products. This potential for a **quantum leap in product innovation** is seen as key to the introduction of much more advanced products much earlier than is possible from Moore's Law alone (figure 1).

4. PERFORMANCE CUBES

The "Performance Cube" [1] provides an objective and consistent method for comparing the cost of implementing Signal Processing applications using different architectural styles. This is a much more general and consistent method for assessing solutions and is adaptable to different and other more commonly used, but much more rigid, performance criteria. It is also applicable across different types of technology including ASIC, DSP processors and FPGA/PLD. The performance cube concept (figure 4) is simple and direct – "How much does it cost to do a specific signal processing computation, measured for example, in terms of processing performance, power dissipation and gate count?" The smaller the volume of the cube the more cost effective the solution!

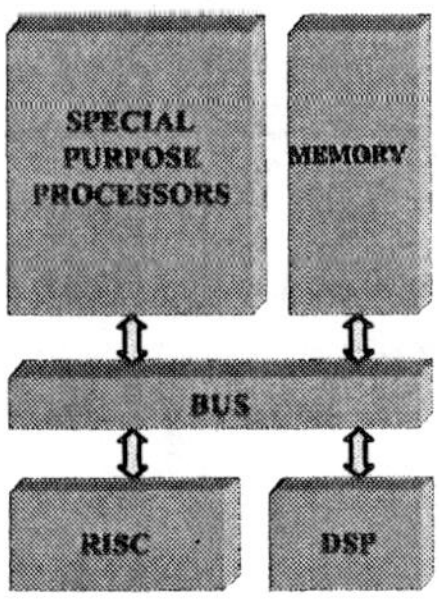

Figure 3. SoC Platform Architecture

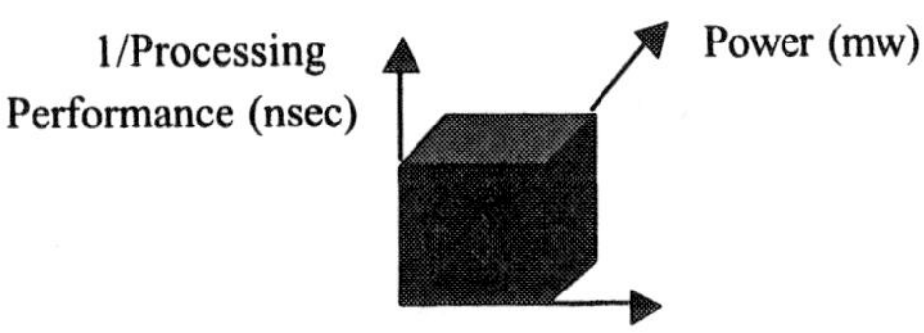

Figure 4. Performance Cube [1]

MPEG-4 Decoder	W$Cycles Frame	Comments
Amphion CS6750	87	
Amphion CS6750	205	Inc. Post-processing
Commercial ASIC	2,880	
RISC/embedded Software solution	24,330	With required cache (11 mm^2)

Table 1 Ultra-low power CS6750 MPEG-4 decoder uses only 4 mW, or less than 8 MW with post-processing

Currently much interest in MPEG4 is focused on low bit and frame rate, QCIF applications e.g. video streaming for mobile wireless handsets or PDAs. However the MPEG4 standard covers a broad spectrum up to top-end studio and digital cinema applications. The processing rates demanded for this are simply not achievable using RISC or DSP solutions alone, and so it is useful to adopt an approach that scales naturally with these requirements.Tables 2 and 3 demonstrate this approach using scaled custom solutions [1]. For example, the performance cubes of custom MPEG2 IP cores are in the region of 30x better than that of other approaches

JPEG		Amphion	Company 1	Company 2	Company 3
TSMC	Encode	27			
	Decode	31			
	Codec	69			Incomplete
Xilinx	Encode	450			
	Decode	645		3,360	
	Codec	1,664	2,207		Incomplete
Altera	Encode	1,920			
	Decode	3,193			
	Codec	5,080			

Table 2 Flexible cores require no additional processor

MPEG2 MP@ML	Amphion	Company 3	Company 4
TSMC	486	16,717	4,199
Xilinx	14,912		
Altera	10,216		

Table 3 MPEG2 performance cube comparisons.

Speech Compression

Compression of lower bandwidth speech signals is usually perceived as very much being in the domain of programmable DSP processors as the computational demands of a single speech channel are not so demanding. However dedicated architectures for implementing ITU standard algorithms such as ADPCM and CELP allow many independent speech channels to be compressed concurrently for the same size silicon and the same processing power. For large multi-channel Voice over IP systems custom IP core solutions can replace expensive racks of equipment with single chips!

CELP	W$/Channel
Programmable DSP – 180nm	11074
Amphion CS4220 – 180 nm	1350

Table 4 CELP comparison between direct mapped solution and implementation based on industry standard programmable DSP

5. COMMUNICATIONS CHANNEL CODING

The same arguments and analysis apply to Communications Channel cores (Turbo coders/decoders, RS and Viterbi decoders, AES -Rijndael encryption/decryption systems). Again very significant benefits exist, the most stark being the 50,000x advantage of the AES cores over Pentium based software solutions (table 5).

AES Encryption	W$cycles Mbit
Rijndael s/w running on Intel Pentium III Xeon	5,764,620
Amphion CS5210 on Xilinx Virtex-II –8	2,789
Amphion CS5210 on TSMC 180nm process	106

Table 5 AES Performance cube comparisons

6. PRODUCTS & MARKETS

The market space for custom IP core solutions range from full SIP platforms to individual core elements. These products are aimed at wireless and broadband communications where highest computational performance/efficiency and lowest power consumption are critical factors for success. Below are some of the market areas for the type of custom solutions discussed in this paper.

6.1.1 SIP Platforms

6.1.2 WLAN baseband processors (IEEE 802.11A, HIPERLAN)

- DVB subsystems for Cable, Satellite and Terrestrial digital television
- Wireless Multimedia systems

6.1.3 Video and Real-time/Streaming Imaging

- MPEG-4 for wireless multimedia, mobile Internet, interpersonal communications (videoconferencing, videophone, etc), networked Database Services
- MPEG-2 - for high-definition digital video, DTV, Games Consoles, Set-Top Box
- JPEG - for digital still cameras, web-cams, office machines, earth observation
- Motion-JPEG - for professional/consumer video editing, and security cameras
- JPEG2000 - digital still cameras, digital documents/digital libraries, medical applications

6.1.4 Telephony/Voice compression

- Kilo-Channel ADPCM Codec - DECT cordless phone, Voice-Over-Networks
- CELP Codec - for multi-channel VoIP Gateway (Internet telephony) applications

6.1.5 Channel coding

- QAM & QPSK modems – for Wireless LAN
- Reed-Solomon & Viterbi - cable DVB, Wireless LAN, Wireless Local LoopTurbocode Codec - 3G telephony (3GPP, CDMA2000)
- Cryptography/Encryption – Rinjdael, IPSec protocols, Secure Socket Layer (SSL) protocol and ATM cell encryption.

6.1.6 Signal Processing Operators

- 2D (9 x 7 and 5 x 3) Wavelet Processors (JPEG2000 compatible).
- FFT Engines (64, 128, 256, 512, 1k, 2k, 8k, parameterized coefficient and data word-lengths)

- FIR/IIR Filters (fully parameterized)
- Arithmetic Operators (fixed- and floating-point, parameterized word-length)

The ability to provide silicon-proven SIP for complex wireless and broadband applications, accelerates customers time-to-market with working silicon ahead of their competitors. An efficient business model is to develop complex cores and platforms in strategic partnerships, and being based on standard algorithms, can be subsequently licensed as standard products - application specific virtual components (ASVC) - fully collateralized with data books, technical briefs, and applications notes. Such cores can be used as hardware accelerators with standard interfacing to a wide range of processors, including industry standard RISC, CISC and DSP processors.

REFERENCE

[1] Amphion Semiconductor Ltd. – Internal research report
[2] "System-on-Chip: What are the limits?" E Roza, IEE Electronics and Communications Journal, January 2002
[3] "Panel weighs hardware and Software design options" EETimes.com, June 20th,2000
[4] "Heterogeneous Reconfigurable Systems" J M Rabaey et al. Proc. IEEE Workshop on Signal Processing Systems – Design and Implementation, SIPS97 1997, pp24-34.

Section 1.3

SYSTEM-ON-CHIP IMPLEMENTATION OF SIGNAL PROCESSORS

Earl E. Swartzlander, Jr.
Schlumberger Centennial Chair in Engineering, Department of Electrical and Computer Engineering, University of Texas at Austin, TX 78712

Abstract: System-on-Chip design is an attractive way to implement many computationally intensive signal processors. System-on-Chip design can achieve high throughput, small size, and low power. This paper examines example signal processors that are reasonable candidates for System-on-Chip implementation, and offers suggestions to facilitate the development. The main concern about System-on-Chip design is the difficulty of producing chips with a mixture of logic and memory on the same chip.

1. INTRODUCTION

Signal processing systems are excellent candidates for implementation via System-on-Chip design. This is due to the high throughput and low power demands of signal processing systems and the well structured nature of the algorithms. Realizing a complete system on a single chip eliminates many chip-to-chip connections that are required for conventional implementations. Replacing a chip-to-chip connection with a connection that stays on chip reduces the capacitive load by two orders of magnitude or more. Since the power consumption of a driver circuit varies in direct proportion with the capacitive load (at a fixed speed), the System-on-Chip design will have much lower power consumption than the conventional version. Typically designers increase the speed so the power reduction is less than the reduction in loading, but still quite significant.

2. FAST FOURIER TRANSFORM SIGNAL PROCESSOR

As a way to explore some of the advantages of System-on-Chip design,an adaptive frequency domain digital filter is examined.The basic system, described in detail in [1], consists of a Fourier transform to transform time domain data into the frequency domain, multiplication of the transformed data in the frequency domain by a filter kernel, and inverse transformation to return to the time domain.This system requires the computation of Fourier transforms and inverse Fourier transforms at the input data rate.The system described in [1] required 4096 point transforms at a data rate of 40 MSPS.In that system radix-4 FFT and inverse FFT were used so that the system clock rate was 10MHz.Implementation today via System-on-Chip would allow a 500 MHz clock rate, for a 2 GSPS data rate and an operation rate of about 200 billion operations per second.With the pipeline FFT, a radix-4 implementation of a 4096 point FFT or inverse FFT requires a cascade arrangement of six radix-4 butterfly processors each of which requires 22 adders and 12 multipliers. The total system (with a 4096 point FFT and a 4096 point inverse FFT) uses 308 arithmetic elements (264 adders and 144 multipliers). If each arithmetic element were realized as a single chip with two 32 pin input ports and one 32 pin output port, there would be nearly 30,000 interconnections between chips. With a load of 50 pf per line, the drivers for the interconnections would consume approximately 3.7kw. A System-on-Chip design with all connections contained on a single chip would consume less than 40 watts for the drivers. Since the pipeline FFT is a cascade implementation, most connections are between adjacent elements. Thus they are very short and the interconnection capacitance may be even less than what was assumed here. From a design perspective the pipeline FFT implementation of the adaptive frequency domain digital fi lter is quite attractive. Only a few relatively high complexity modules (e.g., radix-4 butterfly lement, delay commutator data reordering module, data multiplier, etc.) need to be designed. The data routing is simple and direct. Control is trivial. The calculation of the frequency domain filter kernel is relatively complicated, but this is done at a relatively slow rate, so a standard general-purpose processor core can be used for this function. A potential pitfall for this approach is that the general-purpose processor core will require a sizable memory which may not be easy to produce on the same chip that has a significant amount of logic-circuitry. Traditionally logic and memory are produced with very different processing flows. Other systems may require both analog and digital circuitry which may not be amenable to production with the same fabrication process. Even within the radix-4 butterfly elements, the elements use a simple and consistent combination of adders and multipliers. System compilation sh ould be applicable to this design. This suggests that silicon compilation [2] can be applied (perhaps with higher level ³system² extensions) to create hierarchical ³system compilers^{2} that will facilitate the design and development of System-on-Chip designs.

3. ADAPTIVE EQUALIZER

Another signal processor that illustrates significant ideas about System-on-Chip design is an adaptive equalizer [3]. Time domain data is acquired and formed into fixed size packets. Each packet is analyzed to determine the frequency dependent weighting that it has undergone. Finally the spectra are ^{3}re-scaled^{2} to equalize the frequency dependent weighting. From a computational perspective the adaptive equalizer requires a Fourier transform. Approximately 90% of the computation in the adaptive equalizer is for the Fourier transform. The remaining computation is divided fairly equally between the estimation of the frequency dependent weighting and the re -scaling. This analysis suggests that a pair of custom processor designs be developed, one that implements the FFT and a second unit that does all the rest of the processing. Unfortunately, algorithms for determining the frequency dependent weighting are subject to frequent changes, so a fixed design for the second processor is inappropriate. In fact the second processor is a case where a programmable general-purpose processor core makes a lot of sense. This example illustrates some of the attraction of platform-based design techniques [4] that facilitate creating complex systems with a mixture of IP.A mature FFT processor can be integrated with a well understood general purpose processor to produce a system that achieves a high level of throughput with a measure of programmability

4. CONCLUSION

System-on-Chip design has the potential to achieve high throughput, small size, and low power .It can be expected that system compilation and platform-based design will revolutionize the development of such systems in the same way that silicon compilation and advanced CAD have revolutionized VLSI design.

REFERENCES

[1]Earl E. Swartzlander, Jr., VLSI Signal Processing Systems,Boston:Kluwer Academic Publishers, 1986, Chapter 6, pp. 117-139.

[2]D.D.Gajski,ed, Silicon Compilation, Reading, MA:
Addison-Wesley Publishing Co., 1988.

[3] Earl E.Swartzlander, Jr., VLSI Signal Processing Systems,Boston: Kluwer, Academic Publishers, 1986, Section 4.4, pp. 83-86.

[4]H.Chang, et al.,Surviving the SOC Revolution:A Guide to Platform-Based Design, Norwell, MA, Kluwer Academic Publishers, 1999, p.4

2

DESIGN REUSE

One of the paradigm shifts and also a major strength of the System-on-Chip methodology is associated with design reuse. This has been a particularly major paradigm shift for industry, where the concept has only recently taken a strong hold on the design methodologies employed by industrial design groups. For state-of-the-art chip complexities, design reuse has become a necessity.

Because it is a recent paradigm shift, different methodologies of implementing design reuse are still being explored by reearch and design teams. In this chapter, we present 5 sections associated with various concepts associated with design reuse.

In the first section, "Methodologies and Strategies for Effective Design-Reuse", by Ludovic Loiseau, and Yvon Savaria of École Polytechnique de Montréal, Canada, the authors look at simplifying the "Reuse Methodology Manual" (RMM) from Synopsys and Mentor Graphics (currently a standard in the practice of design reuse) in the context of producing IP for internal use within a design or research group. A strategy is also proposed (planned-redesign-for-reuse) to spread the inevitable additional efforts imposed by reuse methodologies. The concept is based on initially producing a good module, well usable, but not fully reusable, at the first use, and then redesigning it to make it fully reusable. Thus, efforts associated to reuse are distributed between 2 or 3 design stages. This sharply contrasts with the conventional wisdom that redesign-for-reuse is not a good means to produce good quality modules. The method therefore helps in producing better quality modules while requiring moderate efforts. An analysis shows that different quality levels can be obtained for a given module, depending on the dedicated efforts. A layered quality model is proposed, where the quality level of a module can be incrementally improved. Finally, a method is proposed to evaluate the cost of reuse.

In the second section, "A VHDL/SystemC Comparison in Handling Design Reuse", by Luc Charest and El Mostapha Aboulhamid of the Université de Montréal, Canada, the multi-paradigm aspects of SystemC are used to develop hardware libraries, to reuse designs, to accelerate simulation, and to allow efficient design space exploration both at the RTL and architectural levels. Concepts of commonality and variation supported by C++ are used to compare SystemC capabilities to those of VHDL. The use of the methodology is expected to allow the development of resilient and well documents models, and it may also help in developing efficient test-benches.

The third section "Aspect partitioning for Hardware Verification Reuse" by Sébastien Regimbal, Jean-François Lemire, Yvon Savaria, Guy Bois, El Mostapha Aboulhamid and André Baron, from the above institutions and an industrial partner, PMC-Sierra, deals with the problem of verifying large SOC designs by employing partitioning schemes to allow reuse of verification results, thus considerably reducing the complexity of verification. The authors propose a partitioning method based on the hardware verification language, e, which uses aspect-oriented programming to facilitate verification program development. The method is applied to a concrete example; verification of a subset of an SOC protocol converter platform.

The fourth section, "Reconfigurable Combinatorial Accelerators for Real Time Processing" by I. Skliarova and A.B. Ferrari of the University of Aveiro, Portugal, deals with design reuse associated with Field Programmable Gate Arrays (FPGAs); namely, reconfigurability. The paper suggests an architecture for a hardware accelerator for combinatorial computations and a strategy of performing such computations on a reconfigurable hardware/software (RHS) model. The resulting hardware has been implemented using a XCV812E Virtex FPGA, and it can be reused for different combinatorial problems. The complexity of the FPGA allows the implementation of complete systems that include the proposed accelerator linked to a host application-specific embedded controller for real time processing. The paper presents and analyzes a set of experiments with the proposed accelerator on the base of three well-known combinatorial problems: Boolean satisfiability (SAT); a traveling salesman problem (TSP); and a covering problem. The results show that the proposed architecture, together with the respective methods and algorithms, allow significant reduction in computation time.

In the final section, "Tuning Methodologies for Parameterized Systems Design", by Giuseppe Ascia, Vincenzo Catania, Maurizio Palesi of the Università di Catania, Italy, the authors present some new solutions for design space exploration of parameterized systems. Three approaches are presented to perform design space exploration for parameterized systems. The first approach is heuristic and is based on sensitivity analysis to reduce the space of configurations to be explored. The second uses multi-objective genetic algorithms, while the third is a mixture of the previous two, using sensitivity analysis to limit the space of configurations and genetic algorithms to explore the subspace thus obtained. All three approaches are applied to determine the power/performance trade-off of a highly parametric architecture implementing a SOC for digital camera applications. The results obtained show the effectiveness of the purely genetic approach as regards the quality of the solutions obtained, and the mixed approach in terms of efficiency.

Section 2.1

Methodologies and Strategies for Effective Design-Reuse

Ludovic Loiseau, Yvon Savaria
École Polytechnique de Montréal

Abstract: Reuse becomes a necessity to benefit from the growing maximum complexity of integrated circuits. Existing reuse methodologies focus on the development of complex intellectual property (IP) modules. They can be heavy to use when the objective is only to benefit from reuse advantages as an internal tool, to reduce design costs and time to market. This paper proposes a soft methodology that is strongly inspired by the RMM [1], but adapted to a simpler context. A strategy is also proposed to spread the inevitable additional efforts imposed by reuse methodologies. The method helps producing better quality modules while requiring moderate efforts. An analysis shows that different quality levels can be obtained for a given module, depending on the dedicated efforts. A layered quality model is proposed, where the quality level of a module can be incrementally improved. Finally, a method is proposed to evaluate the cost of reuse.

1. INTRODUCTION

Integrated circuits maximum complexity grows rapidly. In spite of the evolution of existing design techniques (more effective tools, better trained engineers...), the design time of complex digital circuits grows.

A pragmatic brute force solution is to put more engineers on a project, but when more than 10 engineers work on a task, the productivity does not grow proportionally any more [5]. Thus, a gap appears between the global productivity and the available number of transistors (see Fig. 1). Without evolution of design methodologies, a small team will simply not be able to design complex tomorrow's chips.

A very popular approach is reusing hardware modules in a context different from their initial development framework. Among existing methodologies, the most popular is the one presented in the Reuse Methodology Manual (RMM) [1], developed for ASIC system-on-chip design. They can be heavy to apply. Many organizations design custom hardware to better compete, without aiming to become integration specialists. For them, FPGAs are quite appealing, thanks to lower design costs and improved flexibility/ reconfigurability. Typically, they do not aspire at becoming suppliers of IP blocks to the general market, but wish to benefit from reuse methods as a means to shorten time-to-market. These organizations need a softer approach to make reuse happen.

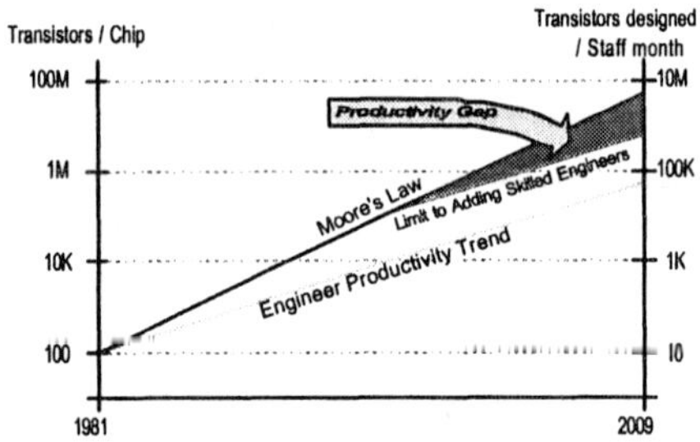

Figure 1. Productivity Gap (source RMM [1])

2. OVERVIEW

The main precepts of design reuse are very simple [1]. A module must be **usable** before it can be made **reusable**. This sets strong requirements for: good documentation, good coding style, thorough and significant comments and thorough verification.

Such modules must be easy to use, integrate or modify. Additionally, to be reusable, a module must be: configurable, thus adapted to a generic use; portable for different technologies or tools; highly verified and fully documented. Ideally, these features imply that the module is adapted for a wide range of applications, and they contribute to raise the user's level of confidence. However, they are generally costly. Thus, strategies and methods to implement them while mitigating risks and additional costs are useful.

Design rules

A complete set of rules, presented in [2], was derived by selecting and enriching relevant rules from [1]. The set of rules is divided in 5 groups: basic coding rules, focusing on lexical conventions or basic constructs, to produce good usable modules; coding for portability, to create technology-independent modules compatibles with various tools; guidelines for clocks and resets, to produce robust designs with good performances; coding for synthesis, to design efficiently synthesizable modules and partitioning for synthesis, to produce optimized blocks.

Configurability

Two techniques can be helpful to produce configurable modules:

- Produce generic architectures, adding optional features in the design. For example, considering a video FIR filter, the following features could be configurable: number of taps, coefficient values and generic data size.
- Facilitate replaceability of sub-blocks in a system. This goal can be reached by supporting a generic integration of modules. Thus, configurability is obtained by interchanging modules with different features. It is often easier to design several sub-blocks, each one being optimal with respect to a specific feature. Considering the example of the FIR filter, it can be difficult to produce an optimal structure for every type of filter.

Verification

Verification has become the main task for "classical" design. It is even more important for reusable components, since its eventual user must be highly confident that it is working. Since 50% or more of the overall development time is often required to perform a good verification, it is critical that it is not re-verified on reuse. If a module is highly configurable, every feature is to be thoroughly verified, which can be highly time consuming. Verification is thus a central task of a good reuse process. Even if a module can never be considered fully verified, good techniques exist to produce highly verified modules [3,4].

3. STRATEGIES

Ad-hoc strategy

The ad-hoc "strategy" discussed in [5] is rather natural. It is not a structured reuse strategy. The first design is carried out without future reuse objectives. It will generally be reused "as-is", or patched. Figure 2 depicts the costs expected for this strategy. It is assumed that each new use is in a new context, requiring different features. Such a process is not very effective, since the code is not easily modifiable, and the architecture does not anticipate future needs. Its internal organization must be understood and reconsidered at each use if new features are necessary. Significant efforts must be invested in verification at each use. This approach does not take advantage of a structured reuse process, thus cost savings are limited, but no additional effort is required on first use.

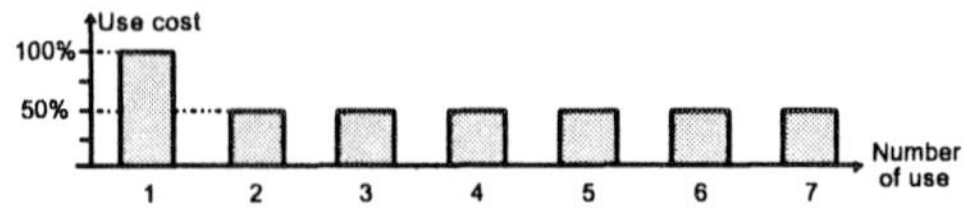

Figure 2. Cost phasing of an ad-hoc reuse strategy

Planned strategy

With this strategy, also proposed in [5], the objective is to produce a reusable block at the first use. Significant additional efforts are required at this first use. The objective is to keep the cost of reuse very low. Nevertheless, significant efforts should be anticipated on the second use to add features not planned initially and to complete verification.

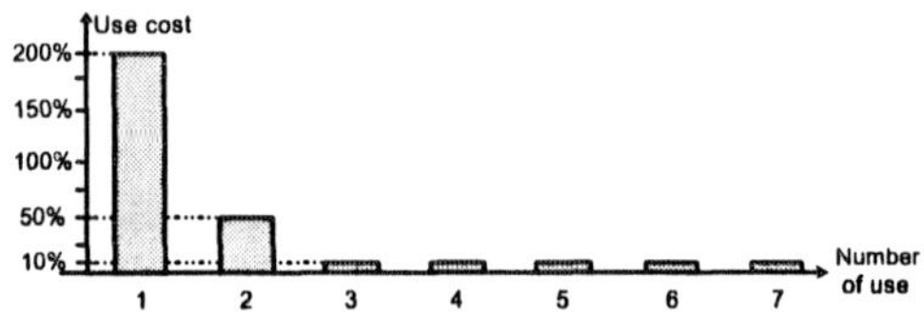

Figure 3. Cost phasing of the planned-reuse strategy

This strategy is considered the best with respect to cost savings, but it is the highest risk approach, due to uncertainties associated with a module's reuse potential. Also, with this strategy, developers should know from the onset how to make a module fully reusable. This assumes the developer knows exactly which features are needed in future uses.

Redesign for Reuse

This strategy, suggested in [1], is generally considered impractical. It was considered in the context of a legacy design that we would try to make reusable. Such a process is indeed inefficient, and generally, the circuit must be remade, because:

- New features are hard to integrate within the current architecture, and/or
- The code is not clearly commented and modifications are complex.

This strategy was used for an MPEG2 video decoder [7]: fewer modules than expected were reused because of inadequate documentation and the lack of flexibility of the architecture. This strategy is also presented in [8], and is called *rework-based reuse*.

Despite similar name, our proposed approach is quite different. It could be qualified of planned-redesign-for-reuse. The principle is to plan the redesign to ensure the initial design is a good one: clear, commented, easy to modify, with a generic architecture, but specifically targeted for its first application. It is not yet fully reusable be fore additional tasks are accomplished. Particularly, the verification plan focuses on the first application, and a limited documentation is produced. That first documentation must support initial integration and document know-how embedded in the module. At the second use, it is thoroughly reworked to make it reusable: add features, verify and document it completely.

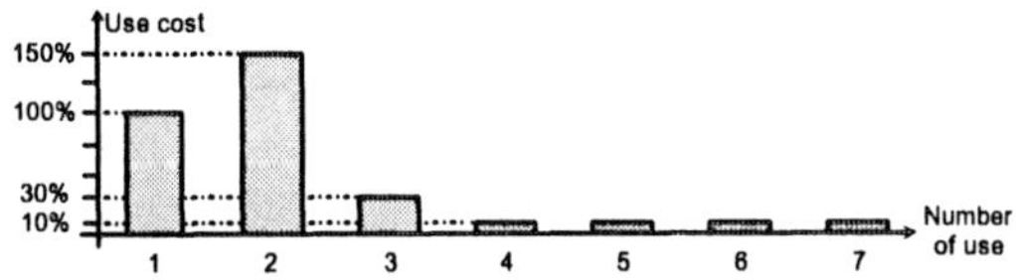

Figure 4. Cost phasing of planned-redesign-for-reuse strategy

The first design follows basic precepts of reuse, but limits the required efforts. Indeed, following good coding style does not really increase cost, but prepares the ground for redesign. Then, more constraining rules are respected in the second use, the first reuse. It means that the overall efforts to obtain a reusable module are divided in two design steps (see Fig.4). When compared to the planned-reuse strategy, our planned-redesign-for-reuse strategy is less constraining, since the efforts are distributed over 2 uses of the module.

Comparisons

Figure 5 shows the cumulative costs for the various strategies. Each new use is supposed to be a new context.

Clearly, the ad-hoc strategy is advantageous only for modules that do not really have a strong reuse potential (typically 3 or 4 uses over the lifetime). A planned-reuse strategy minimizes costs over the long-term. However, it is a more risky approach. Indeed, errors with respect to incorrectly anticipated future uses lead to redesign costs much higher than anticipated. Also, if the expected numerous reuses fail to materialize, a high cost is incurred with not much benefit. A redesign for reuse strategy offers almost the same benefits as a planned-reuse strategy over the long-term, but the initial cost is distributed over the first 2 or 3 uses. The benefits are:

- An opportunity to wait for a demonstrated, as opposed to speculated need for reuse (if a module is used a second time, its potential for reuse is clearer). Thus, we only invest when we are clearer on the reuse potential.
- After an initial design, we have more experience with the modules functionality. We often know better how thing should be done after they have been done once. Having dealt with two use contexts, we certainly know better how to make it reusable. It helps producing better quality reusable modules. That improved quality is hard to quantify.

- Reuse can be considered even for a critical project (where minimum design time is expected from management), since the initial design is not more costly. Experience and common sense show that many industrial projects are critical, since increased design time increases time to market! The proposed approach gives flexibility.

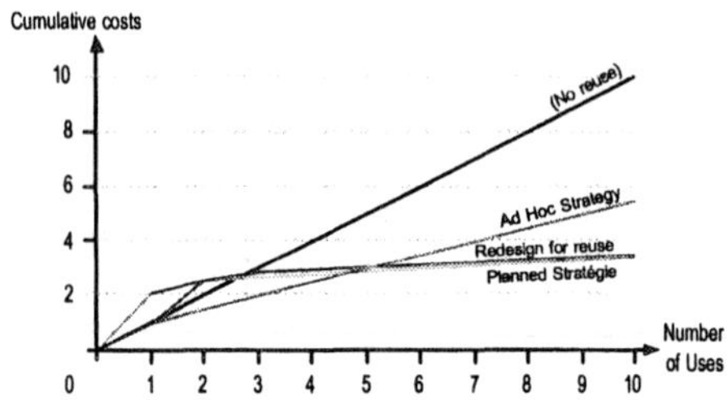

Figure 5. Cumulative costs of various reuse strategies

4. LAYERED MODEL

As shown in Figure 6, modules may fit in one of several quality grades.

Quality grades description

4.1.1 Functional level

It is supposed to be the minimum level for a usable block. The module must be verified for its intended application, but the code is not optimized, the validation is minimum and nothing is documented. Thus, this level is not acceptable even if the module seems to work. Its integration will probably create problems, and any modification will be costly or even impossible.

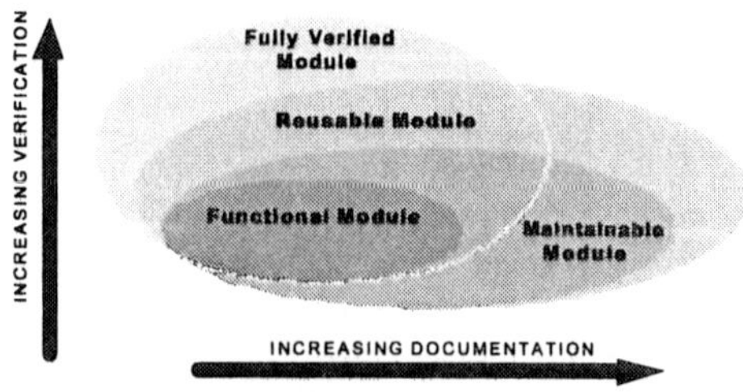

Figure 6. A layered quality grade model

4.1.2 Maintainable level

Compared to a strictly functional module, a maintainable module is not only functional, but it is also well documented (schematics, clear, explicit and commented code...). It is however not necessarily optimized. This level should be the minimum acceptable. Generally, such a module is generic, but may not support a wide scope of features. However, the module is still tested only for a specific application. The main advantage of this quality level is that the module is easier to maintain and modify.

4.1.3 Reusable level

A reusable module is a refinement of the maintainable level. It is a lot more configurable and adapted to a general use. Moreover, it would be verified with a high confidence level for every valid configuration, portable to different tools, and fully documented. The code should also be optimized. Considering our redesign for reuse strategy, the initial design must be **maintainable**, and the redesign for reuse step produces a **reusable** module.

4.1.4 Fully verified level

A fully verified module is an ideal we should aim for in theory, but that may not be practically reachable within limited budgets. This level is not a refinement of the previous levels.

Selective application of design rules

Due to space constraints, the selective application of each rule is only presented in [2], both for the planned-reuse and the planned-redesign-for-reuse strategies. Only a short description is presented in this section.

4.1.5 Redesign-for-reuse strategy

Design rules presented in section 2.1 can be selectively respected for a redesign for reuse strategy.

- Initial design: The rules for this step are numerous, but not really constraining. These are mostly basic rules that should always be applied, as well as guidelines for clocks and resets, and for synthesis.
- Redesign: for this step, all rules voluntarily deferred at the first step (coding for portability and partitioning for synthesis) are supposed to be followed, producing portable, configurable and optimized design.

4.1.6 Relating rules and quality levels

We can also selectively follow the rules depending on the quality level we want to obtain:

- Functional level: only rules dealing with functionality or timing closure should be considered for this level. They mainly consist of guidelines for clocks and resets, and for synthesis.
- Maintainable level: most basic rules should be followed, producing good quality and clear code.
- Reusable level: most constraining rules should be reserved to this level, since it is supposed to produce optimized and highly validated modules.
- Fully verified level: no design rule (in the current state-of-the-art) can guarantee that a complex module is fully verified, since the design process is supposed to be limited to a realistic industrial context.

5. RESULTS

A detailed characterization of the cost of reuse was conducted according to the following methodology. The design process was partitioned according to the tasks composing it. A detailed log of the time spent on each task was maintained. These tasks were classified as non-reuse (that would be necessary even if we did not produce a reusable module) and reuse. Careful accounting of the time spent on each task led to accurate estimates of the cost of reuse, for a planned-reuse strategy and a redesign-for-reuse strategy.

Planned-reuse strategy

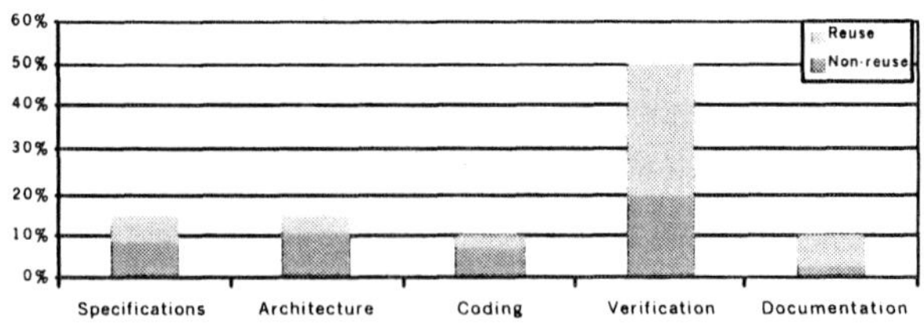

Figure 7. Break down of costs for the planned-reuse strategy

The data we collected confirm that producing a reusable module is two times more costly than a straight design. One of the modules we designed was reused 4 times in an industrial environment. Our analysis showed that the cost of reusing could be as low as 1% of the basic design cost.

Redesign-for-reuse strategy

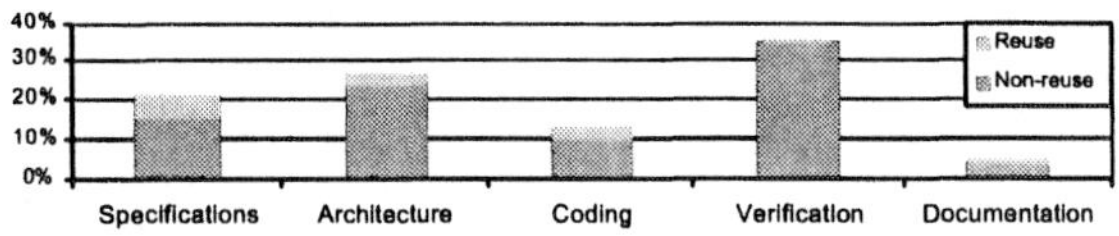

Figure 8. Break down of costs for the planned-redesign-for-reuse strategy (first design)

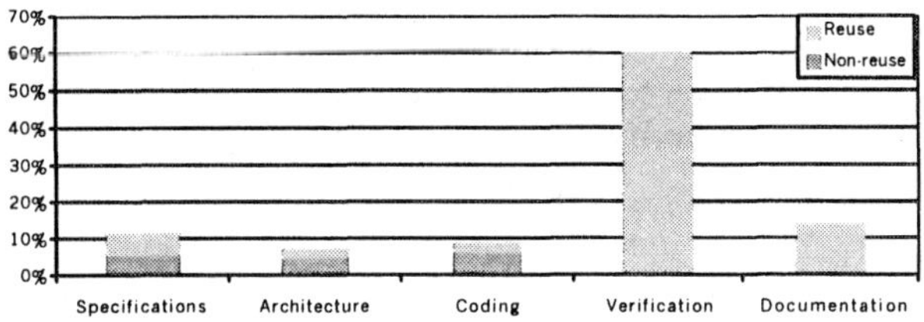

Figure 9. Break down of costs for the planned-redesign-for reuse strategy (redesign)

The cost of reuse in the initial design phase is not significant, and is limited to about 10% of the total cost of that phase. The redesign, at the second (or even third) use focuses on verification, mainly to verify reusable features omitted at first design. It also produces a complete documentation. Globally, the efforts to produce a reusable module with this strategy are similar to those incurred with the planned-reuse strategy, twice the price.

6. CONCLUSION

This paper presented a soft reuse methodology and a strategy to produce good quality reusable modules. The methodology, strongly inspired by the RMM [1] was adapted to better suit the requirements of producing IP for internal use. The strategy called planned-redesign-for-reuse consists in producing a good module, well usable, but not fully reusable, at the first use, and then redesign it to make it fully reusable. Thus, efforts associated to reuse are distributed between 2 or 3 design stages. This sharply contrasts with the conventional wisdom that redesign-for-reuse is not a good means to produce good quality modules. With the proposed strategy, the reuse process can be applied even if a given module is on the critical path of a project. A selective application of the rules helps producing modules of different qualities, by gradually increasing the quality level of a module. The proposed approach mitigates risks associated with designing a reusable module when a function is not yet fully mastered, or when the potential for reuse is still somewhat speculative. An analysis of the costs of reuse also showed that a planned-redesign-for-reuse strategy is not more costly than a planned-reuse strategy, and produces reusable modules as good, or of even better quality.

REFERENCES

[1] M.Keating, P.Bricaud, Reuse Methodology Manual, Second Edition, KAP, 2000
[2] L. Loiseau, Methodologie Design-Reuse, École Polytechnique de Montréal, October 2001, http://www.cours.polymtl.ca/ele4304/references/design_reuse.pdf
[3] J. Bergeron, Writing Testbenches, Functional Verification of HDL Models, KAP, 2000
[4] S. Tasiran, K. Keutzer, Coverage Metrics for Functional Validation of Hardware Designs, IEEE Design & Test of Computers, July 2001, pp.36-45.
[5] Qualis Design Corporation, FPGA Reuse Field Guide, 2000
[6] Xilinx, Xilinx Design Reuse Methodology for ASIC and FPGA Designers
[7] H.J. Brand, R. Siegmund, S. Riedel, K. Hesse, D. Mueller "Redesign of an MPEG-2-HDTV Video Decoder Considering Reuse Aspects, http://www.infotech.tu-chemnitz.de/~microtec/eng/press/annualrep98/mpeg.htm
[8] C. Costi, D.M. Miller, "A VHDL Analysis Environment for Design Reuse", *Virtual Components Design and Reuse*, KAP, 1999, pp. 81-94.

Section 2.2

A VHDL/SystemC Comparison in Handling Design Reuse

Luc Charest and El Mostapha Aboulhamid
Université de Montréal

Abstract: This paper describes the use of the multi-paradigm aspects of SystemC to develop hardware libraries, to reuse designs, to accelerate simulation, and to allow efficient design space exploration both at the RTL and architectural levels. Concepts of commonality and variation supported by C++ are used to compare SystemC capabilities to those of VHDL.

Key words: VHDL, SystemC, RTL Modeling, Design Reuse, Object Oriented Methodology

1. INTRODUCTION

SystemC is a modeling platform consisting of C++ class libraries and a simulation kernel for design at the system-behavioral and register-transfer-levels. Designers create models using SystemC and standard ANSI C++[1]. Different models of computation and design methodologies may be used in conjunction with SystemC. The design libraries and models needed to support these specific design methodologies are considered to be separate from the SystemC core language standard. This work may be considered as a design methodology for RTL and architectural models. By its nature, SystemC inherits the capabilities of C++, like the support of multiple paradigms: classes, overloaded functions, templates, modules, ordinary procedural programming, and others. The freedom afforded by these capabilities may also bring new difficulties to the understanding of the model or the ability to synthesize it. The objective of this paper is to demonstrate the benefits of this multiparadigm environment, and show that the advantages outweigh the risks. We will show that we will be able to develop more resilient specifications and models, express different behaviors and view of the same "entity", facilitating design exploration and ease specification and development of libraries.

We will illustrate this by comparing SystemC capabilities with those of the VHDL hardware description language. The lack of some useful paradigms in VHDL has already been described as early as 1991 as summarized in [2]. Early attempts tried to augment VHDL with other constructs, but since systems have more and more software programmable components, many designers are leaning toward a common language for hardware and software modeling. In order to make a fair comparison, we intentionally restrict ourselves to hardware modeling both at the functional and RTL level. System-level capabilities using C++ and/or SystemC are well described elsewhere [3, 4]. This may serve also as guideline for VHDL designers to understand and potentially adopt the SystemC methodology. In our comparison, we will use the concept of *commonality* and *variation* developed by Coplien [5, 6] to compare the two environments.

In this paper, Section 2 describes the desirable characteristics of a design environment; Section 3 introduces the commonality and variation concepts, deals with design reuse and hardware libraries, and discusses configuration is sues; finally, Section 4 concludes this work.

2. CHARACTERISTICS OF A DESIGN ENVIRONMENT

In our opinion, the desirable characteristics a hardware designer is looking for are as follows:

Reuse: hierarchy is one of the mechanisms that allow reuse. Structural hierarchy is available in VHDL but not behavior hierarchy, as will be illustrated later.

Refinement and automatic path to synthesis.

Families of design: often a design house will be interested in developing a family of products such as processors, controllers, and busses. All express some commonality but inside the same family it is desirable to be able to express variation by changing parameters or other mechanisms.

Express design constraints related to timing, testability, number of processors, etc.

During the design exploration phase it would be very helpful to break the modeling, analysis, elaboration, and simulation cycle by allowing online configuration of a model. This leads to more freedom in comparing results, and a shorter time for the analyzer to evaluate its design. The new cycle would be: model, compile, and examine the execution phase by dynamically reconfiguring the model. If the modules are fully interfaced, implemented and debugged, their construction could be controlled via a GUI at execution time.

Link to the outside world and extension of the capabilities of the environment. In VHDL, this can only be done in a limited fashion via C language interfaces. Since SystemC is an open source environment we have seen some efforts in developing other tools around the SystemC core as well as methodologies to have interoperable environments [7].

Minimize the number of environments and languages to deal with: in the case of systems containing both hardware and software, SystemC seems to have an advantage.

Have delay models that fit the need of the different levels of abstraction. Delay models are very restrictive in VHDL (inertial, delta and transport delays). These seem sufficient for describing hardware components. However they are very inefficient when developing test-benches. Test-benches and abstract architectures necessitate FIFOs, queues and stacks for example. This is very possible in SystemC especially with channels. Moreover, the full power of C++ can be used to develop high-level test generation and analysis tools, and benefit from more abstract datatypes.

Documenting tools for the models: the models would be more usable and more resilient, if well documented. In this case, SystemC seems to have a big advantage compared to VHDL. It is open source and the models are developed using the same paradigms as those used to develop SystemC itself. GNU tools like Doxygen[8] may be used to both document SystemC kernel as well as the models developed using SystemC [9]. Graphical documentation can also be easily obtained using UML [10].

3. COMMONALITY AND VARIATION IN DESCRIPTION LANGUAGES

Reuse in VHDL

Hardware designers are under pressure to develop new design derivatives in less time. In the case of software systems, the benefits of explicitly identifying the common and variable aspects of the different versions of a system are presented in [5]. We will illustrate the ability of SystemC, in comparison to VHDL, to express commonality as well as variation in modeling of hardware systems. VHDL [11] allows the expression of commonality by what is called design units. A design unit is a VHDL construct that may be independently analyzed and inserted in a design library. These design units are:

- Entity declaration: describes the interface view of a component (like a Data Book description). It is implementation independent.

- Architecture body: describes an implementation of an entity (like a single schematic diagram). A single interface may have alternative architectures.
- Package (declaration and body): contains information common to many design units. This information consists of functions, types, signals, and constants. It hides details, simplifies design, and may invoke other packages.
- Configuration: relates local entity and architecture references to actual units in libraries (like a parts reference list).

Commonality can also expressed by generate statements and generic constants for regular structures like a ripple carry adder or an interconnection network.

Variation is obtained by configuration, which allow the designer to choose an architecture among many others during design space exploration; by giving a specific value to a generic parameter; by overloading functions and subprograms in coordination with packages allowing the reuse of models even if we change the basic data types, like going from a bit type to a 9-valued standard logic. Commonality between processes is very limited except if we choose very complex ways like concurrent procedure calls rendering models quite cryptic.

In summary, the main mechanism for design reuse in VHDL is libraries of design units. Structural hierarchy is the way to reuse components declared in a library. On the other side SystemC has the same capability as VHDL and also other mechanisms, based especially on inheritance, templates and overloading. In SystemC 2.0, at the system level, commonality is expressed by abstract classes such as interfaces, then variation is encapsulated from interfaces with a possibility of multiple inheritance. Reuse can also expressed as use of design patterns [12] and other means, which will not be described in this paper.

Module Inheritance in SystemC

In designing a library of gates, latches, etc., we can specify all the properties related to gates in a module `Gate` from which, `And`, `Or`, `Xor`, etc. will be derived later. The purpose of the `Gate` module is to act as an interface. All the properties and methods for Flip-Flops will be captured in a similar module (Setup and hold time, delays, etc.). This is an example of simple inheritance of a module:

```
class Gate : public sc_module {
  public :
  Gate(const sc_module_name& name)
    : sc_module(name) { (...) }
};

class And : public Gate {
  public :
  SC_HAS_PROCESS(And);
  And(const sc_module_name& name)
```

```
    : Gate(name) { (...) }
};
```

Using Inheritance to Insert Tags or Attributes

In VHDL, we can reuse a component with a variation of behavior or semantics based on attributes. We can, for example, use attributes to specify that a process is at the RTL or behavioral level. However, it is impossible to derive processes from previous ones, or specialize their methods. In modern software-oriented methodology, recommendations are made supporting a common root for all objects in the software design. `sc_object` constitutes such a root in the core of the SystemC library. For the purpose of modeling, a library of components and IPs, `sc_module` can be the building block for the models. From this we may derive other specialized modules such as `sc_software_module`, `sc_hardware_module`, `sc_control_module`, etc.

From this point, we can use these basic building blocks to construct our libraries. With this kind of construct, we can now specify how the module should be instantiated. Because all three modules inherit from the same `sc_module`, they have the same behavior (unless we choose otherwise). A designer could decide that a particular software module should be implemented in hardware, by simply changing the ancestor of his design. This kind of construction could be recognized by synthesis analyzers without having to alter SystemC syntax or behavior, but is part of the static configuration and we will see later how to make this kind of choice more dynamic.

Hierarchical Module Construction for SystemC Hardware Libraries

Class hierarchies can be a great help when designing hardware libraries. Here are some proven advantages:

Modularity of structure: hardware properties are reused in derived classes.

Locality of code: when you have something to modify, you know it is in the class or its ancestor(s)

Reusability of code: software implementations in base classes are reused in derived classes. *Figure* 1 represents a basic gate library:

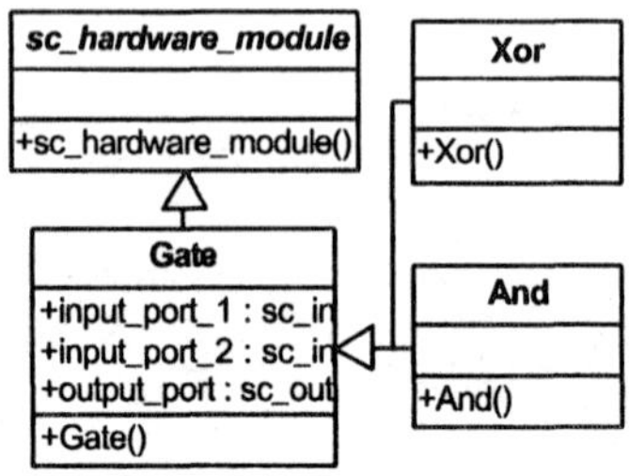

Figure 1 Behavioral hierarchy

There is no need in the "child" class to declare again the port information (this is impossible in VHDL), all we need to do is to implement a sensitive method to these ports and have these method be set sensitive in each of the constructors

```
And::And(...) : public Gate(...) {
  SC_METHOD(main_process)
  sensitive << input_port_1
            << input_port_2;
}
```

This duplication of definition can be avoided. In Figure 2, if the virtual `init()` method defined in the Processor class is sensitive to some signals, then the derived init methods in classes `DLX_Processor`, etc., will inherit this sensitivity. Therefore, a library could be constructed from ground up, by adding and refining the hardware components.

Multiple Architectures

Using VHDL, we can have multiple architectures related to the same entity. We can consider the entity as the abstract class and different architecture as the derived classes from the basic abstract class. Note that in VHDL, this derivation process is limited to two levels of abstractions. In SystemC, this refinement process can continue indefinitely as illustrated in Figure 2.

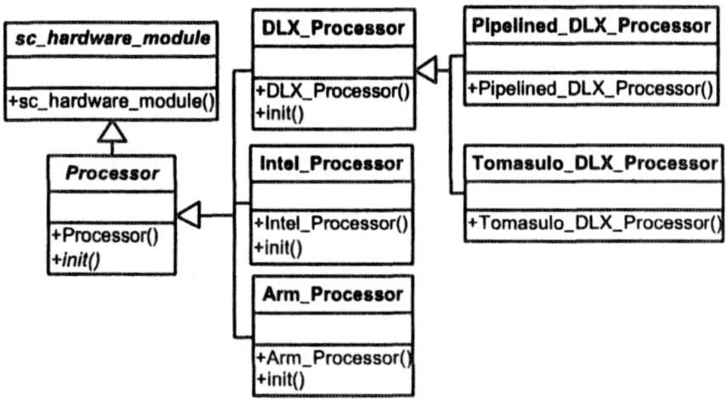

Figure 2: Behavioral refinement

Regular Structures and Their Configuration

In VHDL, generate statements are used to model regular structures composed of processes or components. In SystemC, a simple declaration of an array of components is sufficient, as illustrated below. Configuration is obtained in a very natural way by instantiation.

```
Processor *processor_array[10];
processor_array[0]=new Arm_Processor();
processor_array[1]=new DLX_Processor();
processor_array[2]=new Intel_Processor();
```

The configuration could be put in a file and then read at execution time during elaboration. A switch statement could then be used to instantiate the different processors. This has no correspondence in VHDL.

C++ Polymorphism in SystemC Library

If we reexamine the UML behavioral refinement of, we note that the abstract class Processor has a pure virtual method `init()`. This method is defined in the next level of abstraction in the derived classes `DLX_Processor`, `Intel_processor` and `Arm_processor`. If general enough, these methods are reused in the third level of refinement without having to redefine them. Modularity of code is very important to achieve reusability. A module, which uses these processors, can be written using only the abstract class processor. Depending on the true instance of the processor the right init method will be invoked by polymorphism:

```
for (int i = 0; i < 10; i++)
  //initialize all processors
  processor[i]->init();
```

If we desire the same effect in VHDL we would need a special signal Reset_Processor, which should feed only processor components. This is less elegant and less readable than what we have here. It is also surely more time consuming during simulation because events on signals put a very heavy burden on the VHDL simulator.

Using Overloading Mechanism to Change the Behavior

In VHDL, overloading is limited to functions and procedures. In SystemC, not only methods but also constructors of classes can be overloaded allowing more dynamic configuration of threads and modules. To determine the correct method to call, the compiler only looks at the type of the parameter(s) when the method call is issued.

This is an example of overloaded constructor for a `Processor` object:

```
Processor::Processor(const sc_module_name
                     &name, int bus_format)
 : sc_hardware_module(name) {
  bus = new sc_in<int>();
  (...)
}
Processor::Processor(const sc_module_name
                     &name, char bus_format)
: sc_hardware_module(name) {
  bus = new sc_in<char>(); (...)
}
```

Using Overloading to Speedup Simulation or Specify a Particular Behavior

In VHDL, if we had to specify a generic entity with n ports, we first define a generic and then declare an array of n ports. However, if for a specific number of ports we have a particular behavior or any optimization, the only way is to test the parameter inside the architecture related to the generic entity. In SystemC, by using overloading we can achieve this in a more efficient way as illustrated below:

```
void And::compute(void); {
  output_port = input_port_1
              && input_port_2;
}
void And::compute(int n_ports); {
```

```
  if (n_ports != 0)
    for (int i = 0; i < n_ports; i++) (...)
}
```

Combining Mechanisms

The highest configuration flexibility is achieved by combining all the above mechanisms together. With an overloaded method in a base class, you can choose a constructor of the derived class that best suits your need. With method overloading and polymorphism, the behaviour of the component ca be changed. This can be seen as a *complex design pattern* [12] for hardware/software modules.

In the hierarchy represented by Figure 3, `sc_hierarchical_module` is an abstract class (can not be instantiated) and defines an interface for those later derived classes. Behavioral_process () is an abstract method that simulates the module from a behavioral perspective, while structural_process () is the one that should represent the module by composition from other modules. Because the initialization phase might be different, init_behavioural () and init_structural () methods are provided and the general init () method is responsible for calling the appropriate initialization method.

The constructor of the `sc_hierarchical_module` is the one that builds the module and chooses whether you are going to use the behavioral or the structural definition of the derived class.

The intent of sc_hierarchical_module is to indicate that the module might be composed of other modules. When the hierarchical module is composed of only one module (might be changed at run time during initialization phase), we can use the method `has_subcomponent()` to determines the real nature of the module. This flag hides the `boolean` value that is adjusted by the constructor at compile time. The list of sub-modules is adjusted to allow an easy traversal of the hierarchy.

4. CONCLUSION

In this paper, we described the use of the multi-paradigms aspects of SystemC to develop hardware libraries, to reuse designs, and to allow efficient design space exploration both at the RTL and architectural levels. Concepts of *commonality* and *variation* are used to compare SystemC capabilities to those of VHDL. The use of the presented methodology will allow the development of resilient and well documents models, it may also help in developing efficient test-benches.

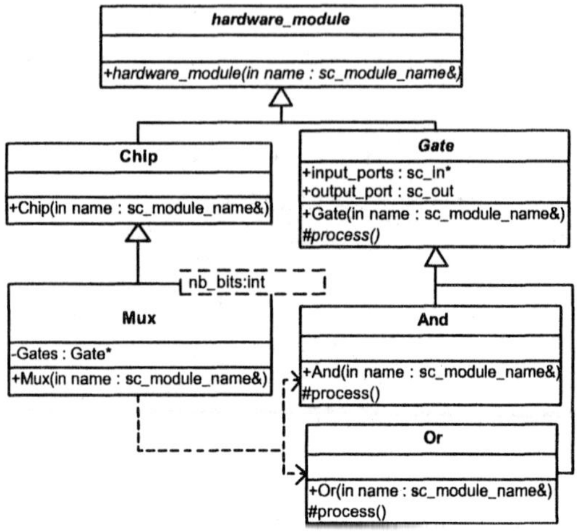

Figure 3: A Multi-Paradigm Module Description

REFERENCES

[1]Open Sy stemC Initiative (OSCI), Functional Specification for SystemC 2.0, http://www.systemc.org, 2001.

[2]S. Swamy, A. Molin, and B. Covnot, "OO -VHDL: Object-Oriented Extensions to VHDL," *Computer*, vol. 28, pp. 18-26, October 1995.

[3]D. Verkest, J. Kunkel, and F. Schirrmeister, "System Level Design using C++," Proceedings of Design, Automation and Test in Europe, Paris, France, 27 - 30 March, 2000

[4]Open SystemC Initiative (OSCI), SystemC, http://www.systemc.org, 1999-2001.

[5]J. Coplien, D. Hoffman, and D. Weiss, "Commonality and Variability in Software Engineering," *IEEE Software*, vol. 15, pp. 37-45, November/December 1998.

[6]J. Coplien, *Multi-Paradigm Design for C++*. Reading, M A: Addison -Wesley, 1999

[7]L. Charest, M. Reid, E. M. Aboulhamid, and G. Bois, "A Methodology for Interfacing Open Source SystemC with a Third Party Software," Proceedings of Design Automation and Test in Europe Conference & Exhibition, Munich, Germany, 13-16 March, 2001, pp. 16-20.

[8]D. van Heesch, Doxygen, http://www.stack.nl/~dimitri/doxygen/index.html, 1997-2001.

[9]L. Charest, SystemC documentation, http://www.iro.umontreal.ca/~chareslu/, 2000-2001.

[10]G. Booch, I. Jacobson, and J. Rumbaugh, *The Unified Modeling Language User Guide 1/e*: Addison Wesley, 1999.

[11]*IEEE Standard VHDL Language Reference Manual*, 1076,2000 ed: IEEE, 2000.

[12]E. Gamma, R. Helm, R. Johnson, and J. Vlissides, *Design Patterns: Elements of Reusable Object-Oriented Software*: Addison Wesley, 1994.

Section 2.3

Aspect partitioning for Hardware Verification Reuse

Sébastien Regimbal[1], Jean-François Lemire[2], Yvon Savaria[1], Guy Bois[2], El Mostapha Aboulhamid[3] and André Baron[4]

[1] *Electrical Engineering Department, École Polytechnique de Montréal, P.O. Box 6079, Succ. Centre-Ville, Montréal, Québec, Canada, H3C 3A7. Computer Engineering Department, École Polytechnique de Montréal, P.O. Box 6079, Succ. Centre-Ville, Montréal, Québec, Canada, H3C 3A7.* [3] *Department of Computer Science and Operational Research, Université de Montréal, C.P. 6128, Succ. Centre-ville, Montréal, Québec, Canada, QC H3C 3J7.* [4] *PMC-Sierra, 3333 Graham Blvd, Suite 500, Ville Mont-Royal, Québec, Canada, H3R 3L5.*

Abstract: Technology advances strongly impact integrated circuits (IC) complexity. Current IC design methods have difficulties to handle this growth. In particular, hardware verification has become the main bottleneck of any major digital design effort. It is thus necessary to develop efficient methodologies for designing verification environments. To deal with this complexity, hardware verification languages (HVL), such as e, allow reducing the verification effort and contribute to raise the level of abstraction at which test benches are described. In addition, aspect-oriented programming (AOP) is an emerging technique, which promotes a better separation of concerns and improves reusability. We propose a partitioning method based on e, which uses AOP to facilitate verification program development, and we apply our method to a concrete example; verification of a subset of a SOC protocol converter platform.

Key words: Functional Verification, Aspect-Oriented Programming, and Object-Oriented Modeling.

1. INTRODUCTION

A difficult part of hardware system design is to ensure functional correctness. The generation of large numbers of high quality stimuli, to uncover design errors with a minimum effort, is one of the key challenges of modern simulation based hardware verification. Another challenge of verification is the need to emulate accurately the system environment with a test bench. Considering that verification can consume over 70% of the overall hardware design effort [1], efficient methods are required for implementing verification environments.

HDL languages such as VHDL or Verilog have improved hardware design by allowing high-level hardware descriptions. However, because hardware verification needs some sophisticated mechanisms that HDL languages do not possess, like an advanced temporal logic or complex data structures, HDL languages are not ideal for verification tasks. The use of Hardware Verification Languages (HVL), such as Verisity's e [5], improves the verification process by addressing the requirements of complex functional verification software development. Furthermore, software reuse is known to be the most important issue for improving the productivity of software development processes. Most reuse approaches are based on an object-oriented methodology. The techniques proposed in [2][3] are guidelines for the creation of object-oriented verification environments.

In [4], some issues have been identified with the object-oriented paradigm in modeling properties of a system that influence several objects. These authors propose new programming techniques called aspect-oriented programming (AOP), which can capture the aspects of a system. Aspect-oriented programming allows efficient software modularization, which facilitates reuse. To our knowledge, the explicit use of an aspect-oriented approach in a concrete hardware verification environment has never been reported, even though [7] observed that a verification language such as Specman's e possesses constructs that can support AOP.

This paper shows how an aspect partitioning of verification environments promote reusability within and between projects. We propose a method that uses the aspect-oriented paradigm to enhance verification. The design under test (DUT) is a subset of a communication protocol converter SOC platform designed at the GRM (Groupe de Recherche en Micro-électronique) of École Polytechnique de Montréal. The paper discusses how a self-checking and configurable test bench was developed in an aspect-oriented framework, to emulate the environment of the DUT. The implementation of this verification environment was done with the e language, and was executed in the Specman Elite™ environment. Thus, the contribution of this work is to present a more formal partitioning methodology that facilitates reuse, through the use of aspects.

The remainder of the paper is organized as follows. In section 2, we review the methodology proposed for the design of our verification environment. In section 3, we briefly describe the DUT and we apply the methodology discussed in section 2 to verify this design. Section 4 discusses the lessons learned from applying our methodology to the verification example. Finally, in section 5, we present a conclusion and suggest possible areas of future investigation.

2. PROPOSED METHODOLOGY

Motivations

Ideally, placing a design in its real operational context is required to perform accurate verification. In general, such a context cannot be completely reproduced by simulation. Usually, it takes the form of a testbench able to:

- Emulate the environment of the design under test (DUT);
- Self-check the response of the DUT to the stimuli generated by the environment;
- Provide a quantitative feedback on the progress of the functional coverage;
- Be easily configurable to allow test writing by users unfamiliar with the verification environment.

In this work, the proposed method facilitates reuse in a verification project. First, we verify a module, and by a bottom-up approach, we integrate the other parts of the design in a verification environment that is gradually constructed. This approach enables reusing some classes of our verification environment during the integration process.

Object-oriented analysis

The object-oriented decomposition methodology for the creation of our verification environment is based on three types of classes: the emulation classes, the verification classes and the utility classes. These class types are described in the Table 1.

Table 1. Different Class Types

Class Type	Definition
Emulation	An emulation class is a class that produces the same behaviour as the real hardware component that should be connected to the interface of the DUT. Moreover, emulation classes must be configurable, to allow the production of all test cases included in the verification plan.
Verification	Verification classes encompass the classes that enable the verification environment to be self-checking.
Utility	Utility classes comprise all other classes needed to complete the verification environment like data structures.

Aspect-Oriented Concepts

Aspect-oriented concepts are based on the principle of the separation of concerns [9]. An aspect defines a behaviour that affects several classes of a system. The object-oriented analysis organizes a system into a clear hierarchy of objects. Because some aspects are not confined in any object boundary, we say that they crosscut the system's modularity. An aspect-oriented methodology resolves that common modularity problem. For example, the inter-object communication elements of a class are usually encapsulated into each class of the objects hierarchy.

However, if we group all inter-object communication elements into a single aspect that affect several classes, we can then pull the communication aspects out of the original class. The reuse potential is then significantly improved, because this approach forces isolating design aspects, to avoid scattering them throughout the code. Consequently, with the aspect-oriented approach, we define objects by slices of behaviour and each slice defines an aspect of the system.

Once we subscribe to the idea of separating aspects, three questions [6] must be answered in order to pursue the development of a complete methodology:

- How to capture the aspects

The e language solves this problem because it provides a construct, modules (e files), which can encapsulate aspects. In a module, we can create the base definition of a class and/or several extensions of other classes.

- What aspect composition mechanisms are to be used

A composition mechanism is embedded in the e language with special extension features [7]. Specman Elite™ is then able to use the different defined aspects to generate the object-oriented framework of the verification environment. The available constructs of the e language needed to achieve this task are presented in [8].

- What issues need to be separated

The answer to that question depends on the kind of applications. In our case, we want to develop a verification environment. Thus, in the next section, we propose an aspect partitioning solution to the problem of designing a verification environment.

2.1.1 Aspect Partitioning

An aspect-oriented analysis allows a concise separation of the verification environment by defining layers in each class of the verification environment. Decomposing the verification environment in aspects is the core of our method. It is important to determine the contribution of aspect partitioning to the verification design effort. Our objective is to enhance the reusability of the verification environment. Thus, we must isolate the aspects of the environment that have a very low reuse potential. We have defined five aspect categories that encompass all the aspects of our verification environment. Each category defines an aspect of existing classes.

As illustrated in Figure 1, the core of our aspect partitioning is the base aspects category. During the object-oriented analysis, we have determined that a certain set of classes was essential to define the core of our environment. The base aspects block includes the definition of all these classes. Each aspect from the other categories extends these base classes.

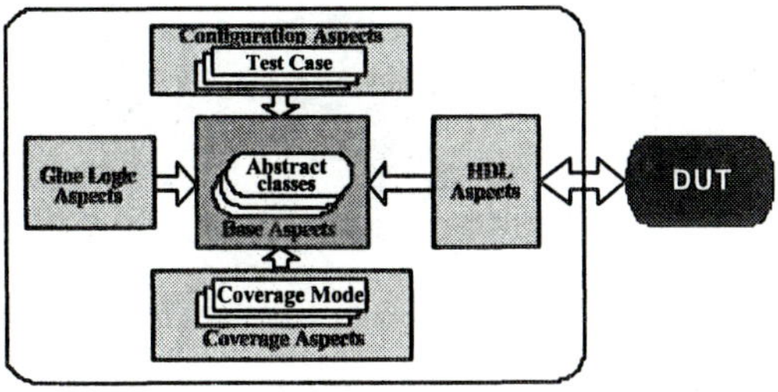

Figure 1. Aspect categories

The extensions are performed by adding features (aspects) to the base classes in other modules. These features have a very low reuse potential, and by defining them as new aspects, we separate them from the base aspects, which have a high reuse potential. Table 2 summarizes the purpose of each category.

Table 2. Different Aspect Categories

Aspect	Definition
Base	Defines the core of each class of the verification environment.
HDL	Define the connections and interactions of the classes with the DUT.
Glue Logic	Define the connections and interactions between classes.
Coverage	Define coverage models for the verification environment.
Configuration	Define each test case necessary to achieve the verification plan of the DUT. Test cases are constraints added on top of the verification environment.

3. APPLICATION

3.1.1 The DUT

The DUT is a module of a SOC protocol conversion platform in development at the GRM of École Polytechnique de Montréal. Its purpose is to convert packets structured according to a network protocol stack into another one. The first version of the design is configured and programmed to convert Firewire (IEEE 1394.b) to Ethernet (IEEE 802.3).

The subsystem on which our verification efforts focus is the block that receives and memorizes the incoming packets (Figure 2). This module also creates packets related tags, identifies the incoming protocols and finally transmits the relevant information to a formatting engine.

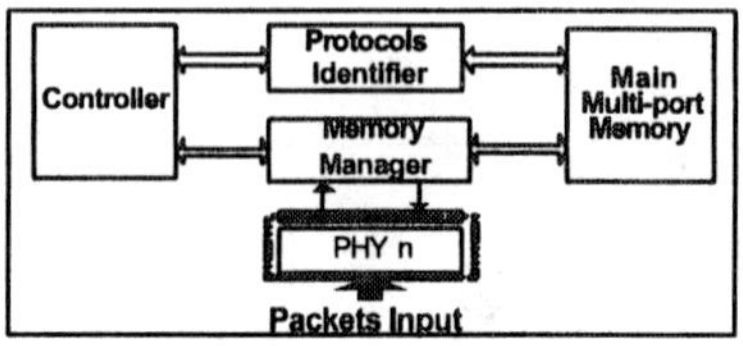

Figure 2. DUT : Part of the Protocols Converter

Partitioning Aspect Results

In this section, we explain how we applied the partitioning methodology developed in section 2 to verify the *Memory Manager* component. We also discuss how to extend the verification environment to cover the *Controller* component, by reusing certain aspects. The discussion is supported by concrete examples from our environment. The main purpose of this environment is to emulate the DUT's functional interface to verify its behavior.

3.1.2 Memory Manager Verification

Figure 3 represents a functional view of the verification environment. It shows the main functional verification components used to stimulate the DUT and to verify its behavior.

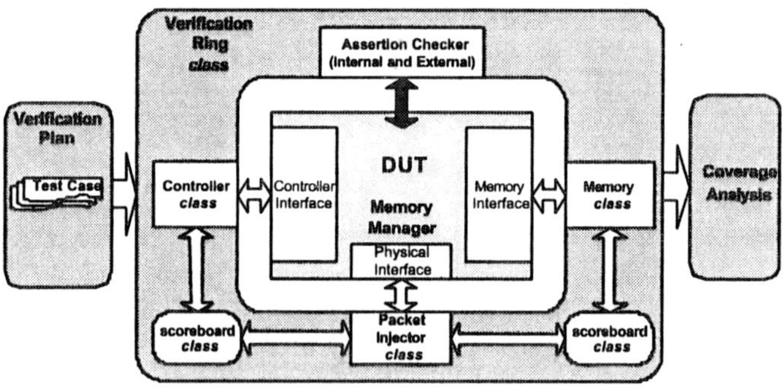

Figure 3. Verification Environment Functional View

In figure 3, we see that each functional verification component is associated with a class within the code structure. The *controller*, *memory* and *packet_injector* classes are part of the emulation type classes. The *assertion_checker* and *scoreboard* classes are part of the verification type classes. We created two*scoreboards* derived from a *scoreboard* super class, to ensure a self-checking process that verifies if the packets injected are transmitted adequately by the DUT, to the *controller* and the *memory.*

The *assertion checker* class is a grey-box and black-box type assertion checker that uses the temporal logic features provided by the e language.

The main characteristic of our methodology resides in the encapsulation of aspects in modules (files) of our verification environment. Figure 4 shows the different aspects of our environment.

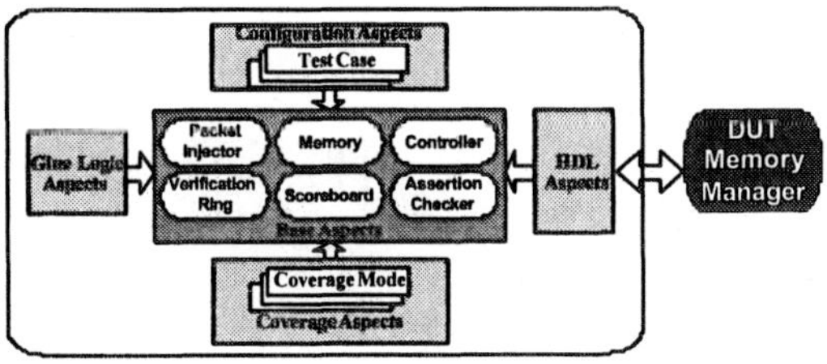

Figure 4. Aspects framework

At the highest-level of the environment, we have the *verification_ring* module aspect. It contains the highest-class definition of our class hierarchy. It is in this specific class that the environment initialization methods are defined. Each aspect in the base aspect group may contain class definitions that are extended in the modules connected by an arrow.

The purpose of the HDL aspect is to create a link between the environment modules and the DUT interface. Each aspect included in the base aspect group may have some of its classes extended to create a connection with the HDL code of the DUT. The glue logic aspect is responsible of connecting together all the other modules in the environment. Thus objects of the base aspect group classes are instantiated in this module. In the coverage aspect, coverage models are defined based on coverage events created in the extensions of the classes encapsulated in the base aspects. Finally, each test case of the configuration aspect category adds few constraints, on top of the existing environment, by extending some class of the environment. These test cases are then loaded on top of the current environment containing all the previously defined modules.

Figure 5 presents the implementation of the base aspect of the *packet_injector* class and a part of the glue logic aspect dedicated to our verification environment

The *packet_injector* class injects by its method a list of packets in the DUT. It is an abstract class because the method *write_to_dut* is undefined. We purposely avoid defining the *write_to_dut* method here, because this method does not belong to the base aspect of this class. It belongs to its HDL aspect.

```
aspect: packet_injector (base)
<'
unit pkt_injector {
   pkts: list of pkt;
   ...
   keep soft pkts.size() == 6;
   inject()@sys.clk is{
      for each pkt (p) in pkts do {
         write_to_dut(p);
      };
   };
   write_to_dut(p: pkt)@sys.clk is empty;
};
'>
```

```
aspect: controller (base)
<'
unit controller {
   ...
   receive_pkt()@sys.clk is{
      var p :pkt;
      wait @new_pkt;
      read_from_dut(p);
      ...
   };
   read_from_dut(p: *pkt)@sys.clk is empty;
};
'>
```

```
aspect: mm_glue_logic (Glue Logic)
<'
extend verification_ring {
   p: packet_injector  is instance;
   c: controller       is instance;
   s: scoreboard       is instance;
   keep p.father == me;
   keep c.father == me;
   keep s.father == me;
   ...
   init_test()@sys.clk is also {
      start p.inject();
   };
};
extend pkt_injector {
   father: verification_ring
   inject()@sys.clk is also{
      father.s.addTransaction(p);
   };
};
extend controller {
   father: verification_ring;
   receave_pkt()@sys.clk is also{
      father.s.matchTransaction(p);
   };
};
'>
```

Figure 5. Example of Aspect implementations

The glue logic aspect, on the right of Figure 5, extends the base definition of the *verification_ring*, the *packet_injector* and the *controller* class. First, a *packet_injector*, a *controller* and a *scoreboard* are instantiated into the *verification_ring* class. The *packet_injector* sends packets into the DUT, the *controller* receives these packets from the DUT and the *scoreboard* verifies the progress of a packet through the DUT. This represents the connection part of the glue logic aspect. For the interaction part, we connect the *scoreboard* to the *packet_injector* and to the *controller* by extending the *inject* and *receive_pkt* methods, with calls to the *scoreboard* methods *addTransaction* and *matchTransaction* respectively. We also start the packet injection with an extension of a method of the *verification_ring*.

3.1.3 Extension of the verification environment

To illustrate reusability within the same project, Figure 6 shows a functional view of the environment with the controller block added as a second design component to be verified. In this case, a version of the *controller* block replaces the class that emulates the controller block from the previous environment. Also, we see in Figure 6 that many classes can be reused to form the new environment. For instance, the base aspects, like *packet_injector* and *verification_ring*, have a high potential of reuse. By contrast, aspects like the glue logic aspect, discussed previously, are dedicated to a specific environment, since they establish the different connections between the environment classes. Then, these types of aspects are not easily reusable.

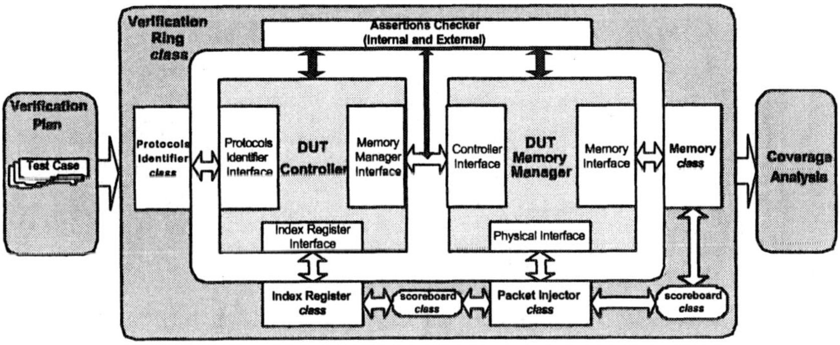

Figure 6. Reusability Realization

4. DISCUSSION

In this section, we discuss the lessons learned from applying the method presented in the previous section.

– Follow the Principles

An aspect-oriented approach combined with an object-oriented approach allows a better modularity that eases the maintenance task. However, coding principles are necessary if we want to benefit from this approach.

– Orthogonal extension

Several people have implemented the verification environment in an orthogonal fashion. More specifically, after an initial definition of the base classes, which is the essence of the object-oriented analysis, the implementation of other aspects was accomplished by extending the base definition. After an experimentation of this technique, we found that the method proposed speed-up the implementation process by allowing an e fficient parallel implementation.

– Automation

An aspect-oriented separation of concerns has clearly demonstrated that some aspects have no reuse potential. Nevertheless, there could be an automation perspective for these aspects, by developing an automatic code generation process. For example, it could be possible to generate e code directly from the HDL description of the DUT to automate the HDL aspect generation.

5. CONCLUSION

In this work, an aspect partitioning applied to a hardware verification environment has been presented An aspect-oriented design methodology offers an efficient separation of concerns within an object-oriented environment. It has been shown that some aspects of our verification environment possess a high reuse potential. By contrast, aspects like the glue logic aspect are dedicated to a particular environment and are not reusable. A good direction to investigate would be to study how the reusable aspects could be transformed into reusable frameworks, or executable patterns, and how the creation of non-reusable aspects could be automated in a functional verification platform.

ACKNOWLEDGEMENTS

We would like to thank PMC-Sierra, Micronet, and the Natural Sciences and Engineering Research Council of Canada for their financial support. This project is made possible by the donation of a license of the Specman Elite™ software by Verisity, and benefits from the technical guidance of PMC-Sierra. Finally, several Polytechnique's students, including Vincent Rocher, Gregory Donzel, and Mame Maria Mbaye, developed the protocol converter code used in this research.

REFERENCES

[1] J. Bergeron, Writing Testbenches, Functional verification of HDL models, Kluwer Academic Publishers, 2000.

[2] P. Whittemore and G. Dearth, Object-Oriented Approach to Verification, Proceeding of SNUG99 Boston, 1999.

[3] F.I. Haque, K.A. Khan, J. Michelson, The Art of Verification with VERA, Verification Central, 2001.

[4] G. Kiczales, J. Lamping, A. Mendhekar, C. Maeda, C.V. Lopes, J. Loingtier and J. Irwin Aspect-Oriented Programming, Proceeding of ECOOP97, pages 220-242, 1997.

[5] http://www.verisity.com

[6] K. Czarnecki and U. Eisenecker. Generative Programming: Methods, Techniques, and Applications. Addison-Wesley, 2000.

[7] Y. Hollander, M. Morley, and A. Noy, The e Language: A Fresh Separation of Concerns. Proceeding of TOOLS-38, pages 42-50, 2001.

[8] S. Regimbal, J.F. Lemire, Y. Savaria, G. Bois, E.M. Aboulhamid, A. Baron, Applying Aspect-Oriented Programming to Hardware Verification with e. accepted to HDLCON 2002.

[9] E.W. Dijkstra, A Discipline of Programming. Prentice Hall, 1976.

Section 2.4

Reconfigurable Combinatorial Accelerators for Real Time Processing

I. Skliarova and A.B. Ferrari
University of Aveiro, Department of Electronics and Telecommunications, IEETA, 3810-193 Aveiro, Portugal

Abstract: Complex combinatorial problems appear in various systems providing real time processing. The majority of combinatorial problems that have to be solved are NP-complete and it limits the use of the respective mathematical models, methods and algorithms for a number of practical applications, for instance, for reactive embedded systems. The paper suggests architecture of hardware accelerator for combinatorial computations and a strategy of such computations based on a reconfigurable hardware/software (RHS) model. The resulting hardware has been implemented on the basis of XCV812E Virtex FPGA, and it can be reused for different combinatorial problems. The complexity of the FPGA allows implementing whole systems that include the proposed accelerator linked to a host application-specific embedded controller for real time processing. The paper presents and analyzes a set of experiments with the proposed accelerator on the base of three well-known combinatorial problems that are a Boolean satisfiability (SAT), a traveling salesman problem (TSP) and a covering problem. The results have shown that the proposed architecture together with the respective methods and algorithms allow to reduce the time of computations significantly.

Key words: FPGA, combinatorial accelerator, combinatorial problems.

1. INTRODUCTION

It is known that many practical problems can be solved with the aid of combinatorial computations using the following technique. At the first step the problem is converted to the respective mathematical model such as logic equation [1], discrete matrix [2], etc. At the second step some combinatorial tasks are formulated in such a way that finding a solution for the considered tasks permits to solve the given problem. This approach has been successfully used for many practical applications and we will give just a few examples. In [3] it was shown that routing of FPGA and estimation of routability can be solved through Boolean satisfiability. There are many other practical problems that can be transformed either to SAT (such as test pattern generation [4], graph coloring [5], etc.) or to another combinatorial task, such as covering [2], TSP [6], etc. For example, consider a set of products and a set of testers to verify some qualities of the products. East tester can just be used in order to check a subset of qualities. The verification has to be done as fast as possible. Thus for each particular product it is necessary to find the minimal subset of testers that can be used in order to verify all the required qualities of the product. This is a typical covering problem, which can be formulated over Boolean matrix [2].

There are many problems that have to be solved within very short period of time and they need real time processing. Many of them might require combinatorial computations and the latter as a rule are time consuming. For example, assembling of microchips and treatment of printed boards include many technological operations (for example drilling [7], soldering, etc.) that can be modeled by the TSP. Combining different operations at the same technological step might change the initial specification and the problem has to be solved during run-time. It should be noted that advances in microchip technology might require to handle very complicated combinatorial problems. For instance, the paper [7] shows that the TSP has to be solved for thousands of nodes. The use of dynamically reconfigurable hardware for reactive embedded systems might require rerouting of FPGA during execution time and if this process is based on SAT for general case it cannot be solved just with the aid of software. Thus hardware accelerators for solving various combinatorial problems are useful and efficient.

It should be noted that it is very difficult to design a universal accelerator targeted to combinatorial problems because they might require absolutely different kinds of input data and computations. For example, SAT and covering problems can be formulated over ternary and Boolean matrices that can be very efficiently handled in FPGA with distributed RAM-blocks, such as XCV812E FPGA [8]. On the other hand TSP requires lots of computations with distances that can be either integers or floating-point numbers. As a result it is very difficult to solve complex instances of TSP just in hardware and an efficiency of such approach is very questionable.

The paper presents two techniques for solving complex combinatorial problems. Both of them assume computations based on a reconfigurable hardware/software (RHS) model. The first technique can be applied to problems formulated over Boolean and ternary matrices. The second technique might be used for combinatorial problems that cannot be formulated over considered matrices but for which evolutionary algorithms (EA) can be efficiently employed.

The remainder of the paper is organized as follows. Section 2 gives an overview of the RHS model of computations. Section 3 presents architecture of a hardware accelerator for combinatorial problems formulated in terms of Boolean and ternary matrices. Section 4 provides the description of the hardware accelerator for evolutionary algorithms. Section 5 contains the results of experiments. Finally, concluding remarks are given in section 6.

2. RHS MODEL

RHS model for combinatorial computations [9] describes partitioning of the problem between reconfigurable hardware (RH) implemented in FPGA and software (S) running on host computer (microcomputer) and specifies the respective communication mechanisms.

For the proposed accelerator we have chosen a matrix representation because of two following reasons. Firstly, discrete matrices are well suited to storing and processing them with the aid of the considered hardware platform because the FPGA employed (which is XCV812E) incorporates a huge amount of distributed memory blocks that can be used in order to store matrices. The second reason is that the accelerator architecture needs to be reconfigurable, and it seems to us that matrix model is well suited for this purpose because we can reserve in an FPGA an extensive area to be used for matrices of different problem instances.

It should be noted that the area reserved for matrix data would not be sufficient for any problem instance that might be considered. That is why we have developed and successfully applied the following software/hardware strategy. It is known that a common way to organize a search process is based on construction of a decision tree [10]. Each problem is firstly formulated in terms of a discrete matrix **M**. During the search process this matrix is modified producing some intermediate matrices **M′** that correspond to various nodes of the decision tree. It is important that the matrix at the root of the tree (which actually is the initial matrix **M**) has always bigger dimensions that the matrices at other nodes of the tree. Thus when we move from the root of the tree to its leaves the initial matrix dimensions are gradually decreased, and during the backtrack process they are increased, as it is shown in fig. 1. This property plays an important role for processing matrices whose dimensions exceed the capacity of hardware platform. If the initial matrix **M** satisfies the imposed constraints, then the respective problem instance can be completely solved in hardware. In the opposite case, if **M** does not satisfy the constraints, a software application implementing the same algorithm will start to handle the problem. The software application treats the problem until it arrives at a point in which an intermediate matrix **M′** that is constructed at the current search step satisfies the constraints. At this moment the sub-problem will be transferred to the hardware. If the hardware finds a solution, the result will be dispatched to the host computer. In the opposite case, if the considered branch of the decision tree does not allow to find a solution, the control will be returned to software application. These steps will be repeated until we either get the solution or traverse the whole decision tree, and, consequently, conclude that the problem does not have any solution. Obviously, the bigger area is reserved for matrices in FPGA, the fewer times it will be necessary to activate the hardware and to transfer the data.

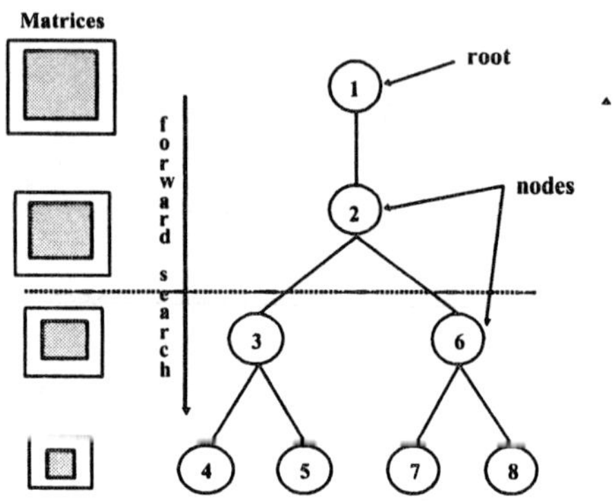

Figure 1. The search process for solving combinatorial problems formulated over discrete matrices. Matrix dimensions are gradually decreased during the motion from the root of the decision tree to its leaves, and are increased in case of a backtrack

The complexity of recent FPGAs, such as Virtex XCV812E, allows implementing a complete system on the same chip, including processor, memory and peripheral controllers. Such a system can be considered as an application-targeted computer. It can be used for example for reactive embedded systems supporting real time processing. The proposed RHS model might be implemented for systems that require solution of combinatorial tasks. In this case software part will be realized in an application-targeted computer on a chip and hardware part will be implemented in a dedicated area of the same chip that was allocated for the respective combinatorial accelerator.

3. HARDWARE ACCELERATOR FOR COMBINATORIAL PROBLEMS FORMULATED OVER BOOLEAN AND TERNARY MATRICES

The proposed architecture of an accelerator that can be used for solving combinatorial problems formulated in terms of Boolean and ternary matrices is presented in fig. 2. The primary architectural blocks for executing operations are a control unit and a reconfigurable core. Both blocks have been constructed from reprogrammable components, such as RAM-based FPGA cells. Thus modifying the contents of the respective RAM might change the functionality of any block. The datapath is controlled by a set of microprograms generated by the control unit for the considered combinatorial problem. As soon as the problem is solved, a new set of microprograms can be loaded. In order to carry out these operations during run-time dual-port RAM capabilities have been used.

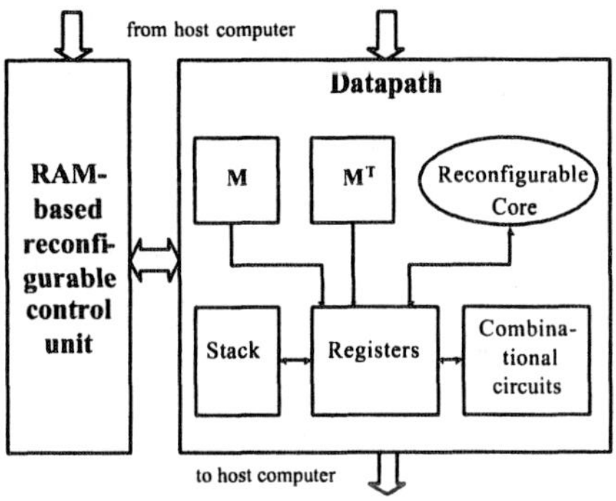

Figure 2. Accelerator architecture for solving combinatorial problems formulated over Boolean and ternary matrices

The reconfigurable core realizes different operations on discrete vectors specified by microprograms of the control unit. It should be noted that we can consider practically infinite number of operations on vectors, but on the other hand any real problem requires just a very limited number of them. Indeed, the majority of combinatorial algorithms are based on a multiple repetition of similar operations that have to be applied to different vectors of a matrix either sequentially or in parallel [11]. This defines the importance of modifications of the reconfigurable core. Since for any practical problem the number of repeated operations is usually very large, the time of reconfiguration is negligible compared to the execution time. Three following types of operations have been considered:

- bitwise binary operations, for example **a** ∨ **b**, where **a** and **b** are discrete vectors and the considered operation will be applied to all pairs of components of the vectors with the same index. The result of a bitwise binary operation is a new vector;
- operations whose result is either *yes* or *no*, for example **a** *ort* **b**?, where *ort* is an operation that checks if vectors **a** and **b** are in the relationship of orthogonality;
- counting operations, for instance, count the number of ones in a given vector.

Combinational circuits in fig. 2 are used for implementing typical operations, which can be reused for majority of combinatorial problems. The registers store the intermediate results. And finally, two memory blocks in fig. 2 keep the initial matrix **M** and its transpose $\mathbf{M}^T$. We use two memory blocks in order to accelerate accessing of columns of **M**. Matrices themselves are not changed during the search process; all possible modifications (such as deleting some rows and columns) are reflected in the registers. The stack memory is used to support the backtracking. In the proposed architecture the result of computations is always a discrete (binary or ternary) vector.

The architecture suggested is well suited to the covering and SAT problems because they can be formulated in terms of a Boolean or ternary matrix [10, 12] and the result might be presented as a discrete vector.

4. HARDWARE ACCELERATOR FOR EVOLUTIONARY ALGORITHMS

Evolutionary algorithms have been shown to be an effective approach for finding near-optimum solutions to various combinatorial problems. It should be noted that EAs successfully applied to many hard optimization problems, are very time consuming. That is why it is worth while to consider and to implement the respective hardware accelerator. However, the hardware implementation of an EA for any practical problem would be very complex and resource consuming. That is why for the considered TSP problem we analyzed which part of EA for the TSP is the most time consuming, and implemented only this critical part in FPGA.

The results of experiments with benchmarks from the TSPLIB [13] have shown that more than 50% of the total CPU time is spent performing the crossover operation. Therefore improving the efficiency of the crossover operator is crucial for improving the overall performance of the algorithm. That is why we propose realizing the crossover in an FPGA. The suggested architecture of the respective circuit is depicted in fig. 3. It includes a central control unit, which activates in the required sequence all the steps of the algorithm needed to be performed. There are four memories that are used in order to keep two parent tours and two resulting offspring. A tour is represented as a path, i.e. a city at the memory address i is visited after the city at the address i-1 and before the city at the address i+1. We realized a PMX crossover, which produces an offspring by choosing a subsequence of a tour from one parent and preserving the order and position of as many cities as possible from the other parent [14]. A subsequence of a tour that passes from a parent to an offspring is selected by picking two random cut points. The cut points used in a crossover are randomly chosen by software application and after that stored in two special registers ("First cut point" and "Second cut point" in fig. 3). The third register ("Max" in fig. 3) stores the actual length of a tour. Taking into account this value, the control unit will only force processing of the required area of parents' and offspring memories. It allows accelerating the crossover of tours of small dimensions. Four additional memories are utilized to help performing the crossover operation. These memories are “CM1”, “CM2”, “SM1” and “SM2” [15].

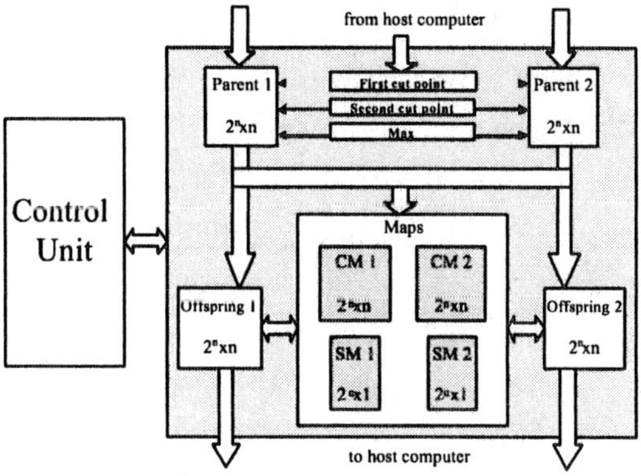

Figure 3. The proposed architecture for realizing the crossover operation

5. EXPERIMENTS

For all the experiments we used an ADM-XRC board [16] containing one XCV812E Virtex FPGA [8]. The first accelerator was analyzed with the aid of randomly generated instances of SAT and covering problems. The results of experiments are shown in Table 1. The maximum matrix dimensions in FPGA are 256x256. Since each matrix element is encoded by two -bit value, the accelerator enables us to process discrete matrices with maximum dimensions of 256x128. The hardware clock frequency was of 40 MHz.

The first column in Table 1 contains the respective instance name. The first five rows correspond to SAT problem, while the last five rows keep the results for covering problem. The second column contains the respective initial matrix dimensions. Thus, instances *sat1-sat3* and *cov1-cov3* can be completely processed by the hardware accelerator, while instances *sat4-sat5* and *cov4-cov5* need to be handled firstly by the software application implementing the same algorithm. The third column in Table 1 stores the execution time when the problem was solved in software on a PIII/800MHz/256MB running Windows2000. The fourth column keeps the execution time when the problem was solved with the aid of RHS model of computations with software part running on PIII/800MHz/256MB and a hardware part implemented in XCV812E FPGA. And finally, the last column presents the speedup obtained with RHS model compared to a pure software realization. The speedups achieved are quite impressive. This can be explained by the following reasons. Firstly, the algorithms employed require just implementing very simple operations, such as counting the number of ones and zeros in discrete vectors, checking vectors for orthogonality, etc. These operations are not well performed by conventional ALUs. The second reason is that since two copies of matrices are kept in FPGA (**M** and $\mathbf{M}^T$), accessing different rows and columns is very fast and requires just one clock cycle. In software the matrix has been constructed as an array of integers. Thus a simple reading of any row or column will require many memory accesses, which will slow down the search process.

Table 1. Experimental results for SAT and covering problems

Instance	Initial matrix dimensions	t_{sw} (ms)	t_{RHS} (ms)	Speedup
Sat1	32 x 32	3.17	0.16	124
Sat2	64 x 64	37.68	0.24	157
Sat3	128 x 128	160.19	0.83	193
Sat4	256 x 256	1993.86	5.83	342
Sat5	512 x 512	17355.4	76.12	228
Cov1	32 x 32	0.336	0.0041	82
Cov2	64 x 64	2.31	0.021	110
Cov3	128 x 128	46.08	0.72	64
cov4	256 x 256	297.4	2.36	126
cov5	512 x 512	1953.7	14.58	134

The second accelerator was analyzed with the aid of benchmarks from TSPLIB [13] that s pecify various instances of TSP problem. A comparison between software and hardware versions of the PMX crossover is presented in Table 2. The first column in Table 2 stores the number of cities in a tour. The column $\mathbf{t_{soft}}$ contains the time of performing the crossover in software. The software version was based on C++ language and executed on a PIII/800MHz/256MB running Windows2000. The column $\mathbf{t_{hard}}$ keeps the time of performing the crossover in hardware. The hardware part was executed on an XCV812E FPGA under clock frequency of 40 MHz. The cut points chosen in software and hardware versions were the same in order to facilitate the comparison. The speedup resulting from our approach is given by t_{soft}/t_{hard}. The results of experiments show that the PMX crossover operator is executed in FPGA 10-50 times faster than in software.

Table 2. Comparison of software and hardware implementations of the PMX crossover operator

Number of cities	t_{soft}(ms)	t_{hard}(ms)	Speedup
280	1.0896	0.08443	12.9
52	0.4293	0.01872	22.9
127	0.6203	0.04040	15.4
657	3.838	0.19486	19.7
51	0.2151	0.01806	11.9

Number of cities	t_{soft}(ms)	t_{hard}(ms)	Speedup
417	2.0605	0.12524	16.5
442	3.8394	0.13045	29.4
575	2.7703	0.17199	16.1
724	10.2145	0.20443	49.9

It should be noted that the time t_{hard} strongly depends on the cut points chosen. The acceleration achieved in FPGA in comparison with software implementation can be explained by the following primary reasons. Firstly, the technique of parallel processing is employed, i.e. a part of the crossover operation is executed in parallel in FPGA. Second, the memory organization in the proposed architecture is tailored to the required data sizes.

6. CONCLUSION

The paper describes two different techniques that can be used in order to accelerate combinatorial computations. The first technique permits traversing a decision tree in such a way that subtasks that exceed a predefined level of complexity are solved in software and the other subtasks are solved in reconfigurable hardware. The paper suggests the respective architecture of a reconfigurable accelerator, which is applicable to combinatorial problems that can be formulated over Boolean and ternary matrices. Two examples of such problems that are SAT and covering have been considered and the results of experiments have been presented. The second technique permits to solve combinatorial problems with the aid of EAs. An example of TSP has been presented with the respective results of experiments. It is shown that the proposed techniques allow achieving a significant speedup comparing with the pure software approaches. The distinctive feature of the proposed architectures for hardware accelerators is an extensive use of such new capabilities of recent FPGA as distributed memory blocks that are very well suited for combinatorial computations.

ACKNOWLEDGMENTS

This work was supported by the Portuguese Foundation of Science and Technology under grants No. FCT-PRAXIS XXI/BD/21353/99 and No. POSI/43140/CHS/2001.

REFERENCES

[1] A.Zakrevskij and I.Vasilkova "Reducing large systems of Boolean equations". - 4th Int. Workshop on Boolean Problems, September 21-22, 2000, Freiberg, Germany, pp. 21-28.

[2] A.Zakrevskij, "Combinatorial Problems over Logical Matrices in Logic Design and Artificial Intelligence", Electronics and Telecommunications, vol. 2, No 2, pp. 261 -268.
[3] R.G.Wood and R.A.Rutenbar, "FPGA Routing and Routability Estimation via Boolean Satisfiability", IEEE Trans. on VLSI Systems, Vol. 6 No. 2, June 1998, pp. 222-231.
[4] P.Stephan, R.K.Brayton, and A.L.Sangiovanni-Vincentelli, "Combinational test generation using satisfiability", IEEE Trans. CAD of Integrated Circuits and Systems, vol. 15, n°. 9, pp. 1167-1176, Sept. 1996.
[5] http://www.intellektik.informatik.tu-darmstadt.de/ SATLIB/benchm.html
[6] Z.Michalewicz and D.B.Fogel, "How to Solve It: Modern Heuristics". Springer, 2000.
[7] J.D.Litke, "An Improved Solution to the Traveling Salesman Problem with Thousands of Nodes", Communications of the ACM, vol. 27, No. 12, pp. 1227-1236, 1984.
[8] Xilinx, "The programmable logic data book", Xilinx, San Jose, 2000.
[9] V.Sklyarov, I.Skliarova and A.B.Ferrari, "Hierarchical Specification and Implementation of Combinatorial Algorithms Based on RHSModel", Proc. of DCIS'2001, Portugal, 2001.
[10] I.Skliarova, A.Ferrari, Design and Implementation of Reconfigurable Processor for Problems of Combinatorial Computations. Proc. of EUROMICRO Symposium on Digital Systems Design, Warsaw, Sept. 4-6, 2001, pp. 112-119.
[11] A.D.Zakrevski, "Logical synthesis of cascade networks", Moscow: Science, 1981.
[12] I.Skliarova, A.B.Ferrari, "A SAT Solver Using Software and Reconfigurable Hardware", Proc. of DATE'2002, Paris, France, Mar. 2002.
[13] http://www.informatik.uni-heidelberg.de/groups/comopt/software/TSPLIB95/index.html.
[14] D.E.Goldberg and R.Lingle, "Alleles, Loci, and the Traveling Salesman Problem", Proc. 1st Int. Conf. on Genetic Algorithms, pp. 154-159, 1985.
[15] I.Skliarova and A.B.Ferrari, "FPGA-based Implementation of Genetic Algorithm for the Traveling Salesman Problem and its Industrial Application", Proc. of the IEA/AIE'2002, Cairns, Australia, Jun. 2002.
[16] http://www.alphadata.co.uk.

Section 2.5

Tuning Methodologies for Parameterized Systems Design

Giuseppe Ascia, Vincenzo Catania, Maurizio Palesi
Dipartimento di Ingegneria Informatica e delle Telecomunicazioni, Università di Catania, ITALY

Abstract: In this paper we present some new solutions for design space exploration of parameterized systems. The approaches use multi-objective optimisation techniques based on the concept of Pareto-optimality to determine the power/performance trade-off front for a highly parametric system-on a chip for digital camera applications. The approaches used are purely heuristic, genetic algorithm-based and a combination of the two. The results obtained demonstrate the effectiveness of the approaches in terms of both validity and efficiency, measured as the number of simulations run.

Key words: Design space exploration, genetic algorithms, system-on-chip architectures, Pareto-optimal configurations.

1. INTRODUCTION

With current technology it is possble to integrate hundreds of millions of transistors on a single silicon die. This large number of transistors means that a single chip can contain a complete processing systems with high-performance CPUs, DSPs, co-processors, memories, and analog blocks that communicate via an extremely complex interconnection network. Unfortunately, the technological gap between the number of transistors that could be used and the number actually used is constantly on the increase [1].

This problem has partially been solved by the IP-based design approach that uses pre-designed and pre-verified cores as building blocks (as gates were previously used). These cores can also be supplied by third parties and can be in different forms (soft-, hard-, firm-cores) [2].

A new approach to IP-based design, called *configure-and-execute* was proposed in [3] and is based on the presence of a highly parametric pre-designed system-on-a-chip or (SOC platform) which is configured according to the application or set of applications it will have to execute.

Once the designer has chosen the SOC platform on which the application is to be mapped, he passes to the *tuning* phase to search for a parameter configuration such as to optimise an objective function that almost always depends on three variables—area, power and performance.

Research in the field of parameterized system design has led to the definition of various approaches to explore the range of configurations. In [4] sensitivity analysis was used to search for the configuration that minimises the power-delay product for a cache memory. In [5] mono-objective genetic algorithms (GAs) were used to search for optimal configurations in terms of area, power and average access time for a memory hierarchy. In [6] a system comprising a CPU, caches and main memory and the interfaces between these cores was analysed to show the power-performance trade-off for various technologies.

This paper presents three new approaches to design space exploration (DSE) for parameterized systems. The approaches use multi-objective optimisation techniques based on the concept of Pareto optimality. The first approach is an extension of sensitivity analysis [4] to multi-objective optimisation, the second uses multi-objective evolutionary programming techniques, and the third is a combination of the previous two. The results obtained in a case study (a highly parametric SOC for digital camera applications) show the effectiveness of using evolutionary programming techniques for DSE in terms of both accuracy and efficiency, measured as the number of simulations run.

The paper is structured as follows. In Section 2 the problem will be stated and a survey of multi-objective optimisation techniques based on genetic algorithms will be given. In Section 3 three new approaches to design space exploration will be presented and then applied to determine the power/performance trade-off for a highly parametric system described in Section 4 together with the results obtained. Finally, Section 5 provides our conclusions and indications as to future developments.

2. PROBLEM FORMULATION

Let there be a parametric system S in which each of the N parameters p_i can take a value belonging to a finite set $V^{(i)}$. We will call a configuration of the system S an N-tuple $(v_1,v_2,\ldots,v_N)$ with $v_i \in V^{(i)}$ such that $p_1=v_1$, $p_2=v_2$ and so on. We will indicate the space of configurations of the system S, i.e. all the configurations the system can possibly have, as $C(S)$. This set will correspond to the cartesian product of the sets of variation of each system parameter and its size will therefore be

$\prod_{i=1}^{N} V^{(i)}$. Let there be M cost functio`ns f_i:$C(S)\rightarrow R$ (i=1,2,...,M) which associate a real value with each possible configuration of the system. The problem is to find an efficient way to search for the Pareto-optimal configurations given the cost functions f_i.

Problems of this kind are particularly felt in VLSI design, especially that of systems-on-chips (SOCs). A system involving the interconnection of several parametric, heterogeneous functional blocks has to be configured in such a way as to find the Pareto-optimal configurations which are typically in terms of area, power and execution time. Beside the basic problem of the enormous size of the space of possible system configurations, another great problem is that of evaluating the costs functions for a given configuration. This is a computationally onerous task, especially for complex systems like SOCs. First of all, in fact, it requires the system to be configured according to the given configuration, then execution of the application, and finally measurement (or estimation) of the variables of interest. It is therefore clear that a technique to search for the Pareto-optimal configurations of a parameterised SOC based on an exhaustive approach is unfeasible.

The use of heuristic techniques can help to reduce the space of configurations that have to be analysed by identifying and discarding any Pareto-dominated [1] configurations. Another approach presented in [5] proposes the use of genetic algorithms as an efficient technique for DSE.

Both techniques reduce the problem of multi-objective optimisation to one of scalar optimisation by aggregation of the objective functions [7],[8]. The main disadvantage to aggregation functions is that they do not generate proper Pareto-optimal solutions in the presence of non-convex search spaces, which is a serious drawback in most real-world applications. These problems are solved by Pareto-based approaches that select Pareto non-dominated individuals from the rest of the population. These individuals are then assigned the hightest rank and eliminated from further contention. Another set of Pareto non-dominated individuals are determined from the remaning population and are assigned the next hightest rank. The procedure is repeated until the whole population is suitably ranked.

3. NEW MULTI-OBJECTIVE APPROACHES FOR SYSTEM LEVEL EXPLORATION

In the following subsections, the approaches based on sensitivity analysis and GAs will be extended to conduct a proper multi-criteria analysis of the notion of Pareto optimum.

Pareto-Based Sensitivity Analysis

The sensitivity analysis (SA) presented in [4] reduces the space of possible configurations in two phases. The first phase identifies the parameters which most influence the objective function to be optimised (sensitivity analysis phase (SAP)). For a system with P parameters, determination of the degree of sensitivity of each parameter consists of fixing $P-1$ parameters and varying one of them, determining the maximum range of variation of the objective finction. One way to fix the parameters is to consider the mean value of their variation set.

The next phase, design space exploration phase (DSEP), identifies the optimal value for each parameter, from the most to the least sensitive. If $V^{(i)}=\{v_1^{(i)},...,v_{Ni}^{(i)}\}$ is the set of values the parameter p_i can take, the number of configurations to be evaluated goes down from $\prod_{i=1}^{P} N_i$ to $\sum_{i=1}^{P} N_i$.

To understand how this approach works, let us consider the following example. Let us assume that we want to find the configuration of a cache memory in terms of size (s), block size (b) and associativity (a) that minimises the power-delay product (pd). Let S, B and A respectively be the set of possible values of the parameters s, b and a. The SAP is performed as follows: we fix $b=b_0$ and $a=a_0$ and s is made to vary in S, obtaining the set $PD=\{PD_1,PD_2...,PD_{|S|}\}$ of values of pd. If pd_{max}=max(PD) and pd_{min}=min(PD), the sensitivity of the parameter s is $s_S=pd_{max} - pd_{min}$. The same procedure is repeated for the remaining parameters s_B and s_A.

Under the hypothesis that $s_S>s_A>s_B$, we set $a=a_0$, $b=b_0$ and vary $s\in S$, obtaining $PD=\{PD_1,..,PD_{|S|}\}$ values of pd. If s_{opt}=min(PD), we fix $s=s_{opt}$ and $b=b_0$ and vary $a\in A$. Proceeding as previously, a_{opt} is determined. In short, having fixed $s=s_{opt}$, $a=a_{opt}$ and varying $b\in B$ we determine b_{opt}. The configuration $\langle s_{opt}, b_{opt}, a_{opt}\rangle$ will determine a pd value close to pd_{min}. Algorithm 1 shows the generalization of the DSEP assuming all parameters sorted by decreasing values of sensitivity.

Algorithm 1: Monono-objective sensitivity analysis (DSEP)
for all i$\in$ {1,2,...,P}
  for all j$\in$ {1,2,...,i-1}
    $v_j = v_j^{opt}$
  end for
 for all j$\in$ {i+1,i+2,...,P}
    $v_j = v_{j0}$
 end for
 $v_i^{opt} = v_1^{(i)}$
  for all j$\in$ {2,3,...,N_i}
    if $f(v_1,..., v_j^{(i)},...,v_P) < f(v_1,..., v_i^{(opt)},..., v_P)$
      $v_i^{(opt)} = v_j^{(i)}$
    end if
  end for
end for

To overcome the limits of a mono-objective approach we extended the technique based on sensitivity analysis to perform multi-objective optimisation based on the notion of Pareto optimum (PBSA).

Defining a new metric to measure the sensitivity of a parameter modified the SAP. We have defined the sensitivity s_i of the i-th parameter as:

$$s_i = \max_{h,k \subset \{1,2,\dots,N_i\}} dist(\underline{o}_h^{(i)}, \underline{o}_k^{(i)})$$

where dist is the Euclidean distance and $\underline{o}_j^{(i)}$ are the N_i points in the n-dimensional space obtained by fixing $P-1$ parameters and varying $p_i \in V^{(i)}$.

Indicating with S_i a parameter order by decreasing degrees of sensitivity, i.e. such that $s_{Si} > s_{Si+1}$, we defined the DSEP as follows.

Having fixed $p_{S2},\dots, p_{SP}$ parameters, $p_{S1} \in V^{(S1)}$ is made to vary. From the N_{S1} points obtained, whose components represent the objective values, the non-dominated configurations are extracted and accumulated in a set P. At the second iteration for each configuration in the set P, $p_{S2} \in V^{(S2)}$ is made to vary. From the $N_{S1} \times |P|$ obtained, the non-dominated configurations are extracted and accumulated in the set P. The procedure is repeated for all the parameters p_{Si} with $i<v$, where v is the first index such that $s_{Si} < T$ and T is a *sensitivity threshold*. Parameters whose sensitivity is less than T influence in a limited way the value of the objective functions. So they are fixed to a reference value. At the end of the algorithm the configurations in P will represent the trade-off surface identified. The Algorithm 2 gives the pseudo-code of the DSE procedure.

Algorithm 2: Pareto-based objective sensitivity analysis (PBSA)
Require $S_1, S_2, \dots, S_m$ //sorted by sensitivity parameter's index
ND = $\underline{p}^*$ // initialize non-dominated set
i = 1 // high sensitive parameter index
Repeat
  C={}
  for all $\underline{c} \in$ ND
    for all $v \in V^{(Si)}$
      $\underline{c}[S_i]$=v
      C = C ∪ $\underline{c}$
  end for
  end for
  ND =ND ∪ C
  ND = ND \ Dominated(ND) //*remove dominated solutions*
  i = i + 1 // next high sensitive parameter

constraints (eg. power < x. area<y

until (i>P **OR** Sensitivity(S_i) < T)

Pareto-based Genetic Algorithms

In this paper we consider multiobjective optimisation techniques that use Pareto-based GAs. More specifically, we chose the *Strength Pareto EA* (SPEA) [9] approach, which is very effective in sampling from along the entire Pareto-optimal front and distributing the solutions generated over the trade-off surface. The flow proposed is shown in *Figure 1* An initial population of configurations is simulated to obtain an estimate of the parameters to be optimised. Together with the constraint specifications, these parameters will be used by the genetic exploration algorithm to generate the next population to be evaluated. The cycle is repeated until a stop condition is met and Pareto-optimal solutions are provided.

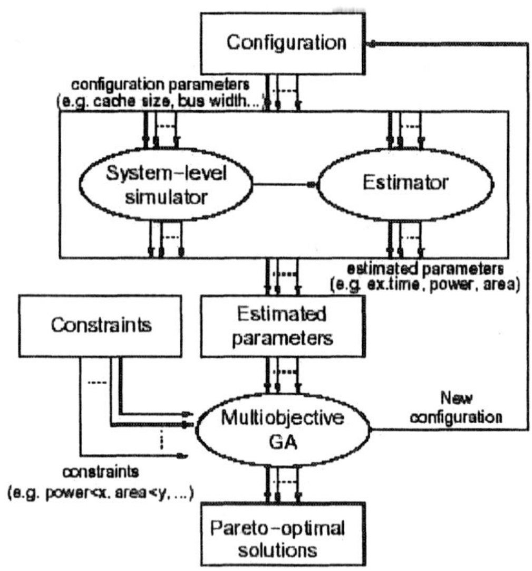

Figure 1 . GA Design flow.

The algorithm was implemented using GAlib [10] (a C++ library of genetic algorithm components). A configuration is represented by an individual of the population whose genome defines its parameters. Each gene represents a system parameter (defined by means of an allele) that only codes values defined within the range that is admissible for the parameter involved. Impossible configurations were excluded by using the approach classified in [11] as rejection of unfeasible individuals.

A Mixed Approach

A mixed approach can be used to exploit the potential of the two approaches presented in the previous subsections. The advantage of sensitivity analysis lies in its reduction of the configuration space by neglecting parameters that have less effect on the objective functions. The genetic approach, on the other hand, allows for rapid exploration of the space of configurations. We could therefore envisage defining the chromosome, using only the most sensitive parameters determined by the sensitivity analysis, to define a new mixed approach which we will call SAGA (Sensitivity Analysis Genetic Algorithm).

Obviously the trade-off front obtained by SAGA will not be better than that obtained by the pure GA approach, given that the space of configurations on which SAGA operates is a subspace of the space on which the GA approach operates. As compared with PBSA, on the other hand, the solutions found will be better, as the parameter tuning is not constrained: any combination of parameter values is admissible.

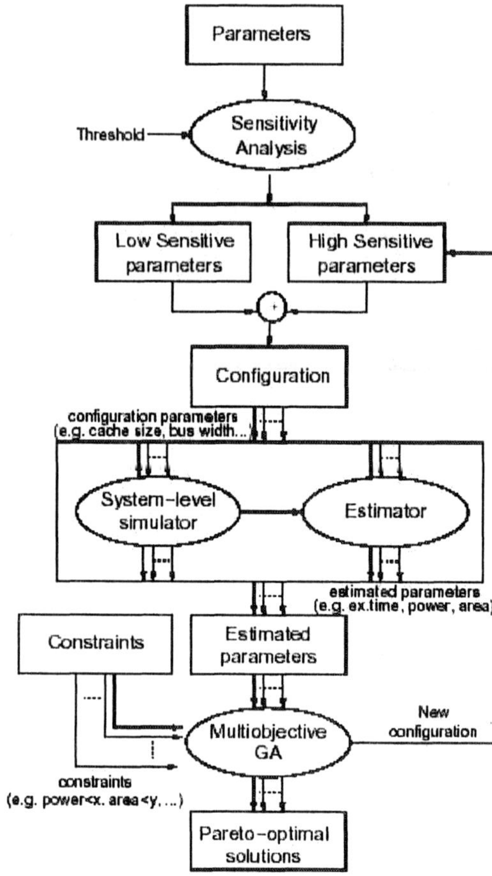

Figure2. Mixed approach design flow.

The flow of operations performed by SAGA, *Figure 2* is divided into two stages. In the first, the sensitivity analysis is performed and the parameters that exceed a certain sensitivity threshold are determined. These parameters will define the chromosome of the genetic algorithm that will be applied in the second stage. The remaining parameters that remain below the threshold are set to a reference value (e.g. the mean value of their range of variation).

4. EXPERIMENTAL RESULTS

In this section we will first describe the parametric reference architecture and then apply the approaches described in the previous section.

Reference Architecture

To test the methodologies proposed in Section 3 we used the architecture shown in *Figure*. It is a highly parametric SOC for digital camera applications developed under the Dalton Project [12]. The project is an open source one and comprises a parameterized simulation model of a system-on-a-chip composed of an MIPS R3000 processor core, instruction cache (I$), data cache (D$), memory, MIPS to instruction cache bus, MIPS to data cache bus, instruction/data cache to memory bus, bus bridge, peripheral bus, uart and codec. Each core is parametric. For each bus (data bus or address bus) it is possible to configure the number of lines and the encoding scheme to minimise the switching activity. The caches can be configured in size, line size and associativity. For the UART it is possible to define the transmission and reception buffer sizes, and for the JPEG Codec the pixel width can be varied. In all there are 26 separate parameters, giving a total of 9.7×10^{15} possible configurations.

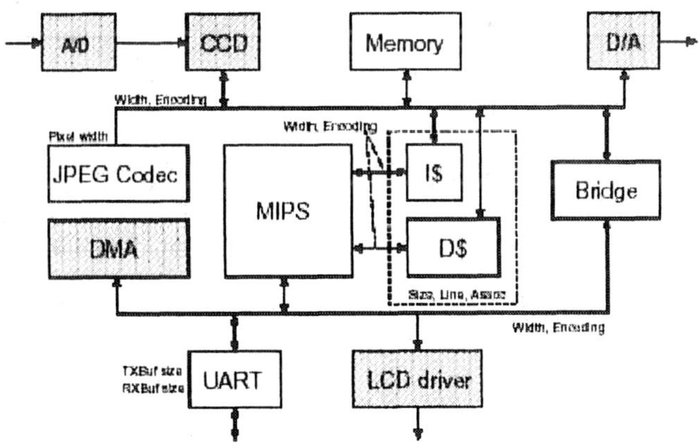

Figure 3 Reference architecture.

There are two versions of the system: both a synthesisable VHDL version and a high-level model written in C++. With this model it is possible to perform rapid simulations of the system when it is executing an applications, as well as estimating the execution time and power consumption by using the estimation model described in [13].

Experiments

The three approaches were compared considering three different applications. The first, *image*, copies a bitmap from one memory region to another. The second, *key*, works on large-size matrices. The third, *matrix*, performs arithmetical operations on two 10x10 matrices of integers.

Figure gives the power/execution-time trade-off for the *image* application, obtained using the three approaches. PBSA was executed with different threshold values. As was to be expected, when the threshold decreases more parameters are taken into consideration, thus leading to an improvement in the solutions obtained but an increase in the number of simulations required (respectively 1780, 4958 and 8633 for thresholds of 10%, 5% and 1%). SAGA gives the same results as PBSA with a 1% threshold but after only 30 generations with internal and external populations of 50 individuals, and a total of 2238 simulations. With GA, after 50 generations with internal and external populations of 50 individuals, a total of 4581 simulations gives dominant solutions as compared to those obtained with PBSA and SAGA.

The same qualitative results are obtained in the other two applications -- *key* and *matrix* and *Table 1* gives the number of simulations run by each approach and the percent gain over PBSA. The results were obtained using a 1% threshold for both PBSA and SAGA. For SAGA and GA we used an internal and external population of 50, a crossover probability of 0.9 and a mutation probability of 0.01. GA and SAGA respectively converged after 50 and 30 generations for *image* and *key* and after 40 and 20 generations for *matrix*.

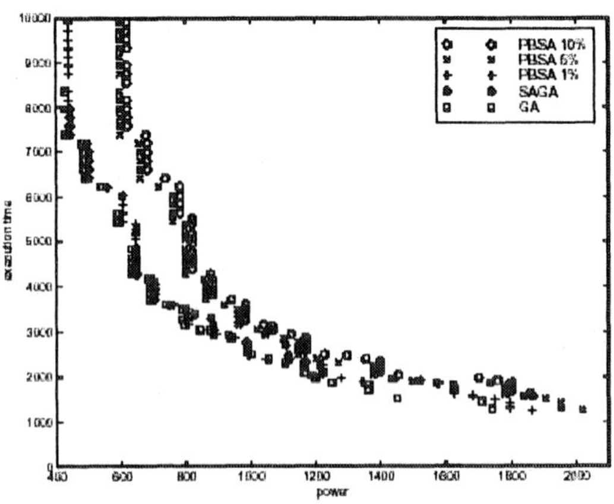

Figure 4. Power/execution time trade-off for the *image* application

Benchmark	PBSA	GA		SAGA	
	simulations	Simulations	% saving	simulations	% saving
Image	8633	4581	47.5	2238	74.1
Key	11019	4587	58.4	2642	76
Matrix	4372	3751	14.2	1714	60.8
Total	24024	2919	46.2	6594	72.6

Table1. Number of simulations run by each approach and the percent gain over PBSA.

In short, the solutions obtained with GA dominate those obtained with PBSA and are achieved with on average 46% fewer simulations. If we are willing to sacrifice the quality of the solutions to obtain an increase in efficiency, we can use the mixed SAGA approach, which gives solutions close to those obtained with GA but with 72% fewer simulations than PBSA.

5. CONCLUSIONS

In this paper we have presented three approaches to design space exploration for parameterized systems. The first approach is heuristic and is based on sensitivity analysis to reduce the space of configurations to be explored. The second uses multi-objective genetic algorithms, while the third is a mixture of the previous two, using sensitivity analysis to limit the space of configurations and genetic algorithms to explore the subspace thus obtained. All three approaches were applied to determine the power/performance trade-off of a highly parametric architecture implementing a SOC for digital camera applications during the execution of different applications. The approaches were evaluated in terms of both the quality of the solutions obtained (using the concept of Pareto dominance) and efficiency, measured as the number of simulations required to determine the trade-off front. The results obtained show the effectiveness of the pure genetic approach as regards the quality of the solutions obtained, and the mixed approach in terms of efficiency.

REFERENCES

[1] International technology roadmap for semiconductors", SIA, 1999.
[2] R. Rajsuman, System-on-a-Chip Design and Test, Artech House, 2000.
[3] F. Vahid and T. Givargis, "The case for a configure-and-execute paradigm" Int. Workshop on Hardware/Software Codesign (CODES), May 1999, pp. 59-63.
[4] W. Fornaciari, D. Sciuto, C. Silvano, and V. Zaccaria, "A design framework to efficiently explore energy-delay tradeoffs", 9th. International Symposium on Hardware/Software Co-Design, Copenhagen, Denmark, Apr. 25-27 2001, pp. 260-265.

[5] G. Ascia, V. Catania, and M. Palesi, "Parameterized system design based on genetic algorithms", 9th. International Symposium on Hardware/Software Co-Design, Copenhagen, Denmark, Apr. 25-27 2001, pp. 177-182.
[6] T. Givargis, J. Henkel, and F. Vahid, "Interface and cache power exploration for core-based embedded system design", Int. Conference on Computer-Aided Design (ICCAD), Nov. 1999, pp. 270--273.
[7] D. Van Veldhuizen and G. B. Lamont, "Multiobjective evolutionary algorithms: Analyzing the state-of-the-art", Evolutionary Computation, vol. 8, no. 2, pp. 125-147, 2000.
[8] J. D. Schaffer, "Multiple objective optimization with vector evaluated genetic algorithms", First International Conference on Genetic Algorithms, 1985, pp. 93-100.
[9] E. Zitzler and L. Thiele, "Multiobjective evolutionary algorithms: A comparative case study and the strength pareto approach", IEEE transactions on Evolutionary Computation, vol. 4, no. 3, pp. 257--271, Nov. 1999.
[10] M. Wall, GAlib: A C++ Library of Genetic Algorithm Components, Mechanical Engineering Department, Massachusetts Institute of Technology, Aug. 1996.
[11] C. A. Coello Coello, "Treating constraints as objectives for single-objective evolutionary optimisation", Tech. Rep., Laboratorio Nacional de Informatica Avanzada, Rebsamen 80, Xalapa, Veracruz 91090, Mexico, 2000.
[12] "The UCR Dalton Project IP-Based Embedded System Design", http://www.cs.ucr.edu/~dalton/.
[13] T. Givargis, F. Vahid, and J. Henkel, "A hybrid approach for core-based system-level power modelling", Asia and South Pacific Design Automation Conference, 2000.

3

TEST AND VERIFICATION

Verification and test is a critical component in the SOC design process. With SOC level complexities, the problems of verifying designs and then testing the designs is formidable. The four sections of this chapter deal with various issues associated with test and verification of SOC complexity designs.

In section 3.1 we start with an overview of formal verification in "Formal Verifications of Systems on Chips: Current and Future Directions", by Khaled M. Elleithy of the University of Bridgeport, CT, USA. Formal verification is a departure from the traditional 'exhaustive simulation' approach and this survey work deals with a) design of formal languages; b) design and implementation of formal frameworks for verification at different architectural levels and c) design and implementation of different cell class libraries that support formal verification. The survey examines different approaches using formal logic (first order logic, high order logic, temporal logic), and a case study of an object-oriented paradigm is presented. The section concludes with a brief discussion on future directions.

Following on from this survey, in section 3.2, "A Practical Approach to the Formal Verification of SoC's with Symbolic Model-Checking" by Emil Dumitrerscu of the TIMA Laboratory, Grenoble, France, addresses the issues of defining a minimal set of efficiency criteria. The work is illustrated by the linear-time model-checking verification of an instruction cache controller designed by ST Microelectronics. The set of model simplification guidelines presented in this section allow a reasonable functional coverage for the verification of the cache controller design. All the functional blocks have been addressed, despite the lack of hiera rchy in the RTL description of the design. The final conclusion is that, given the available tools, it is essential that verification engineers be aware of the implementation details of the design they verify. The blind application of automatic model-checking fails in most cases, and knowledge about the underlying theory of this technique is vital.

The third section of this chapter, "High Performance Verification Solutions for SOC Designs", by Makarand Joshi and Ashok Kumar of Quickturn, a Cadence Company, introduces a set of integrated, processor-based simulation acceleration solutions for verification. Hardware accelerators (Palladium and CoBALT), feature software-like compile times, scalable gate capacities, and functional congruence across all engines with a flexible upgrade path on the performance curve. Processor-based hardware acceleration engines allow users to verify large designs in unprecedentedly fast turn-around times. In SOC design workspaces, verification strategies are instrumental in determining the success of design projects, and simulation acceleration solutions enable designers to perform more verification earlier in the design cycle.

Finally, in section 3.4, "Novel Test Methodologies for SoC/IP Design: Implementation and Comparison", by L. Hong, M. Nahvi, R. Fung, A. Ivanov and R. Saleh, University of British Columbia, Canada, issues associated with System on Chip designs are addressed. The authors describe a network-processing engine SoC, designed using reusable Intellectual Property (IP) and a test harness based on the emerging IEEE P1500 standard. Three different test access mechanisms (TAMs) are compared on this platform – a serial threading approach analogous to the IEEE1149.1 standard, a test-rail approach, and a packet-switching test network (NIMA). The NIMA architecture is the most effective when there is a limited number of I/O pins and could potentially become the fastest architecture when it is allowed to operate at higher clock speed than the core DFT. However, the test rail approach is the most effective of the three TAMs if the test width is not restricted or if the area overhead is critical.

Section 3.1

Formal Verifications of Systems on Chips:

Current and Future Directions

Khaled M. Elleithy
Department of Computer Science and Engineering
University of Bridgeport
Bridgeport, CT 06611

Abstract: In this tutorial we present the area of formal verification of systems on Chips. The paper discuses the following topics: different approaches of formal logic such as first order bgic, high order logic, temporal logic. A case study of object-oriented paradigm is presented. A survey of the current research status is presented. The paper concludes with a section on the future directions.

Keywords: Formal verification, first order logic, high order logic, temporal logic, automated theorem provers, object oriented paradigm

1. INTRODUCTION

The design process is a transformation between different specifications (Figure 1). An input algorithm may be specified using a specific algorithmic specification language. Architecture may be specified using a realization specification language. The role of the design process can be viewed as a transformation between the algorithm specification language and the realization specification language. The objective of any design procedure is to produce an architecture that correctly implements the required behavior subject to a given set of constraints on area and timing. It is very expensive to fabricate a design before verifying the functional correctness of the design. There are two approaches for verification; simulation and formal verification. Simulation is efficient with small size architectures where it is possible to exhaustively run the simulator. Formal verification is suitable for large size architectures.

A verification methodology is formal if it satisfies the following characteristics [1]: There is a formal framework to describe the architecture.

There is a formal technique to prove the equivalency of the implementation and the specifications with out physically construct or simulate the design. It is possible to manipulate and study the design's performance without the physical implementation.

The heart of any formal verification methodology, then, is the availability of a formal specification language where formal proofs can be driven. Logic is one of the most widely used specification languages for verification. First order logic has been used in a number of systems [2-4]. Higher order logic has been used in a number of applications [5-8]. Windely [7] and Joyce[8] used the HOL system to verify a microprocessor. Temporal logic is an appropriate approach for specifying timing characteristics of a design. Temporal logic has been successfully used for verification in [9-10]. Production systems have been used in formal verification [11].

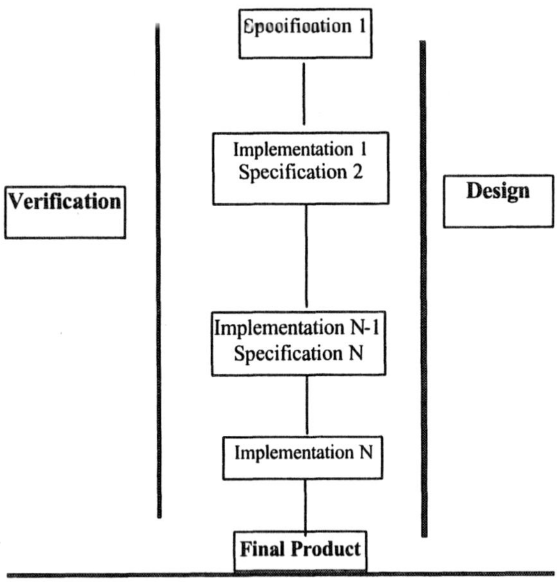

Figure 1: A Hierarchical Representation of Design and Verification.

Object Oriented Paradigm (OOP) has been proven successful in software engineering due to its reusability, which increases design productivity. Object oriented techniques provide a number of features that fits hardware verification. In [12], Nebel argued the suitability of object-oriented techniques in hardware system design. Several proposals were introduced to adopt OOP in hardware description languages such as VHDL [13-15]. In [16], Schumacher discussed the problems in theses approaches.

2. FORMAL VERIFICATION

With the current advances in Very Large Scale Integration (VLSI) it is impossible to produce correct designs using ad-hoc methods. It is essential to detect as many errors as possible before any fabrication stage. There are two approaches for verification; simulation and formal verification. The following steps are considered in any simulation based verification approach: (a) modeling the design using a suitable language, (b) testing input stimuli generation, (c) running the simulator, (d) extracting results from simulation runs, (e) comparing extracted and expected data, (f) redesigning circuit/system. The first two steps of the simulation are very critical to the validity of the simulation. If the modeling of the design is not correct or the input test patterns are not exhaustive, then there is no proof that the verification process is complete.

In formal verification a mathematical framework is used to prove that the specification is an isomorphism of the design. Two approaches may be used in formal verification: *total* verification and *partial* verification. In total verification all *safety* and *liveness* properties of the system are verified. Safety properties are used to assure that undesired properties will never occur. Liveness properties imply that desirable properties will occur in the future. In partial verification, some of *safety* and *liveness* properties of the system are verified.

Research in the formal hardware verification area started in the early eighties. The first approaches have been adopted from software verification techniques. For example Floyd's assertion method for program verification has been modified for hardware verification by Shostak[17]. In this section we are not trying to give a comprehensive study of available approaches, rather some of the pioneering and most important works.

3. FIRST ORDER LOGIC

In first order logic, thc following connectives are used:
negation ($\sim$),
conjunction ($\cap$),
disjunction ($\cup$),
and then ($\Rightarrow$).
The existential ($\exists$) and universal ($\forall$) quantifiers are used.

First order logic has been used in a number of systems[2-4]. In [2] a system called DDL verifier developed in Fujitsu Laboratories has been applied to synchronous systems. In [3] Fensel discussed knowledge acquisition and representation. . The Booyer-Moore logic has been used by Hunt[4] of the University of Te xas at Austin to verify the FM8501 microprocessor.

4. HIGHER ORDER LOGIC

In higher order logic the domains of variables range over functions and predicates, and functions can take functions as their arguments.

Higher order logic has been used in a number of applications [5-8]. Hanna introduced a system called Vertias[5]. In Vertias low-level timing issues can be taken into considerations. Gorden introduced a system called HOL which is used in [6] for automatically analyzing cryptographic protocols. Examples to illustrate specification and verification techniques include a CMOS inverter, a CMOS full adder, an n-bit ripple carry adder, and a sequential multiplier have been given by Gordon. Windley [7] and Joyce [8] used the HOL system to verify a microprocessor originally verified using the LCF-LSM system by Gordon. Joyce described several related examples of hardware verification based on Gordon's computer and other microprocessor examples.

5. TEMPORAL LOGIC

Temporal logic is used to specify properties of all possble execution times that may evolve from the present state. The following operators are defined:

Henceforth(◊): means that an assertion is true in the present state and all future states.

Eventually(∇): means that an assertion will be true at some state in the future (possibly the current) state, but not necessarily remains true.

Next(O): means that an assertion will be true in the next instant of time. This operator introduces the idea of discrete time and a transition that occurs between subsequent time instants.

While(Ξ): a statement like A Ξ B means that if B is true at the present, then A is true at the present and remains true as long as B remains true.

Temporal operators are not required in the verification of combinational circuits since there is no time dependency in this case. Memory elements have two aspects, setting of a new value and keeping it for some period. While combinational circuits and memory elements are considered passive components, processes are active components that initiate system changes. Safeness properties of processes can be expressed using predicate logic and *While* operator. Liveness assertions can be expressed using the ◊ and ∇operators.

The first attempt to use temporal logic for specification and verification has been given by Bochmann [9]. Bochmann used a linear and discrete time model to verify an arbiter. Fujita used linear temporal logic and prolog for hardware verification [10]. This system is applicable to finite state machines verification. Clarke used a branching temporal logic approach for verification of synchronous and asynchronous control paths of digital systems[18]. Clarke presented a language called Compositional State Machine Language (CSML). CSML is intended to allow modular design of finite state controllers. Moszkowski introduced an approach based on the interval temporal logic (ITL) for hardware verification [19]. Moszkowski has adopted a subset of ITL as the programming language Tempura [20].

6. AUTOMATED THEOREM PROVERS

In the case of large-scale devices, e.g., microprocessors, the applicability of automated theorem provers to prove the mathematical theorem of equivalency between design and specification is of considerable interest. The formal theory developed by Booyer and Moore [21] is one of the early pioneering works. The HOL introduced by Gordon is based on the Boyer-Moore logic. The theorem prover has been used by Hunt [4] to verify the hypothetical microprocessor, the FM8501. The FM8501 is similar in size and complexity to a PDP-11. Its design is described at the gate level, and its specification at the instruction set level.

The Royal Signals and Radar Establishment (RSRE) of the British Ministry of Defense applied formal verification techniques to design a microprocessor suitable for real-time control of safety-critical systems. The Viper microprocessor has been specified, designed and verified by RSRE using automated theorem provers [22]. The industrial manufacturers of the VIPER have endorsed the formal methods developed by RSRE.

German [23] has developed a technique for proving functional properties of parameterized hardware designs. Examples include recursively defined designs and synchronous systolic arrays. Chandrasekhar [24] used a hardware verification approach based on the term rewriting techniques. This system is capable of proving the correctness of combinational circuits.

7. OBJECT ORIENTED PARADIGM

Object Oriented Paradigm (OOP) has been proven successful in software engineering due to its reusability, which increases design productivity. Object oriented techniques provide a number of features that fits hardware verification. In [12], Nebel argued the suitability of object-oriented techniques in hardware system design. Several proposals where introduced to adopt OOP in hardware description languages such as VHDL[13-15]. In [16], Schumacher discussed the problems in theses approaches.

The primary characteristics of any object-oriented framework are:

abstraction: is a higher level, more intuitive representation for a complex concept;

encapsulation: is the process whereby the implementation details of an object are masked by a well-defined external interface;

inheritance: where classes may be described in terms of other classes by use of inheritance;

polymorphism: is the ability of different objects to respond to the same message in a specialized manner; and

dynamic binding: is the ability to defer the selection of which specific message-handlers will be called for a message until run-time.

Hoover is hardware verification tool based on OOP. In HOOVER, a Cell Class Library is build out of the description of hardware components. These components are organized in a hierarchy that allows **inheritance** of common attributes between different components. The behavior description is the only accessible attribute of these components (i.e. **encapsulation**). The hierarchy structure of the Cell Class Library allows the component of common attributes to be on the top level (i.e. **abstraction**). Dealing with these components description would be only through messages. The same message may be passed to two different components that results two different responses (i.e. **polymorphism**).

8. CASE STUDY

HOOVER Environment

HOOVER is hardware object oriented verification. HOOVER gets its inputs as a circuit description and the behavioral description of that circuit. The circuit description would be one or a combination of different hardware descriptions. These descriptions include transistors, gates, logical, functional, and module descriptions. HOOVER has a class hierarchy contains Cell Class Library. The Cell Class Library contains a predefined set of hardware components. It consists of five subclass libraries represent the five level of hardware descriptions. These subclass libraries are Transistor-level Class Library (**TCL**), Gate-level Class Library (**GCL**), Logic-level Class Library (**LCL**), Function-level Class Library (**FCL**), Module-level Class Library (**MCL**). The block diagram of HOOVER is shown in Figure 2.

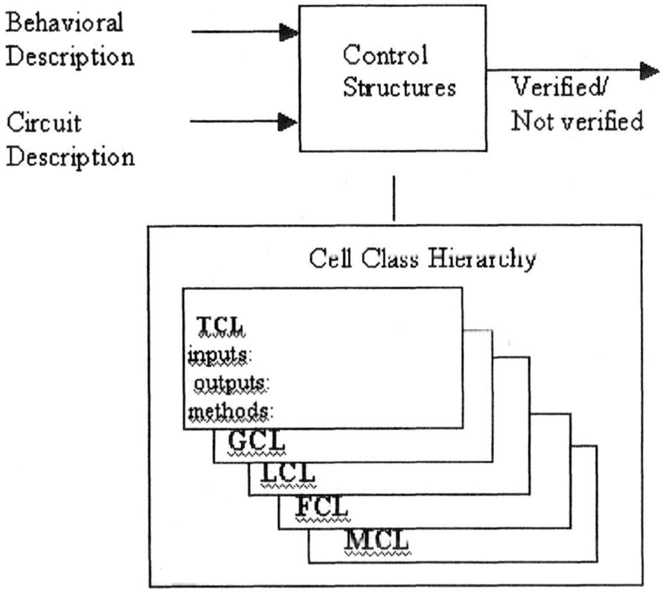

Figure 2: The Block Diagram of HOOVER

Each cell class contains several properties describe its inputs, outputs and behavior. The behavior of the cells is described through methods that may be inherited to (or override by) the sub cells. Some examples of these classes are shown in the following subsection.

Examples of Classes

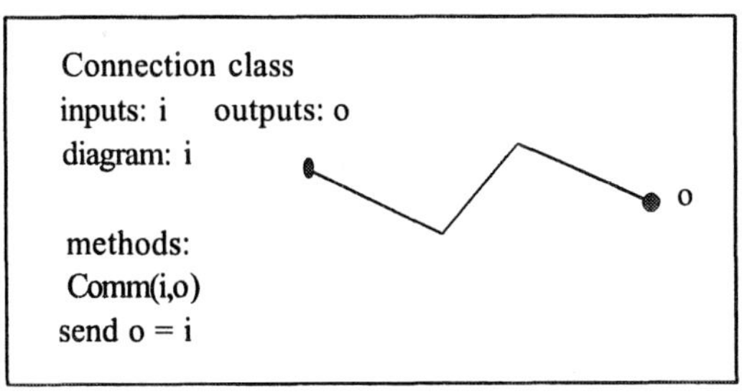

Transistor class
inputs: b, s outputs: d
diagram:

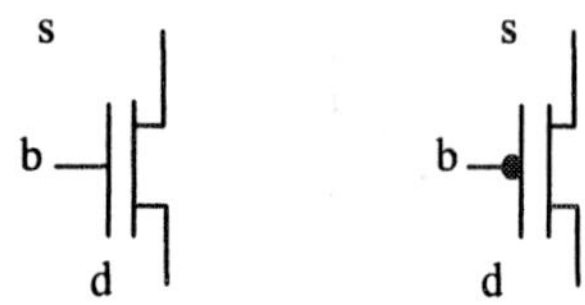

methods:
Trans(**b**,**s**,**d**)

send **d** depends on the logical level of the base **b** and **s**

An n-type transistor is an instance of the transistor class. The method is overridden by:

NTrans(**b**,**s**,**d**)
send **d** = **s** if logical level of the base **b** is 0

An p-type transistor is an instance of the transistor class. The method is overridden by:

PTrans(**b**,**s**,**d**)
send **d** = **s** if logical level of the base **b** is 1

Inverter class
inputs: i outputs: o
diagram:

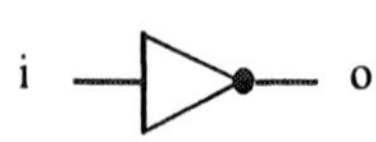

methods:
Inv(**i**,**o**)
send **o** = not **i**

OR class
inputs: $i_1, i_2, \ldots i_n$ outputs: o
diagram

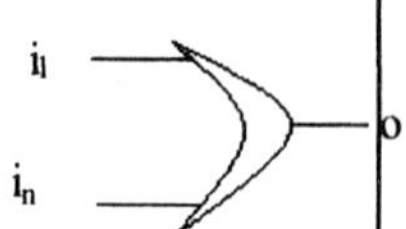

methods:

$Or(i_1, i_2, \ldots i_n, o)$
send $o = i_1 \vee i_2 \vee \ldots \vee i_n$

AND class
inputs: $i_1, i_2, \ldots i_n$ output: o
diagram:

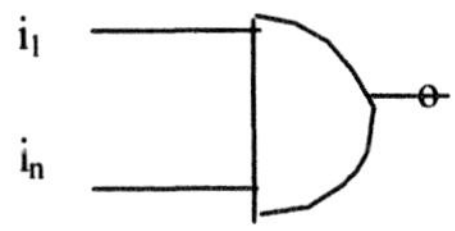

methods:
$And(i_1, i_2, \ldots i_n, o)$
send $o = i_1 \wedge i_2 \wedge \ldots \wedge i_n$

The cell class library is a hierarchy structure. The top class represents a simple hardware component; connection class. In the same cell library, there exists a different transistor level component. As we go down the hierarchy, more specific hardware components are described. These components may inherit some characteristic from the upper ones.

EXAMPLES

Several circuit examples are used as input to HOOVER. These examples include transistor/gate circuit (e.g. inverter), transistor/logic circuit (e.g. 1-bit full adder) and logic/function circuit (n-bit full adder). Here, two examples are presented. The first one is a CMOS inverter. The second one is a 1-bit full adder.

Example 1:
A CMOS inverter consists of power, ground, p-transistor and n-transistor components as shown in Figure 3.

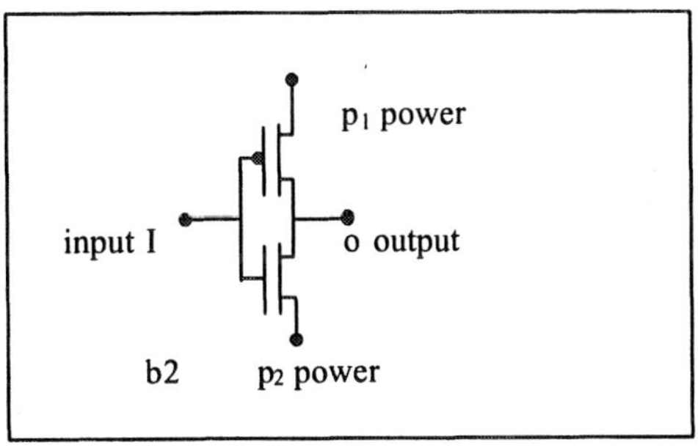

Figure 3: A CMOS Inverter

The circuit description of the inverter is described as a set of instances from the exiting classes.

$Ntransistor_1 (b_1, p_1, o)$
$Ptransistor_1 (b_2, p_2, o)$
$Connection_1 (i, b_1)$
$Connection_2 (i, b_2)$

The behavioral description of the inverter is as follows.

Output = invert (input)

For i = 0, HOOVER sends 0 to $Connection_1$ and $Connection_2$. Their methods Comm (i, b_1) and Comm (i, b_2) send b_1 and b_2 equal 0 to $Ntransistor_1$ and $Ptransistor_1$. The method of $Ntransistor_1$ Ntrans (b_1, p_1, o) send $o = p_1$. The output is equal 1.

Similarly, for i = 1, HOOVER sends 1 to $Connection_1$ and $Connection_2$. Their methods Comm (i, b_1) and Comm (i, b_2) send b_1 and b_2 equal 1 to $Ntransistor_1$ and $Ptransistor_1$. The method of $Ptransistor_1$ Ptrans (b_2, p_2,o) send $o = p_2$. The output is equal 0.

This shows that the circuit description verifies the behavioral description.

Example 2:

A 1-bit CMOS full adder consists of power, ground, 12 p -type transistors and 12 n-type transistors, as shown in Figure 4.

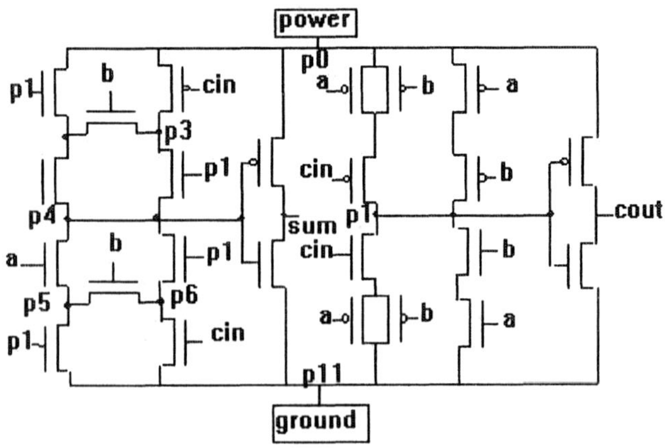

Figure 4: A 1-bit CMOS Full Adder

The circuit description of the adder circuit is as follows.

$Ptransistor_1(p_1,p_0,p_2)$ $Ptransistor_2(c_{in},p_0,p_3)$ $Ptransistor_3(b,p_2,p_3)$
$Ptransistor_4(a,p_2,p_4)$ $Ptransistor_5(p_1,p_3,p_4)$ $Ptransistor_6(a,p_0,p_7)$
$Ptransistor_7(b,p_0,p_7)$ $Ptransistor_8(p_4,p_0,sum)$ $Ptransistor_9(a,p_0,p_8)$
$Ptransistor_{10}(c_{in},p_7,p_1)$ $Ptransistor_{11}(p_1,p_0,c_{out})$ $Ptransistor_{12}(b,p_8,p_1)$

$Ntransistor_1(p_1,p_4,p_6)$ $Ntransistor_2(b,p_5,p_6)$ $Ntransistor_3(p_1,c_{out},p_{11})$
$Ntransistor_4(p_1,p_5,p_{11})$ $Ntransistor_5(c_{in},p_6,p_{11})$ $Ntransistor_6(c_{in},p_1,p_9)$
$Ntransistor_7(a,p_4,p_5)$ $Ntransistor_8(b,p_1,p_{10})$ $Ntransistor_9(p_4,sum,p_{11})$
$Ntransistor_{10}(a,p_9,p_{11})$ $Ntransistor_{11}(b,p_9,p_{11})$ $Ntransistor_{12}(a,p_{10},p_{11})$

The behavioral description of the adder circuit is given in a logical level as follows.

$$sum = a \oplus b \oplus c_{in}$$

$$c_{out} = (a \cap b) \cup (a \cap c_{in}) \cup (b \cap c_{in})$$

For $a = 0, b = 0, c_{in} = 0$

The following classes receive inputs: $Ntransistor_2(b,p_5,p_6)$, $Ntransistor_5(c_{in},p_6,p_{11})$, $Ntransistor_6(c_{in},p_1,p_9)$, $Ntransistor_7(a,p_4,p_5)$, $Ntransistor_8(b,p_1,p_{10})$, $Ntransistor_{10}(a,p_9,p_{11})$, $Ntransistor_{11}(b,p_9,p_{11})$ and $Ntransistor_{12}(a,p_{10},p_{11})$

They send the following:

$p_5 = p_6$ $p_6 = p_{11}$ $p_1 = p_9$ $p_4 = p_5$ $p_1 = p_{10}$,
$p_9 = p_{11} p_9 = p_{11}$, $p_{10} = p_{11}$

which means that $p_4 = 0$, $p_1 = 0$, $p_9 = 0$. Then the class $Ntransistor_3(p_1,c_{out},p_{11})$ send $c_{out} = p_{11}$ which means c_{out} equals 0.

The final HOOVER's output indicates that the behavioral description is equivalent to the circuit description.

9. CURRENT AND FUTURE DIRECTIONS

Currently, the area of formal hardware verification is of intensive research and development. Although major activities started in academia, a number of industries are entering this domain. It is noteworthy that some industries have incorporated formal verification tools into their Computer Aided Design (CAD) systems.

Regularly held conferences devote part of their programs to the active research in this area. We mention here: Symposium on Computer Hardware Description Languages (CHDL), International Conference on Computer Design (ICCD), International Conference on Computer Aided Design (ICCAD), Design Automation Conference (DAC), and European Design Automation Conference (EDAC). In addition, the major part of a number of international workshops is devoted to formal hardware verification research.

Important research areas in this field:

- Design of formal languages for specification at different levels that support formal verification.
- Design and implementation of formal frameworks for verification at different architectural levels.
- Design and implementation of different cell class libraries that support the formal verification.

REFERENCES

1. Elleithy, K. M. "Formal Hardware Verification of VLSI Architecture Current Status and Future Directions," Fifth International Conference on Microelectronics, Dhahran, pp. 197-201, Dec. 1993.
2. Uehara, T., et al., "DDL Verifier and Temporal Logic," Proc. CHDL 83: IFIP 6th Int'l Symp. Computer Hardware Description Languages and their Applications, Pittsburgh, May 1983, pp. 91.
3. Dieter Fensel, Jürgen Angele, and Rudi Studer, "The Knowledge Acquisition and Representation, KARL," IEEE Transactions and on Knowledge and Data Engineering, July-August 1998, pp. 527-550.
4. Hunt, W. A., "FM8501: A verified Microprocessor," IFIP WG 10.2 Workshop, From HDL Descriptions to Guaranteed Correct Circuits Design, North Holland Publishing, Amsterdam, Sept. 1986, pp. 85-114.
5. Hanna, F. K. and Daeche, "Specification and Verification of Digital Systems Using Higher order Logic," IEE proc., Vol. 133, Pt. E, No. 5, Sept. 1986, pp. 242-254.
6. Brackin, S.H., "A HOL extension of GNY for automatically analyzing cryptographic protocols," Proceedings of the Ninth IEEE Computer Security Foundations Workshop, 1996.
7. Phillip J. Windley, "Formal Modeling and Verification of Microprocessors," IEEE Transactions on Computers, 1995 IEEE Vol. 44, No. 1; January 1995, pp. 54-72.
8. Joyce, J., "Formal Verification and Implementation of a Microprocessor," VLSI Specification, Verification, and Synthesis, Birtwistle, G. and Subrahmanyam, P.A., eds., North Holland, Amsterdam, The Netherlands, 1988, pp. 371-378.
9. Bochmann, G. V., "Hardware Specification with Temporal Logic: An Example," IEEE Trans. Computers, Mar. 1982, pp. 223-231.
10. Fujita, M., et al., "Logic Design Assistance with Temporal Logic," Proc. CHDL 85: IFIP 7th Int'l Symp. Computer Hardware Description Languages and their Applications, Aug. 1985, pp. 129-137.
11. Elleithy, K. M. and Aref, M., "A Production Based System for Formal Verification of Digital Signal Processing Architectures," Twenty-Seventh Annual Asilomar Conference on Signals, Systems and Computers, Pacific Grove, California, pp. 1618-1622, Nov. 1-3, 1993.

12. Nebel, W. and Schumacher, G., "Object-Oriented Hardware Modelling - Where to Apply and what are the objects?," Proc. of the Euro-Dac 1996 with Euro-VHDL 96, IEEE Computer Society Press 1996.
13. Glunz, W., et al., "System Level Synthesis, "in Michel, P. and et al. (eds): The Synthesis Approach to Digital System Design, Kluwer, pp. 221-260, 1992.
14. Zippelius, R. et al. "An Object Oriented Extension to VHDL," Proceedings of the VHDL-Forum, Spring'92 Meeting, 1992.
15. Willis, J., et al."A Proposal for Minimally Extending VHDL to Achieve Data Encapsulation and Multiple Inheritance, " Proceedings of the VHDL International User's Forum, 1994.
16. Schumacher, G. and Nebel, W., "Inheritance Concept For Signals in Object-Oriented Extensions to VHDL," Proc. of the Euro-Dac 1995 with Euro-VHDL 95, IEEE Computer Society Press 1995.
17. Shostak, R. E., "Formal Verification of Circuit Designs," *Proc. CHDL '83: IFIP 6th Int'l Symp. Computer Hardware Description Languages and their Applications,* Pittsburgh, May 1983, pp. 13-30.
18. Clarke, E. M., Long, D. E., McMillan, K. L., "A Language for Compositional Specification and Verification of Finite State Hardware Controllers," *Proc. Nith Symp. on Computer Hardware Description Languages and their Applications,* Darringer, J. A. and Rammig, F. J., eds., North Holland, Amsterdam, The Netherlands, 1989, Participants Edition, pp. 281-295,
19. Moszkowski, B., "A Temporal Logic for Multilevel Reasoning about Hardware," *Computer,* Feb. 1985, pp. 10-19.
20. Moszkowski, B., *Executing Temporal Logic Programs,* Cambridge University Press, Cambridge, England, 1986.
21. Boyer, R. S.and Moore, J. S., *A Computational Logic Handbook, Academic Press,* San Diego, Cali, 1988.
22. Cohn, A., "A Proof of Correctness of the Viper Microprocessor: The First Level," *VLSI Specification, Verification and Synthesis,* Birtwistle, G. and Subrahmanyam, P.A., eds., North Holland, Amsterdam, The Netherlands, 1988, pp. 27-71.
23. German S. M. and Wang Y., "Formal Verification of Parameterized Hardware Designs," *Proc. of the Int'l Conf. on Computer Design: VLSI in Computers,* 1985, pp. 549-522.
24. Chandrasekhar, M. S., Privitera, J. P. and Conradt., K. W., "Application of Term Rewriting Techniques to Hardware design Verification," *Proc. of the 24th Design Automation Conference,* IEEE Computer Society Press, Los Alamos, Cali, 1987, pp. 27-71.

Section 3.2

A Practical Approach to the Formal Verification of SoC's with Symbolic Model-Checking

Emil DUMITRESCU
TIMA Laboratory, Grenoble, FRANCE

Abstract: The successful application of model-checking to industrial designs calls for a minimal set of efficiency criteria. This work addresses these issues, based on the linear-time model-checking verification of an instruction cache controller designed by ST Microelectronics.

1. INTRODUCTION

We present a formal verification approach concerning the main modules of a SoC. This approach has been experimented on a system designed by ST Microelectronics: the digital core of a cellular phone, which implements two main features: one real-time component performs digital signal processing, and implements the communication protocol; the second component is in charge of all user-interface functionalities.

Given its high degree of complexity, the elaboration of this system involves a large design team. The verification task is separated from the actual design and performed by verification engineers who are not involved in the design process. This paper presents, from the point of view of a verification engineer, a set of model checking verification strategies applying to the different parts of a SoC. The presentation is illustrated using, as a running example, one SoC module: an instruction cache controller.

Model Checking state of the art

Temporal logic and symbolic model checking present considerable advantages for the specification, analysis and verification of real-life circuits. It is now common practice to use this technique and traditional simulation jointly, as two complementary tools. Unfortunately, the use of model checking on an industrial design often requires non-negligible efforts, despite the "pushbutton" advantage advertised for this technique. The fully automated approach of symbolic model checking is limited by combinational explosion of the BDD model representation [4], [1]. To solve the exponential space complexity problem, two main directions are being explored.

Various alternative representations have been successfully tried. Word-level operations have shown efficient symbolic representations [2]. As for the general representation of a controller state space, ZBDDs [6] seem to bring a good spatial improvement compared to traditional BDDs.

The second important research direction is the enhancement of symbolic model-checking algorithms. Several algorithms are very efficient for a particular class of circuits, but can give bad results if applied to another class [9]. In [7], a heuristic analysis of the circuit representation, based on variable dependency matrices, is presented, and can be used to choose the appropriate verification algorithm, if it exists.

Despite these improvements, the asymptotic exponential complexity has not been eliminated. Besides, some of them are not implemented inside industrial verification tools accepting a standard HDL entry. Hence, care must be taken when using this technique for verifying an arbitrary circuit. Some essential guidelines for model checking a design already exist. For instance, the proof of arithmetic operations is very inefficient. As for control parts, they cannot have an arbitrary size. Deciding whether a circuit is reasonably sized depends on several aspects: number of lines, number of state variables, complexity of the transition and output functions, the importance and the size of the data part, as well as the possible depth of the corresponding finite state machine. All these aspects require a good knowledge of the circuit.

On the other hand, classical efficiency guidelines concerning model checking include design decomposition according to assume-guarantee techniques [3], data-path abstraction, case splitting, and data type reduction [5]. These techniques successfully apply to most common RT-level designs but they require a good knowledge of the existing temporal logics, and also of the symbolic model checking mechanism.

In this paper we show the practical application of several of the above techniques on a SoC design. Section II presents the sub-module taken as running-illustration together with the characteristics of the VHDL model. Section III is the core of this paper, in which we formalize the design specification and explain the various simplifications that we applied to the model to check the requested properties. Section IV gives our results: the figures showing that a simplification strategy cannot be avoided. The last section concludes on the task of the verification engineer.

2. THE ST-MICROELECTRONICS CACHE CONTROLLER

Overview of the architecture

The cache controller is actually a memory interface connecting a DSP on the one side, and internal (on-chip) and external memory banks on the other side, as shown in Figure 1. At each clock cycle, the DSP can issue a fetch request to the cache controller. If the requested word is present in the on-chip memory, it is delivered with a three clock cycles constant latency. Requests are pipelined towards the internal memory. Each word delivery corresponds to a request received three clocks earlier.

In case the requested word is not in the on-chip memory, the fetch request is transmitted to the *external RAM fetch* engine. A read request is issued through the **external fetch port**. This request may be answered within an arbitrary amount of time, which is usually very long.

Unlike classical mechanisms, cache misses are not followed by an automatic download of a new memory page. It is the responsibility of the DSP to order the refreshment of one particular memory bank. To do so, it transmits a download command to the *DMA* engine, via the **command port,** specifying which memory bank is to be loaded, and which memory page is needed. In parallel, it updates the *internal status register file* to indicate that the bank being loaded is no longer valid. Fetch requests to an invalid bank are answered with an error. The DMA engine controls the external RAM fetch engine, which will perform as many word transfers as needed to fill the bank. When the download completes, the bank is validated.

During the DMA transfer of a bank, using the external fetch port, other valid banks must remain accessible for DSP fetch requests. Moreover, if a fetch request corresponds to an external address, it must be issued. Hence, a DMA transfer and an external fetch must share the port in a fair manner.

However, if a transaction arrives on the command port, internal fetches are interrupted. This happens because the transaction may invalidate a bank, and hence allowing further fetches may be unsafe.

The **internal status register file** records the current status of the cache controller: which banks are valid or invalid, the status of current and pending DMA transfers, and the base addresses of each page recorded inside the internal banks.

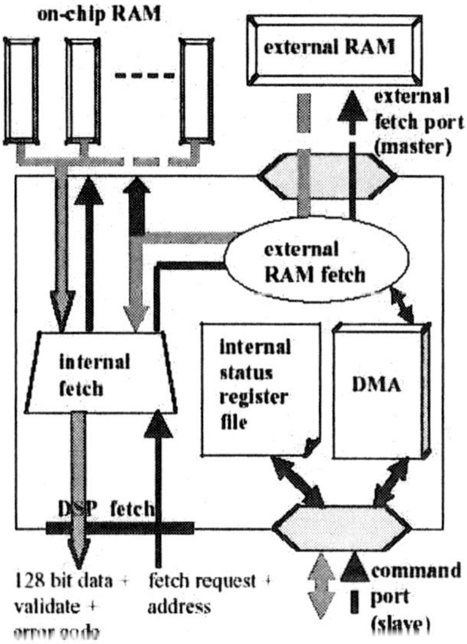

Figure 1 - Architecture of the cache controller

The VHDL model

The whole implementation of the cache controller is described using RT-level VHDL. A simple code inspection has revealed that this design does not present any explicit hierarchy indications. The separation between the different engines represented above (internal fetch, external fetch, DMA, etc.) is not clearly done by the designer, who has preferred to keep all these behaviors inside the same VHDL architecture. Thus, these engines are implemented as collections of interconnected clock synchronized and combinational processes. Hence, it is difficult to reason about the different parts of the design in an independent manner, by applying hierarchical design decomposition.

The control part of this design is predominant. However, transition functions often use chained comparisons between bit vector objects, for instance each time an address is analyzed. This complicates their symbolic representation.

A small arithmetic computation is implemented inside the DMA engine: a counter indicates the index of the last word transferred from the external memory to an internal bank. The upper limit of the counter reflects the size of a bank, which currently equals 1024 words. The reachability analysis step may become extremely inefficient, because the state space exploration of this counter requires 1024 iterations, which are added to the exploration of the global system.

Besides, reasoning about aggregate objects can generally not benefit from symmetry considerations, because the different parts of a bit vector most often follow different paths.

The most representative quantitative figures concerning this design are the following: it is about 1500 lines long. The VHDL elaboration gives a total number of 1000 flip-flops. Address busses are 30 bits wide, and data buses are 32 bits wide. The width of the fetch port is 128 bits.

Given the features of this design, it is impossible to use the "push-button" approach of model checking. A number of preliminary manipulations are necessary, which are presented in the following section.

3. THE VERIFICATION TASK

The formal specification

The specification document of a circuit implicitly defines the properties that need be verified. Ideally, a verification engineer should not even have to look inside the HDL code, as it should be enough to formalize and verify all the assertions given by the specification. If all these assertions are proved correct, then the circuit may be considered "correct". However, a specification is rarely clear and explicit. Many aspects are often left unspecified; hence, the first important step to achieve is to set up a verification plan addressing the key behaviors of the design. The correctness of the design is assessed with respect to this plan. Obviously, it is important to define a large number of unrelated assertions.

The set of assertions written define the formal specification of the design. They are split into two classes: environment assumptions and design properties.

3.1.1 Environment assumptions

Constraints on the environment are most often required, as designs usually work under the assumption that it never exhibits an unexpected behavior. An environment constraint is a collection of invariants declared either as Boolean formulae, or as simple state machines. For example, a Boolean formula could say: "it is never true that the DSP sends a request while the cache controller has its DSP fetch stalled". A simple state machine environment would say: "after a request signal is risen, it must remain high, until an acknowledgement is received".

The environment assumptions are composed with the finite state machine description of the design. Unfortunately, the size of the result is usually much bigger than the design alone, except for very simple assumptions that assign constants to primary inputs, allowing obvious simplifications on the design.

3.1.2 Design properties

The design properties are temporal formulae that the design should satisfy. Safety and liveness can usually be expressed. In our experience, response times are always constant, or at least bounded. Hence, instead of expressing liveness requirements, which are usually very hard to prove, similar safety properties can be written.

The concept of monitor is a very interesting alternative to temporal logic for writing properties [8]. Figure 2 displays the principle that we have applied. Let A and B be two components communicating with a protocol on their interconnection signals. The monitor takes as inputs the signals of interest that must be observed, and provides as outputs one bit for each Boolean formula that must evaluate to true. It is written as a component, which is instantiated together with A and B in the enclosing architecture. The body of the monitor itself can be written as one or more finite state machines whose state transitions are triggered by events on the observed signals.

For instance, in Figure 2, one output displayed says that an error occurs if a "validate" is asserted without a previous request. It is verified like an invariant. Thus, reasoning about the past is done in a natural way using this specification method, unlike most classical temporal logics available.

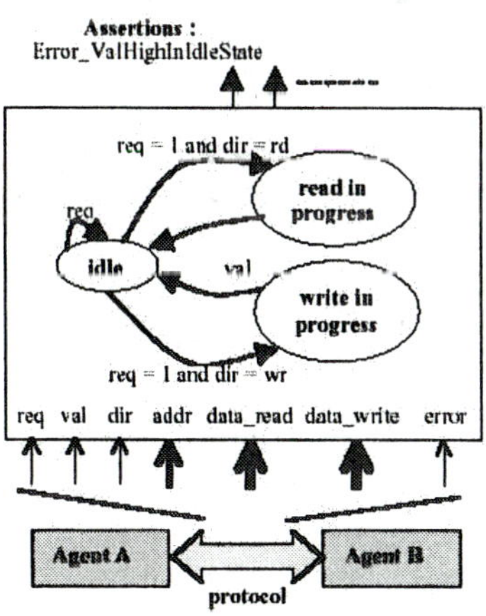

Figure 2 - A monitor specification for the command and external fetch ports

Writing a monitor as a component has no side effects on the description of the design itself. It allows easy plug-and-play of the added VHDL text for verification purposes, which can easily be removed for synthesis.

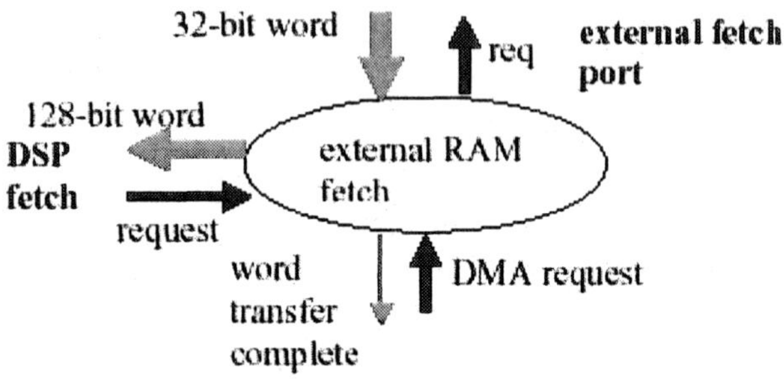

Figure 3 – Isolating one functional block

The Model Simplification

3.1.3 Isolating functional blocks

The most natural verification approach consists in writing a set of properties relating the inputs and the outputs of the cache controller. However, given the size of this design, only a small part of them can be expected to give a result when run by a model checker, as their symbolic representation probably uses an important part of the circuit description. In order to be able to go further, decomposition is vital.

After a detailed code inspection, the different functionalities of the cache controller have been isolated as shown in Figure°1. It is now possible to reason about each part of the design independently, and write a set of local properties for each one of them. Consider for instance the external RAM fetch engine (Figure 3).

A DMA or a fetch request can trigger the transfer of an external word. Thus, the *req* signal of the external fetch port is asserted together with the address that is required. The following properties should be satisfied:

P1. the *req* signal cannot be asserted spontaneously. At least one incoming request should be present, either on the DSP fetch side or on the DMA side;

P2. conversely, the *req* signal cannot be de-asserted unless a response has been received on the external fetch port. However, it is also true that this request holds as long as the incoming request that triggered it.

The first property states that a *req* rising event can only be triggered by one of the two incoming DSP fetch or DMA requests. The Boolean function defining the *req* signal depends on the values of these two incoming requests. But as the external RAM fetch engine remains connected to the rest of the circuit, these two signals have their own transition functions. They depend on other intermediate signals, and so on, until a set of primary inputs is reached. Thus, the verification of this property is inefficient, due to complicated transition functions that involve both the DSP fetch port and the command port.

3.1.4 Eliminating useless parts

At the local level of the external fetch engine, the behaviors associated to the incoming requests can be abstracted away. Indeed, a request is issued on *req* **when** an incoming request arrives. Hence, we may consider both incoming requests as *primary inputs*. Thus, for verifying these two particular properties, both the DSP fetch and the DMA functional blocks may be eliminated. The transition function of *req* becomes much more simple and the verification of property P1 is done at the local level of the external fetch engine. This simplification is done manually. The verification engineer directs the model-checker to override the transition functions of the incoming requests.

According to these simplifications, the proof of properties P1 and P2 runs on a new model consisting of the external RAM fetch separated from its environment. The interpretation of the verification results is the following. If the property is true at the local level, this result has been obtained under the assumption that **any** behavior can occur on the incoming request inputs. The original implementations of these inputs define more restrictive behaviors, which are necessarily included in the set of all possible behaviors. Hence, the property is also true in the original implementation. However, in case the property is proven false, it is not safe anymore to draw a conclusion at the global level of the circuit. Erroneous behaviors of the incoming request signals may cause the property failure, while they are not allowed by the initial description. This case is illustrated by property P2, which says that *req* may fall if the incoming requests fall. The same simplified model remains interesting, but it is obvious that the property will fail: the scenario displaying an incoming request rising and then falling can occur, because any behavior is now allowed for the incoming requests. The model must be complemented with the assumption that incoming requests remain high once they are asserted, until a transfer is completed. Thus, the erroneous behavior is eliminated and the property passes.

3.1.5 Using non-determinism

Eliminating a part of a design for simplification purposes means deleting all the behaviors it implements. However, some of the behaviors deleted may still be needed so that the model obtained after simplification remains realistic. In other terms, a simplified model focusing on the desired behaviors must replace the eliminated part of the initial description, while still allowing all other. When an appropriate scenario occurs, the behaviors needed are exhibited. All other scenarios correspond to an unspecified behavior. Thus, the simplified model obtained is both *abstract* and *non-deterministic*.

In our example, a model must implement the assumption required for proving property P2. It must ensure that when a request rises it must remain high until a transfer is completed. However, the request may rise at an arbitrary moment.

The abstract non-deterministic model replacing the initial description is usually written by the verification engineer. Hence, it must be proved correct before it can be used. The correctness criterion states that, while focusing on desired behaviors, the model allows all the behaviors defined by the original transition functions it replaces. This is synonymous with *refinement checking*[5]. Replacing a part of the design (in this example, the definitions of the two incoming requests) by a simpler, non-deterministic, abstract definition, together with performing the subsequent checks is known as *assume-guarantee* reasoning°[3].

3.1.6 Behavioral partitioning

The different functional blocks that are running in parallel may execute independently, or they can influence the execution of each other. Hence, different execution scenarios are possible. Any property a design should satisfy must be valid for all scenarios that may occur.

The verification of a property can be split into a set of sub-goals, one for each interesting scenario. Each sub-goal asserts that the initial property is true under the assumption that only one scenario may occur. If all sub-goals are verified, then the initial property may also be true.

For instance, a verification goal asserts that transactions on the command port of the cache controller are correct. This goal can be split into two sub-goals, according to the following scenarios: (1) only read transactions are allowed and (2) only write transactions are allowed. If both read and write transactions are proved correct, we may conclude that all transaction on this port are correct. The two sub-goals constrain the signal indicating the direction (read or write) to constant values. Thus, the size of their underlying model becomes considerably lower due to BDD simplifications.

Behavioral partitioning acts both as a powerful simplification method and as an environment constraining method. However, its application needs care: if all sub-goals are true, the initial goal is not necessarily true. For instance, assume that a bug occurs only when a read transaction is followed by a write transaction. If behavioral partitioning is employed, this bug shall not be uncovered.

Classical simplification methods that are inapplicable

When a design manipulates bit-vector objects, a natural simplification consists in reducing their size to only a few bits wide. This is especially appropriate for data. However, control bit-vectors are not always symmetric, as individual bits or bit fields may correspond to different meanings. This is the case for bit-vector objects implemented inside the cache controller. In that case, most properties have to be split into as many sub-goals, as there are bit fields in the vector (up to the number of bits in the worst case).

Concerning the refinement-checking technique as a support for the assume guarantee reasoning, most industrial verification tools do not implement the mechanical check. Thus, before using an abstract model, a manual assessment with respect to the initial design must be performed. In practice, this limits the method to abstract models that are very simple.

4. RESULTS

For verification purposes, the initial model has been slightly modified. In order to solve the reachability problem stated at 2.2 the verification has run under the assumption that memory banks contain only 4 words instead of 1024. It is important to note that in order to set up a property a "design cycle" must be followed. As it is impossible to know in advance whether a verification task ends within a reasonable amount of time and resources, it is important to establish a limit on the memory size to be used, and kill the verification if this limit is exceeded. If this happens, a model simplification should be attempted before re-trying a new verification.

The cache controller has been verified using Cadence FormalCheck v2.3 on a Sparc Station Ultra 2-60, running at 400 MHz with 2 Gbytes of memory.

A total number of 50 properties have been written. We highlight a few interesting classes of properties that we have written, as well as the importance of the model simplification used in achieving good verification times (Table 1).

Behavioral partitioning is the only model simplification needed for verifying the *internal fetch (IF)* functional block: to make sure that fetches are not interrupted, the incoming requests on the command port are deactivated. Two properties are representative:

IF1. a fetch request is acknowledged within 3 clock cycles;

IF2. a rising acknowledge corresponds to a fetch received 3 cycles earlier.

The verification of the *external RAM fetch (ERF)* functional block has needed more important model simplification efforts, as it had to be cut from the rest of the design. We highlight two properties for this block

ERF1. external fetch requests are never retracted;

ERF2. DMA and internal fetch incoming requests are treated fairly.

The *DMA* functional block must comply with a very important requirement (DMA): it must not write into the internal RAM if it is in use. The model is simplified by cutting the functional block from the remaining design, and by using behavioral partitioning.

In Table 1, the first two columns list the properties and the way they are written; the third column gives the number of state variables and the number of iterations necessary to check a property. The fourth column tells which simplification technique, if any, has been applied. The last column gives the performance in computation time and memory.

Functional block	Temporal logic (TL) or monitor (M)	State variables/ model depth	Model simplification technique	Time (sec)/ Memory (MB)
IF-1	TL	153/19	behav.part.	63/16,5
IF-2	M	163/19	behav.part.	52/16,5
ERF-1	M	210/ -	-	out of memory
	M	10/11	cut	49/16.5
ERF-2	TL	300/-	-	time limit exceeded
	TL	256/33	cut	65/31
DMA	TL	310/ -	-	out of memory
	TL	147/57	cut + behav.part.	62/16

Table 1 — Different properties and verification results

For properties ERF1,2 and DMA, we see that simplification by cutting away a functional block is essential. If this technique is not applied, either the proof runs out of memory or it lasts several days, which is impractical. One third of the 50 properties written have been verified using functional cut. For the remaining properties, behavioral partitioning and environment constraining assumptions were sufficient to obtain verification answers within useful time.

5. CONCLUSION

The set of model simplification guidelines presented here allowed a reasonable functional coverage for the verification of the cache controller design. All the functional blocks have been addressed, despite the lack of hierarchy in the RTL description of the design.

A few minor incompatibilities have been faced, concerning the synthesizable HDL code used by the designer team, which is not always recognized by the verification tool.

This work also raises an interesting question: would the verification have been quicker if the designers had partitioned the design? In theory, the answer should be "yes"; in practice, it is probably "no". In order to avoid looking inside a component, its documentation should be very detailed, clear, and maintained up to date as the design evolves, which is very hard to achieve. Using the existing documentation as an aid to understanding the behavior of a component seems more realistic.

Thus, we believe that in the current context, given the available tools, it is essential that verification engineers are aware of the implementation details of the design they verify. On the other hand, the ignorant application of automatic model-checking fails in most cases. Knowledge about the underlying theory of this technique is vital.

REFERENCE

[1]. R. E. Bryant, Graph-based algorithms for Boolean function manipulation, IEEE Transactions on Computing, vol. C-35, no. 8, Aug. 1986, pp. 677-691

[2]. R. E. Bryant, Y. Chen, Verification of arithmetic circuits using binary moment diagrams, International Journal o n Software Tools for Technology Transfer, vol. 3, no. 2, Springer, pp. 137-155, Mar. 2001

[3]. T. A. Henzinger, S. Qadeer, and S. K. Rajamani, You assume, we guarantee: methodology and case studies Proceedings of the Tenth International Conference on Computer-aided Verification (CAV 1998), Lecture Notes in Computer Science 1427, pp. 440-451

[4]. K.L. McMillan, Symbolic Model Checking, Kluwer Academic Publishers, Boston, MA, 1994

[5]. K.L. McMillan, Verification of infinite state systems by compositional model checking, in L. Pierre and T. Kropf, editors, CHARME99, vol. 1703 of LNCS, pp. 219-233

[6]. S. Minato, Zero-suppressed BDDs and their applications, International Journal on Software Tools for Technology Transfer, vol. 3, no. 2, Springer, pp. 156-170, Mar. 2001

[7]. Moon, J. H. Kukula, K. Ravi and F. Somenzi, To Split or to Conjoin: The question in Image Computation, in Proceedings of DAC, Los Angeles, CA., June 2000, pp. 23-28
[8]. K. Shimizu, D. Dill and A. J. Hu, Monitor Based Specification of PCI - Proceedings of the Third International Conference of FMCAD, November 2000
[9]. R.K. Ranjan, A. Aziz, R.K. Brayton, B.F. Plessier and C. Pixley, Efficient BDD algorithms for FSM synthesis and verification Presented at IWLS95, Lake Tahoe, CA, May 1995

Section 3.3

High Performance Verification Solutions for SOC Designs

Makarand Joshi & Ashok Kumar
Advanced Simulation Division, Quickturn, a Cadence Company

Abstract: Growing design complexity, sizes and shrinking turn-around times are representative of impending veri.cation crises for the electronics industry. To address the void in productivity from veri.cation solutions of today, Quickturn introduces a set of integrated, processor-based simulation acceleration solutions. These hardware accelerators _ Palladium and CoBALT, feature software-like compile times, scalable gate capacities, and functional congruence across all engines with a .exible upgrade path on the performance curve. The use model for these engines is simple_a single design model is compiled to target the multiple accelerator engines. Users can perform high performance simulation or debug interactively with complete design visibility on the Palladium and CoBALT accelerator engines. A design bug identi.ed during simulation acceleration on a Palladium or CoBALT engine can be debugged off-line on the software engine, thus freeing the accelerator for other users. Performance statistics from customer benchmarks validate the claims of high simulation throughputs offered by these unique and powerful high performance verification solutions.

Key words: Debug, hardware accelerators, save and restore, simulation acceleration, test bench, turn-around time, verification.

1. INTRODUCTION AND MOTIVATION

Increased verification throughput is a primary concern of design and verification houses wrestling an ver-losing battle against design size, complexity and increased time-to-market pressures. Amidst the erpetual quest for simulation speed, verification teams have resorted to numerous solutions that stretch from low-cost simulation solutions, such as event-based simulators, cycle-based simulators and formal verification tools, to high-end, high performance emulation engines. However, many of these solutions deliver limited capacity, slow simulation throughput, huge set-up or turn-around times and/or a prohibitively high price tag.

Clearly, the electronics design industry is in need of a robust solution with optimal performance, fast turnaround time and sound return on investment. A fast, functional verification environment that uses hardware acceleration engines in conjunction with software simulation provides a set of high performance verification solutions that effectively address this requirement of the design industry.

2. THE SOLUTION

We discuss a set of HDL simulation tools spanning both hardware and software solutions. The family of products includes three execution engines: a compiled code software engine and two processor-based hardware acceleration platforms viz., Palladium and CoBALT (Concurrent Broadcast Array Logic Technology). The toolset includes a single compiler that converts design descriptions into a form targeted to execute at high speed on any and all of the supported execution engines.

Simulation acceleration engines feature fast compile times (around 6 Mgates/hour) and hardware like simulation speeds ranging from 500 Hz to 600 KHz for different engines, resulting in fast turn-around times throughout the design and verification stages. A single run-time environment (RTE) manages execution of the compiled design on all the engines. This run-time environment provides a simulation interface with full visibility and a uniform look-and-feel across all execution engines.

A user can trade-off between price and performance and use the appropriate simulation acceleration engine at any particular stage in the overall design process. The combination of the software engine, Palladium and CoBALT truly offers users a simple upgrade path for performance. The typical use model allows for a user to perform high-speed simu lation on any hardware engine, identify a design bug, save the state and restore it on another engine (typically the software engine) for off-line interactive debugging. Further sections will provide details on the commonality of the different engines and features.

3. SIMULATION ACCELERATION ENGINES

The tools discussed here are a combination of software developed at Quickturn and hardware acceleration platforms developed by IBM.

CoBALT High-end, High-performance Hardware Engine

The first engine — CoBALT is a high-performance hardware platform manufactured by IBM. The architecture of this system is comprised of an array of emulation processors interconnected by a high-performance broadcast network. The compiler automatically converts and partitions a design net list for execution on this array of concurrent processors. The CoBALT solution provides speeds ranging from 500 Hz (i.e. simulation cycles per second) to 400 kHz with capacities up to 100 million gates [4].

Palladium Mid-range, High-performance Hardware Engine

The second engine — Palladium is also manufactured by IBM. Palladium [5] uses an array of high-speed custom logic processors to accelerate simulation and provides simulation speeds ranging from 500 Hz to over 600 kHz. The compiler automatically converts and partitions a design net list for execution on the array of concurrent processors. The Palladium engine can support capacities up to 16 million-gates.

Compiled Code Software Engine

The compiled code software engine utilizes a standard workstation to execute a circuit described in machine code and represents the lowest-cost and best debug option in the family of engines. After a design bug is identified, a user can off-load the design onto a workstation using the software engine and free the hardware accelerator for other simulations. When a user is running the software engine, the host workstation manages both the user interface (UI) functions and the circuit execution.

In summary, the three engines mentioned above comprise the solution and can be used interchangeably at any stage in the design process with functional congruence guaranteed across all the engines — CoBALT, Palladium and the software engine.

4. LANGUAGE SUPPORT

The simulation acceleration solutions allow stimulus to be applied to the design-under-test via test vectors or a software test bench (modelled in Verilog, VHDL, C or an abstract verification language such as *e* or *VERA*)[1]. The designs may be synchronous (yield high performance) or asynchronous in nature. A wide range of Verilog and VHDL language constructs are supported. In addition to the synthesizeable subset, the following constructs are supported.

– A subset of the Verilog and VHDL constructs used to define memories.

– Verilog system tasks, user-defined tasks and functions, C mo dels and portions of the Verilog PLI.

– A limited number of constructs required for test benches, including initial blocks, force/release, and some simulation-related compiler directives.

1. e and VERA are registered trademarks of Verisity Design Inc., and Synopsys Inc. respectively.

5. USE MODEL

The salient features of the different engines and the design styles supported have been discussed in the earlier sections. Let us now turn our attention to the process flow from a user¡¯s perspective for compilation, simulation and debug of designs using one or more simulation acceleration engines. Figure 1 illustrates the use model for the simulation acceleration engine(s). A user specifies Palladium or CoBALT to be the target hardware platform and software engine to be the target software platform during the invocation of the compilation process. This specification enables compilation of the design into a set of simulation and execution models optimized for providing high performance on the target verification engine. The compile process is discussed in detail in Section 6.: Compile Process for Simulation Acceleration. Simulation acceleration is performed on the corresponding target engine and results from simulation can be recorded in a variety of formats. As mentioned in earlier sections, the hardware engines serve as high-speed simulation platforms and the less expensive software engine serves as a debug engine.

Off-line debug leads to improved ROI: When simulation is being done off-line (on the software engine), th e hardware engine can continue to be used to perform other simulations. This represents a recommended use scenario where the hardware engines are used for high speed functional verification to identify a bug and then the design is off-loaded onto one or mo re software engines for isolation of the bug via interactive debugging. This mode of operation is illustrated in Figure 1.

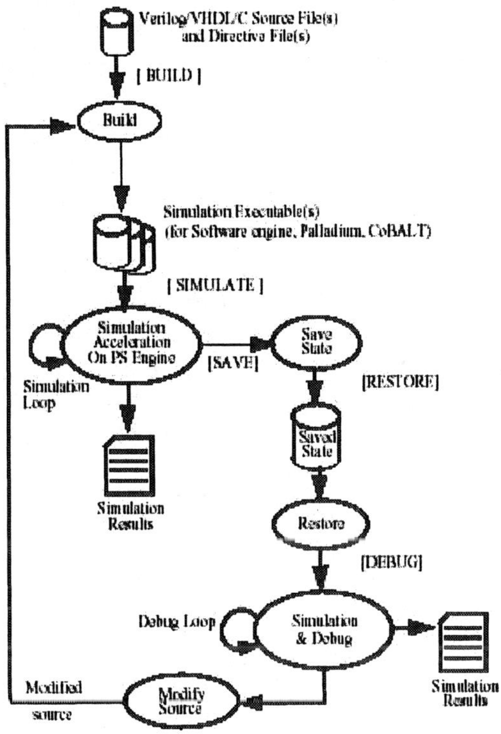

Figure 1: user Module

6. COMPILE PROCESS FOR SIMULATION ACCELERATION

A key benefit of the compile process for simulation acceleration is its fast compile time. The compile process resembles that of a traditional simulator and is illustrated in Figure 2.

The initial portion of the compile process is common for Palladium, CoBALT and the software simulator and is followed by accelerator-specific and software simulator-specific compile phases. The compile phase produces an intermediate dump of the design data and the net lists for the hardware accelerators and software simulator formats.

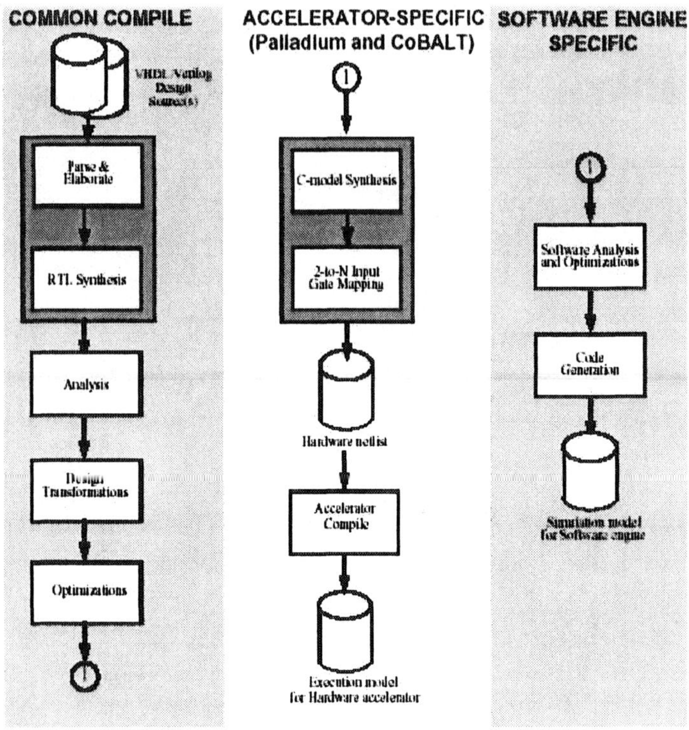

Figure 2. Compile Process for Simulation Acceleration.

Common Compile Phase

The compilation phase utilizes a traditional simulator front-end that includes parse, elaborate and RTL synthesis stages.

Parse, Elaborate The parser reads the Verilog and VHDL source files and directives files. The elaborator uses the parsed information as it elaborates the design hierarchy and flattens it for use by subsequent stages of the compile process.

RTL Synthesis In this phase, the elaborated design representation is synthesized into a gate-level net list. Also, construction optimisations for accelerated simulation are performed during gate-creation. The non-synthesizeable constructs and modular constructs like arithmetic multipliers, shifters and dividers are mapped onto abstract models and are processed at a later stage of the compile process for optimal performance on the target simulation engine.

Analysis and Optimisations A number of analysis passes are performed on the synthesized design. These include combinational loop analysis and memory ordering. The design is als o subjected to optimisations such as constant propagation, common sub-expression removal and DeMorgan¡¯ s theorem.

Design Transformations Subsequent to RTL synthesis, special state device transformations are performed on the design to produce a net list that can be mapped onto the Palladium and CoBALT hardware accelerator(s). These transforms allow the simulation acceleration engines to support both asynchronous designs and synchronous designs with multiple clock domains and including edge and level sensitive devices. Note that the transforms preserve the logical and functional structure of the design.

In addition to the state device transformations, other transformations are made on the design to enable support for evaluation of initial blocks, reset logic, clock-enabling for memories, evaluation of C-models, and assertions.

Accelerator-specific Compile Phase

The accelerator specific compilation phase comprises of the following stages.

Generic Phase This phase is identical for both the hardware accelerators: Palladium and CoBALT.

Modular constructs in the design like adders, multipliers, shifters, etc., execute faster as gates on the Palladium and CoBALT accelerators and are synthesized with 2-input logic gates using algorithms that reduce the critical path of the generated logic. Reduction of net list critical path enhances acceleration performance.

A generalized mapping algorithm is applied to the net list (which at this stage is a low-level structure consisting of 2-input gate primitives) to map it to primitives on the hardware accelerators— Palladium and CoBALT. A number of optimisation passes precede and succeed the mapping process. The entire accelerator net lists and the original software net list are congruent with respect to the state points and signals in the design.

Hardware mapping This stage is specific to the individual hardware engines, Palladium or CoBALT. As mentioned earlier, Palladium and CoBALT are processor-based hardware accelerators. Hardware specific analysis and optimisations passes perform resource mapping and allocation to ensure efficient distribution and scheduling across the processors. This stage maps the design into the wirelist format required by the accelerator and inserts any additional data required for that accelerator-like memory execution sequences. The compile process ends with the dumping of the accelerator design files.

Software-specific Compile Phase

The internal representation resulting from the common compile process is subjected to some optimisation and analysis phases and a simulation model to be executed on the software engine is generated.

Abstract model synthesis As stated in the description for common compile phase, the modular constructs such as arithmetic multipliers, shifters and dividers are mapped into special C-models tuned for performance on the software engine.

Analysis and Optimisations This stage precedes the generation of native code for the design. In preparation to use the rich instruction set available for the host workstation, the compiler performs software specific design optimisations and organizes data for enabling effective generation of code.

Code Generation Just before writing the software simulation file, the signal indices and names in the software net list are annotated back to the accelerator net list for synchronization and user debug. Finally, the machine language code for the design is generated and written along with the symbol table into the software design file.

In summary, it can be seen that most of the compile process resembles a traditional software simulator. The extra compilation phases for the target accelerator such as hardware mapping involve linear traversal of the design and do not significantly increase the compilation time. The simplicity and efficiency of the compile flow results in fast compile times.

7. SIMULATION AND DEBUG

The Simulation Process The compiled design is now in the form of software simulation and acceleration executable models. As stated earlier, a design compiled by the compiler can be simulated on any of the different supported engines. Using different command line switches applied to the same simulation executable causes the design to simulate on the software engine, or on either the Palladium or CoBALT acceleration engine. A key enabling feature is that the simulation output and any tool-generated messages are identical (congruent) across all engines, the only difference being the speed of simulation.

Simulation and Debug Modes

A user has a choice of two main simulation flows: High-performance simulation on the accelerator or the interactive debug mode on the workstation. Simulation is performed in either of the modes so as to maximize ease of debug and efficient use of resources. The different modes of operation for Palladium, CoBALT and the software engine are explained here.

High-Performance Acceleration Mode

Synthesizeable Designs and test benches A design and test bench having only synthesizeable constructs can be run on the accelerator without having to interact with the software engine. In this case, the design is automatically loaded on the accelerator and simulates the required number of cycles as determined by the user command or the test bench. The result state or data is directly transferred from the accelerator to the user through the high-speed interface (between the accelerator and the host workstation).

Simulation results are obtained at high speed, which allows an increased number of tests to be run on the accelerator. Figure 3, is an illustration of simulation of a synthesizeable design and testbench.

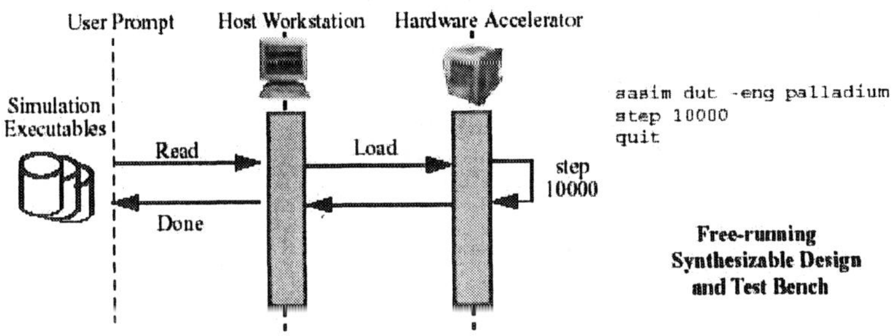

Figure 3. Simulation of Synthesizeable Design and Test bench.

Handling Non-synthesizeable Constructs A control module is added to the design (shown in Figure 4) during the compile process to control, stimulate and monitor the design under test with no user intervention. It controls the design clocks and controls the accelerator execution. For accelerator interactive mode and for execution of non-synthesizeable C-models, the control module determines when to stop the accelerator and pass control to the run-time core on the host workstation

Figure 5, is an illustration for simulation using a C test bench. In this example, a C-model for the system task $display gets fired after 4096 cycles and a Cmodel for the user-defined task $doMyTask() gets fired after 8192 cycles. The sequence diagram in Figure 5 illustrates the sequence of actions.

Interactive Debug Modes

Software Mode A design is run in software mode for debugging or if the accelerator is unavailable or in use. The simulation performance here is comparable to a stand-alone software simulator and is orders of magnitude less than the accelerators. However, the software mode has exactly the same user interface, interactive commands and simulation output as the accelerator modes.

The following scenario illustrates the value of interoperability across different platforms: (The sequence of actions for this scenario is illustrated in Figure 6.) A verification engineer is debugging a design that takes 400,000 cycles to complete simulation. Let us assume that it takes 12 hours to simulate on a software simulator and simulates in 2 minutes on the Palladium hardware accelerator. Running the design on Palladium, the user detects an error at the 300,053rd cycle (~2 mins). He then re-runs the simulation at accelerator speed, saves the state of the design at the 300,000th cycle (~2mins), thus freeing the accelerator for another user. The debug engineer can restore the state on the software engine and start interactive debugging from the 300,000th cycle, thus saving 9 hours of simulation time.

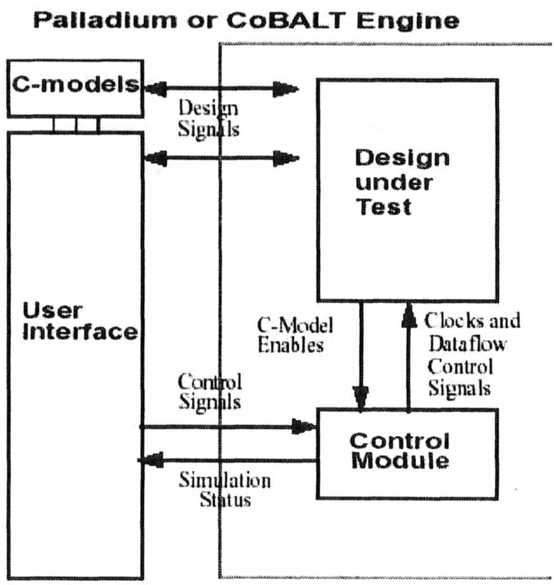

Figure 4. Design and Test bench with C-models and Control Module.

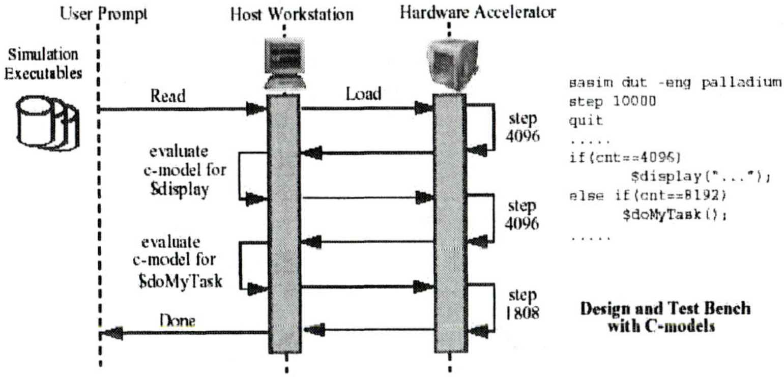

Figure 5. Simulation of Design and Test bench with C-models.

Accelerator Interactive Mode Debugging problems in the software engine is convenient because it represents an optimum and efficient use of resources. However, the Palladium and CoBALT engines are also equipped with the same debug capabilities as the software engine.

Simulating the design in the accelerator (Palladium or CoBALT) in interactive mode offers complete visibility into the design and supports interactive user tools like the ¡¯ watch¡¯ command. However, the accelerator needs to stop each cycle and have the software engine execute the user commands. Hence, this process does not match the performance of the high-performance Palladium or CoBALT mode. However, it is still faster than running interactive debug on the software engine.

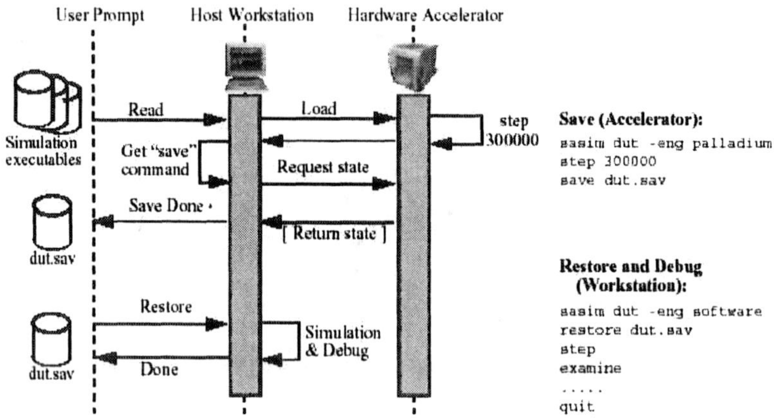

Figure 6. Save on Accelerator and Restore, Debug on Workstation.

8. PERFORMANCE MEASUREMENTS

The simulation acceleration engines discussed here feature fast, simulation-like compile times and high simulation throughputs in addition to the support for interoperability and functional congruence between different engines. In turn, the users can achieve unprecedented, low turn-around times.

Multiple customer benchmarks were used to conduct performance measurements. These benchmarks included several different types of designs and test benches. Simulation of the benchmarks resulted in different performance characteristics based on the nature of the verification strategy. We will discuss here the different types of test benches seen during measurement of simulation performance numbers for the different engines.

Types of Test Benches

Some of the test bench styles that have been seen during the performance measurements are outlined here.

Synthesizeable Test Benches These test benches are well suited for Simulation Acceleration. Here, the design-under-test, stimulus generator and the results checker are mapped onto the hardware accelerator. These test benches exhibit high simulation performance (comparable to the theoretical maximum for the accelerator) and require minimal interaction with the external environment.

Embedded Test Benches For efficiency, embedded test benches typically use a memory to hold stimulus and capture simulation results. The memories and the design-under-test are mapped onto the hardware accelerator (Palladium or CoBALT) for performance. Simulation is initiated by loading the stimulus memory and the results¡¯ memory is filled up on completion of simulation.

C-model Test Benches In C test benches, one or more C-models running on the workstation act as a stimulus generator(s) and results verifiers as they interact with the hardware accelerator on a clock cycle or an as-needed basis. Amdahl¡¯s law prevails here. The overhead of running the C-model on a slow workstation, the communication events between the accelerator and the host workstation during simulation, and the starting and stopping of the accelerator can significantly limit simulation performance.

Test Benches with Vectors Vector playback mode test benches are treated similar to an embedded test bench. Typically, the test bench loads in a vector, runs simulation, and dumps vector results that are verified by a results¡¯ checker on the workstation. The inputs vectors are loaded onto the hardware every cycle, but outputs from the box are obtained only when necessary, i.e. when there is a test failure.

Transaction based Test benches In this mode of test bench the interaction between the accelerator and the host workstation are infrequent and limited to trigger events from the host workstation to the transactors on the accelerator and the response events from the accelerator. The result is a very low overhead from communication events and yields high overall simulation performance.

Results

The results from compilation and simulation of multiple customer design benchmarks are summarized here. The designs used for benchmarking varied in sizes from 15,000 gates to 8,100,000 gates. The test benches for the different designs included synthesizeable test benches, embedded test benches, C test benches, VCD test benches and transaction based test benches.

8.1.1 Palladium Engine

Palladium exhibited very fast compilation times resulting in build rates averaging 6 Mgates per hour. The nature of test bench had a considerable influence on the simulation performance for a particular design. (The different types of test benches are explained in Section 8.1). For the different designs simulated on Palladium, cycles simulated per second (performance) ranged from 5000 cycles/sec on test-benches with considerable communication events¡¯ overhead to cycles/sec in excess of 350K for transaction based test benches. The fast compilation times and high simulation performance from Palladium result in excellent turn-around times.

8.1.2 CoBALT Engine

CoBALT also featured good compilation times resulting in build rates over 6 Mgates per hour. As explained in Section 8.2.1, the nature of test bench had a considerable influence on the simulation performance for a particular design. For the different types of designs (and types of test benches explained in the earlier section) simulated on CoBALT, cycles simulated per second (performance) ranged from 2000 cycles/sec (for C test benches) to cycles/sec in excess of 250K (for embedded test benches). In summary, the fast compilation times and very high simulation performance from CoBALT result in excellent turn-around times for the user.

9. CONCLUSIONS

We have presented simulation acceleration as a solution that combines the ease of use and fast compile times of software simulation with the high performance processor-based hardware acceleration engines to allow users to verify large designs in unprecedentedly fast turn-around times. In today¡¯s design workspace where verification strategies are instrumental in determining the success of design projects, simulation acceleration solutions enable designers to perform more verification earlier in the design cycle. Designs simulated in a comprehensive and thorough verification environment offer a competitive advantage to customers by meeting industry¡¯s time -to-market needs.

ACKNOWLEDGMENTS

The authors would like to express their sincere thanks to customers and members of the Advanced Simulation Division at Quickturn for their support and reviews.

REFERENCES

[1]"Cycle-Based-Verification", http://www.quickturn.com/tech/white.paper.cby.htm
[2]"Reducing Logic Verification Time with Cycle Simulation", Keith Westgate, Don McInnis, http://www.quickturn.com/tech/cbs.htm.
[3] CoBALT Technical Data Sheet, http://www.quickturn.com/products/cobaltplus_data_sheet.htm
[4] Press release: "QUICKTURN¡¯S CoBALT ULTRA BREAKS THE 100-MILLION-GATE CAPACITY BARRIER", http://www.quickturn.com/w_new/CoBALTUltra.htm.
[5] Press release: "QUICKTURN ANNOUNCES PALLADIUM, THE MOST ADVANCED SIMULATION ACCELERATION AND IN-CIRCUIT EMULATION SYSTEM'http://www.quickturn.com/w_new/QT%20P alladium.htm

Section 3.4

Novel Test Methodologies for SoC/IP Design

Implementation and Comparison

L. Hong, M. Nahvi, R. Fung, A. Ivanov, R. Saleh
Dept. of Electrical and Computer Engineering, University of British Columbia, Vancouver, BC.

Abstract: This paper addresses testing issues associated with System-on-Chip (SoC) designs. A network-processing engine SoC was constructed using reusable Intellectual Property (IP) and a test harness based on the emerging IEEE P1500 standard. Three different test access mechanisms (TAMs) are compared on this platform – a serial threading approach analogous to the IEEE1149.1 standard, a test-rail approach, and a packet-switching test network (NIMA). The NIMA architecture is the most effective when there is a limited number of I/O pins and could potentially become the fastest architecture when it is allowed to operate at higher clock speed than the core DFT. However, the test rail approach is the most effective of the three TAMs if the test width is not restricted or if the area overhead is critical.

1. INTRODUCTION

Many experts believe that testing Systems-on-Chip (SoC) designs will be the most critical bottleneck in future designs if the looming Design-for-Test (DFT) issues are not addressed [1]. One important challenge in testing SoCs is accessing the embedded cores from the chip boundary. Zorian et al. [2] propose generic conceptual test access architecture for embedded cores as shown in Figure 1. The components of this architecture are: *Source/Sink*, *Wrapper*, and *Test Access Mechanism* (TAM). The TAM is the physical mechanism that connects the source and the sink with the core. It determines how efficiently information is transmitted and received from the outside world, and thus controls the total test time. Therefore, the optimal design of a TAM is critical in terms of the overall test strategy. There are a number of challenges associated with the design of a TAM. These include using the limited number of primary I/O pins for test; testing different types of cores with compatible methodologies; integrating different core test schemes into a system

DFT; limiting test time, power, area overhead; and integrating new cores, with or without TAM support, into a previous design.

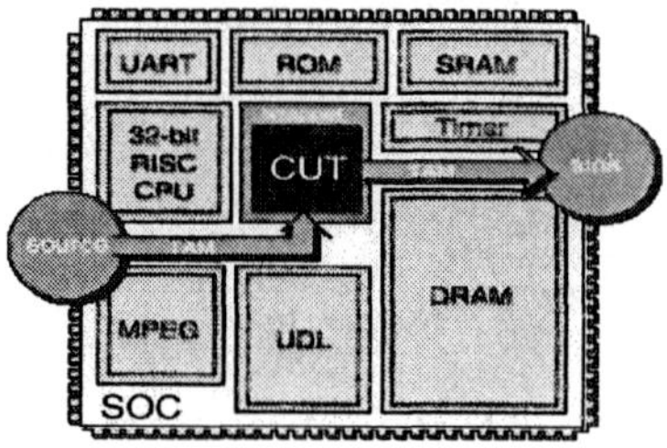

Figure 1. A Generic Conceptual Test Architecture [2]

There are many proposed TAM architectures in the literature. These can all be grouped into three main categories: multiplexer-based, serial-based, and bus-based connections. In the first category, multiplexers are used to access the cores; e.g., [3] modifies the cores and inserts multiplexers such that each core has a transparent mode for testing. TAMs in the group of serial-based connection adopt the established IEEE 1149.1 standard. References [4,5] are examples of serial-based TAMs. A number of different variations of the bus-based TAM have been reported. References [6,7] are examples of this method of TAM connection.

A *Novel Indirect and Modular Architecture* (NIMA) for TAM was proposed in [8], where emphasis is placed on modularity, generality, and configurability of the architecture to exploit the advantages offered by the re-use paradigm. The key idea in NIMA is to establish an indirect digital communication path through the use of a packet-switching connection. The notion of a network on a chip has been the subject of numerous research efforts [9]. We use the on-chip network for test purposes. In this method, test vectors are converted to packets before they find their way from the source to their destination core. Test results can also be transferred in the same way from the core to the sink. On- and off-chip sources and sinks are used to provide/receive data to/from the cores. Each core has a specified DFT which may be scan-based, or full or partial Built-in-Self-Test (BIST). All test-related communications destined/originated to/from the chip need to pass through the on-chip network.

NIMA is designed such that the communication between source/sink and cores is achieved in a simple 3-layer hierarchical model, i.e., *Application Layer*, *Network Layer*, and *Physical Layer*. Each layer, except the *Physical Layer*, deals directly with its own peer through a virtual link. Moreover, the idea of source routing is used, so the routing path is pre-determined. To simplify the *Network layer*, packets are sent in order and assumed to reach their destinations in order.

The new packet definition for the forward path is illustrated in Figure 2, the *Sync Word* signals the beginning of a packet, and the *Data Length* and *Address*

S	L_D	L_A	A	D
Sync Word	*Data Length*	*Address Length*	*Address*	*Data*

Figure2. Packet structure for the forward path

Length define the length of the *Data* and *Address* fields respectively. Finally, the *Address* field is a variable length of address bits, and the *Data* field holds a variable number of data bits [8].

This paper compares three of the TAM architecture for SoC design. The following sections describe the platform SoC, the implementation of the three different TAM architectures, and the results of our experiments.

2. PLATFORM IMPLEMENTATION

In order to evaluate the proposed test access mechanisms, a network-processing engine has been developed as the test vehicle. The network-processing engine is an OSI Layer 3 device that forwards IPv4 packets

The architecture of the network-processing engine is based on information gathered from the public domain [10]. The major blocks of the engine include a pre-processing unit, a classifier, an embedded processor, several post-processing units, and various memory components. The blocks communicate with each other either through an AMBA™-compliant high-speed bus. Point-to-point connection is also used for control signals and interrupts.

The development of the blocks follows the reuse guidelines such as I/O buffering and sub-block partitioning [11]. The embedded processor is a modified Motorola HC11™ micro-controller. The AMBA-compliant bus was developed using the AMBA AHB specifications available from ARM Ltd. [12]. The bus uses a master-slave request-based architecture with multiple clock cycles per transfer. The pipelined bus design allows for efficient implementation that satisfies the high-bandwidth requirements of the network-processing engine.

For testing at the core level, full-scan testing is deployed for the pre-processing unit and the post-processing units. One single scan chain is created in each block, and scan vectors generated by an ATPG tool are provided externally. To create a heterogeneous testing environment, the classifier and the embedded processor uses logic BIST. The logic BIST uses a 32-bit linear-feedback shift register (LFSR) to generate pseudo-random test vectors and uses a signature analyzer to compact the test result. The memory modules use memory BIST that runs a Marching C algorithm for testing. All the blocks and the associated test structures are encapsulated with P1500-compliant wrappers. P1500 is a standard under development by the IEEE targeted specifically at SoC core testing [13]. Its purpose is to provide a uniform interface between the cores and the chip-wide TAM, analogous to how JTAG facilitates board-level testing.

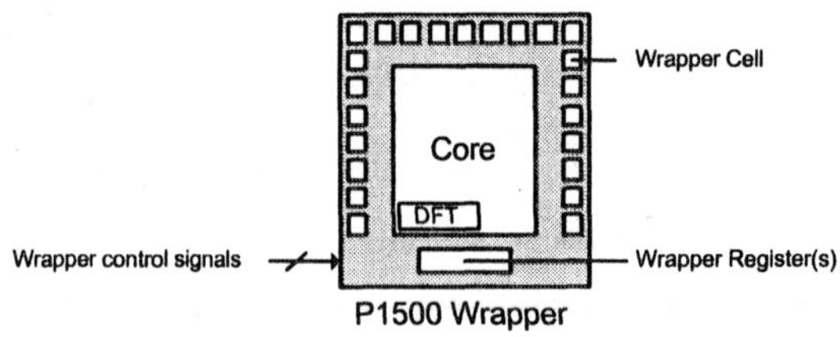

Figure3. Conceptual diagram of a core, the core DFT, and the 1500 Wrapper

3. TAM IMPLEMENTATION

Three different TAM architectures were compared on the network processor platform as shown in Figure4. The first TAM investigated leverages the new P1500-compliant wrappers. The P1500 standard is very similar to the legacy JTAG boundary scan; the most noticeable difference is the absence of TAP controller and the addition of a parallel test port. By detaching the TAP controller and providing more access ports, the serial-input constraint of JTAG is removed and a wide variety of test access mechanisms are supported. It is possible to serially thread the P1500 wrappers to create a simple TAM for a SoC. This was the first TAM architecture in the comparison.

The second TAM is the NIMA as described above. In this case, we varied the number of test pins to allow test packets of varying widths to be used to determine the optimal number for this design. By increasing the width, the bandwidth of the test access mechanism scales with the increasing number of cores so that the test time is within an acceptable threshold.

A third TAM used in this comparison is a test-rail approach [7]. Our specific test rail TAM uses time-division multiplexing for sharing the data bus. Each core automatically assumes control over the bus when an internal counter reaches some predetermined time slots. This eliminates the need for an address bus.

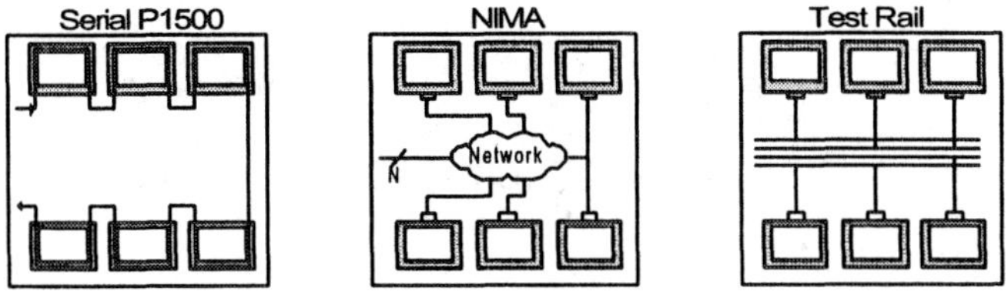

Figure4. Competing TAM architectures

While all TAMs described above use P1500 wrappers, the NIMA TAM requires that test data be pre-processed into packets. Therefore, before testing begins, test data (either test vectors for scan-based modules or commands for the BIST) is formatted into test packets and sent to the designated block. In order to translate the test packets into information understandable to the P1500 wrapper, a NIMA interface module must be developed and attached to each SoC block. A controller within the NIMA interface module generates the P1500 wrapper control signals. Since P1500 does not specify the implementation of such a controller, the TAP controller design from the JTAG standard was used because of its robustness and similarity between the standards. This controller is responsible for sequencing the P1500 control signals such that the P1500 wrapper can perform the operations outlined in the standard. In addition, the NIMA test network parallelizes the test data, so a data funnel is required between the test network and the NIMA interface module in order to down-convert the data width. This enables the packet switching capability of NIMA test network by allowing the network to have a higher bandwidth than required by a single core. Since the test data from the NIMA test network arrives at the NIMA interface in bursts, a buffering scheme is incorporated to ensure the P1500 wrapper gets the data only when it is ready to accept it.

4. RESULTS

There were two modes of DFT used in the network processor design: full scan and BIST. The BIST approach has the advantage of lower test traffic than full scan. In addition, a BIST approach avoids off-chip vector storage and management, thus simplifying the ATE setup. However, the current BIST algorithm is based on pseudo-random pattern generation, which typically suffers from reduced fault coverage. To mitigate the problem, we require BIST to use more vectors (2 to 10 times more) than full-scan test. This of course results in longer test time. Consequently, it is important to explore the impact of various BIST and full-scan DFT combinations when considering the choice of TAM architectures, as shown below.

The NIMA TAM is first compared to the serially threaded TAM in six different scenarios. In each scenario, the number of cores that use full-scan increases from the previous scenario, which results in a larger volume of test data that need to be transferred. Power simulation indicates that the power budget is not exceeded even when all DFT structures are operating concurrently. As a result, the DFT structures are allowed to run in parallel whenever possible. Figure 5 shows the test time of exercising full-scan DFT using NIMA of different widths (from 1 bit to 32 bits) and the serial TAM. The horizontal axis indicates the total test data (in bits) transferred. The BIST execution time is approximately 530,000 cycles for all six scenarios, and it is chosen as the threshold for total test time.

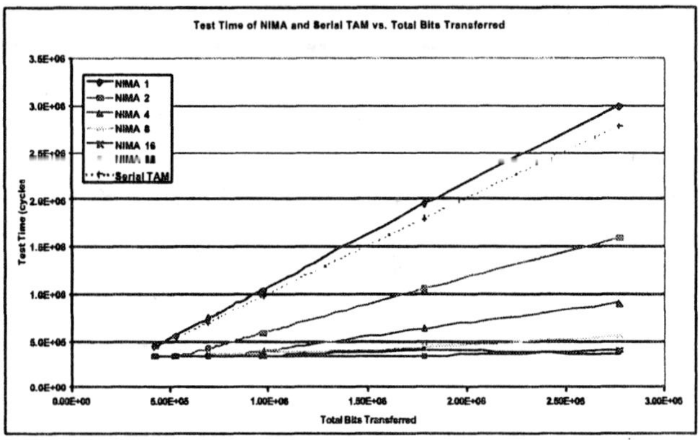

Figure 5. Test Time Comparison between NIMA and serial TAM

As seen in Figure 5, NIMA with a bit width of 2 or above performs better than the serial TAM in all scenarios. NIMA with a bit width of 1 is performing worse than the serial TAM due to the overhead introduced by the packet header and the delay through the NIMA interface module. As the number of cores using full scan increases, a larger NIMA width is required to keep the total test time below threshold. For this particular SoC platform, a 8-bit NIMA is sufficient to keep the test time close to the chosen threshold, even in the worst-case scenario where all logic cores use full scan.

The performance of NIMA is partially dictated by the scheduler, which determines when to send a particular test packet. The current scheduling algorithm sends the packet for the first-ready core. The time for a core to be ready is calculated in the algorithm based on the time to deliver the packet and the time to perform the test. If more than one core is ready at the same time, the core with larger scan chain has higher priority. The algorithm also looks ahead in time so that a non-ready core with a longer scan chain may preempt a ready core with a shorter scan chain. This heuristic sometimes reduces the idle time of the core DFT and thus decreases overall test time. An optimal scheduling algorithm is the subject of future research.

In Figure 6, the test-rail TAM is compared to the NIMA TAM using the same six scenarios in NIMA-serial comparison. Here the test width is varied from 1 bit to 5 bits. The results suggest that the test rail performs better or comparable to NIMA, except for the case where the test width is 1 or 2. The low performance of test rail with a small test width is caused by the limitation of our time-division multiplexing scheme.

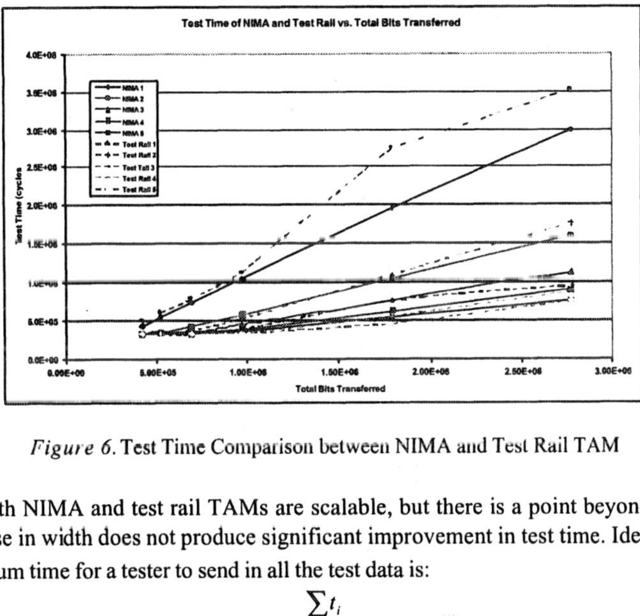

Figure 6. Test Time Comparison between NIMA and Test Rail TAM

Both NIMA and test rail TAMs are scalable, but there is a point beyond which increase in width does not produce significant improvement in test time. Ideally, the minimum time for a tester to send in all the test data is:

$$T_{opt} = \frac{\sum_i t_i}{W} \qquad (1)$$

where t_i is the total test cycles of core i, and W denotes the width of the TAM. However, if there is one core whose total test cycle time is larger than T_{opt}, than it implies that the core cannot consume all the data that is sent to it. Either the tester has to slow down or a buffer is required. In this case, the actual test time is t_{max}. Then, it becomes apparent that the optimal width is the one that makes $T_{opt} = t_{max}$. Thus,

$$W_{opt} = round\left(\frac{\sum_i t_i}{t_{\max}}\right) \qquad (2)$$

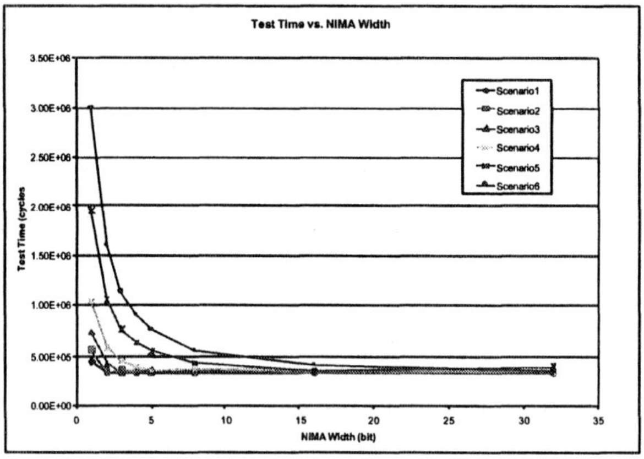

Figure 7. NIMA test Time versus test width

Figure 7 shows the plot of NIMA test time versus different test widths in the six scenarios previously discussed. The optimal width calculated using (2) is labelled in the figure. As shown by the graph, no significant improvement in test time is achieved when the width is raised above the optimal value. Note that (2) does not take the TAM overhead into account, but it still serves as a reasonable estimate of the optimal test width.

	Optimal Width
Scenario1	1
Scenario2	2
Scenario3	2
Scenario4	3
Scenario5	5
Scenario6	8

Aside from test time, there are many other factors that need to be evaluated for thorough comparison. Table 1 summarizes several factors that SoC designers should consider when choosing TAM architecture. Despite its simplicity, the serial TAM is not a scalable solution and is not suitable for complicated SoCs where large test data traffics could occur. However, we still include it in the comparison with NIMA and test rail to illustrate some of the trade-offs that designers could make. Judging from the merits that are presented, test rail is a better TAM if area overhead is important, while NIMA is preferred if reducing I/O pin count is paramount. NIMA is able to save pin count because the control signals are embedded in the packets; once a packet arrived at the core, an interface module decodes the packet and generates appropriate control signals. By incorporating the technique of embedding control signals into the test-rail architecture, a scalable TAM with low-test time and low pin count could be devised.

Table 3. TAM Comparison (++: strong advantage, +: advantage, -: disadvantage)

	Serial P1500		NIMA		Test Rail	
Test Time	-	Not scalable	++	Scalable	++	Scalable
Logic Area	++	No overhead	-	Overhead due to routers, interface logic, and funnels that act as buffers	+	Less than 2% of the SoC area
Wire Area	+	Small	-	Routing overhead due to aggregated channel lines	-	For large width, there is routing overhead
Total Test Pins	+	5 fixed pins (1 for data, 4 for control signals)	+	2x width of NIMA network	-	3x width of test rail
Synthesis automation	++	No issue	+	Minor issue	+	Minor issue
Pre and post processing of test data	+	ATPG	-	ATPG, test packet generation, packet scheduling, and extract test results from returned packets	-	ATPG, distributing test data according to optimized test width and core assignment, and re-organizing test outputs for analysis

The control signals of the test rail are a hidden cost that is not emphasized in previous research papers. The number of I/O pins available for testing and the extra routing resource required by TAM control signals are important considerations. The proposed NIMA architecture reduces the pin count and wires for control signals at the expense of more silicon area for additional logic. One additional advantage of NIMA is that it allows the test bandwidth to be re-allocated dynamically. For instance, during diagnostic test engineers can use the full bandwidth of the network to test a single core if so desired. For the test rail, the bandwidth allocated to a particular core is fixed during implementation of the TAM. Furthermore, it appears that to achieve any significant improvement in performance, the TAM should be decoupled from the core DFT. Our current architectures of NIMA and test rail both use the same clock frequency as the core DFT. If the number of I/O pins available for testing is less than optimal, clocking the TAM at higher speed makes up for the lower bandwidth and thus reduces test time. The test rail is an extension to the scan chain concept and is difficult to decouple from the core DFT. NIMA on the other hand, is conceived as a system completely independent of the core DFT, and thus allow such decoupling to be implemented quite easily. A similar decoupling scheme has already been proposed for the functional communication between the cores [14], and it is believed to be the solution for developing high-speed SoCs.

5. CONCLUSIONS AND FUTURE WORK

We believe that the concept of an on-chip network will be an important topic for research in SoC design. Our focus is on the use of this type of network for test purposes. The results in this paper suggest that the packet-switching network-based TAM is comparable to the test-rail TAM in terms of test time when applied to the same SoC platform. Both architectures also exhibit the same degree of flexibility to allow trade-offs between test time and test pins. While the test rail allows for slightly better performance and lower area overhead, NIMA yields significant savings in pin count and is independent of the core DFT structures.

Further analysis is required to quantify the area overhead introduced by the TAM structures. Moreover, more work is needed to develop optimal scheduling for NIMA and to improve the time and space overhead of the TAM. Finally, it is important to understand the trade-offs between BIST and scan in the NIMA-based DFT methodology. These issues are topics of ongoing research.

ACKNOWLEDGEMENTS

The authors wish to thank NSERC, CMC, PMC-Sierra and Micronet for providing support for this project. A number of students at UBC helped compile the results in this paper: Zahra Ebadi, Julien Lamoureux, Gary Lim, Simon Chan, Martin Ma. We gratefully acknowledge their contributions.

REFERENCES

[1] E.J. Marinissen, Y. Zorian, "Challenges in Testing Core-Based System ICs", *IEEE Communication Magazine*, June 1999, pp. 104 –109.
[2] Y. Zorian, E.J. Marinissen, S. Dey, "Testing Embedded-Core Based System Chips", *Proc. IEEE ITC*, 1998, pp. 130-143.
[3] I. Ghosh, N.K. Jha, S. Dey, "A Low Overhead Design for Testability & Test Generation Technique for Core-Based Systems", *Proc. IEEE ITC*, 1999, pp. 50-59.
[4] L. Whestsel, "An IEEE 1149.1 Based Test Access Architecture for ICs with Embedded Cores", *Proc. IEEE ITC*, 1997, pp. 69-78.
[5] Debashis Bhattacharya, "Hierarchical Test Access Architecture for Embedded Cores in an Integrated Circuit", *Proc. VLSI Test Symposium*, 1998, pp. 8-14.
[6] A. Benso et al, "HD2BIST: Architectural Framework for BIST Scheduling, Data Patterns delivering & Diagnosis in SoCs", *Proc. IEEE ITC*, 2000, pp. 892 –901.
[7] E.J. Marinissen, R. Arendsen, G. Bos et al, "A structured & Scalable Mechanism for Test Access to Embedded Reusable Cores", *Proc. IEEE ITC*, 1998, pp. 284-293.

[8] Mohsen Nahvi, Andre Ivanov, "A Packet Switching Communication-Based Test Access Mechanism for System Chips", *Proc. IEEE European Test Workshop*, 2001, pp. 81-86.
[9] William J. Dally, Brian Towles, "Route Packets, Not Wires: On-Chip Interconnection Networks", *Design Automation Conference (DAC)*, 2001
[10] Stephen J. Sheafor, "Network Processors: Ushering in a New Era of Performance and Flexibility", www.sitera.com.
[11] M. Keating, P. Bricaud, Reuse Methodology Manual, 2nd Edition, Kluwer Academic Publishers, 1999.
[12] "AMBA Specification Rev 2.0", www.arm.com.
[13] http://grouper.ieee.org/groups/1500/
[14] Drew Wingard, "MicroNetwork-Based Integration for SOCs", *Design Automation Conference (DAC)*, 2001

4

MODELING

Modeling is an important component in any design technique. In the case of SOC complexity designs, modeling can take a variety of forms. In this chapter we present four different aspects of modeling.

In section 4.1, "SOC Modeling and Simulation Based on Java", by Henry Fu and Richard F. Hobson of Simon Fraser University, Canada, an instruction level SOC simulator is proposed to speed up the functional verification and optimization of SOC hardware-software specification. The simulator is written in the Java programming language, and a Java hardware interface model is proposed to interface the hardware to the simulator. Clock synchronization techniques are used in the simulator to model real time processing. A comparison with VHDL simulation shows that the proposed instruction level simulation approach is around two to three orders of magnitude faster than the VHDL simulation on the same platform.

Section 4.2, "RTOS Modeling Using SystemC", by Mohamed Abd El-Salam, Ashraf Salem and Gamal Aly of Mentor Graphics and Ain Shams University, Egypt, presents a methodology for modeling a priority-based preemptive real time operating system (RTOS) kernel in SystemC. The current modeling constructs of SystemC 2.0 are used with additional constructs, that can be used in RTOS modeling, being proposed throughout the development of the kernel's system calls. The authors then demonstrate the interaction of the embedded software module, consisting of the RTOS kernel and the running application tasks, with a hardware module representing a bus functional model (BFM) of a generic microcontroller. By applying appropriate design techniques for the RTOS and the BFM, performance estimates can be obtained for software execution, which can be used to drive the co-design process and speedup system integration phase early in the design cycle.

Section 4.3, "Modeling, Synthesis and Implementation of Communicating Hierarchical FSM", by Valery Sklyarov, of the University of Aveiro, Portugal, deals with the specification and synthesis of fast control systems implemented on an FPGA chip, and a technique is suggested for the design of high-speed communicating hierarchical finite state machines (HFSM). The specification method for a HFSM provides support for modularity, hierarchy and parallelism. It is shown that an implementation of a HFSM can be provided on the basis of a reusable circuit with reprogrammable RAM-blocks. Hierarchical mechanisms have been realized with the aid of a stack memory. The proposed synthesis method allows the given specification to be formally translated into bitstreams for the RAM-blocks. Using RAM-blocks with dual port capabilities allows run-time modifications of the hierarchical control algorithms, making these algorithms flexible and extensible.

Finally, in section 4.4, "A Modeling Method for Reconfigurable Architectures", by Lilian Bossuet, Guy Gogniat, Jean-Philippe Diguet and Jean-Luc Philippe of the University of South Brittany, France, a reconfigurable FPGA architecture modeling method is presented. The proposed model is applied to a domain representative component in order to show its ability to describe a complex architecture. The modeling method, which is based on a functional view, is able to describe complex, heterogeneous and multi-granularity architectures. The approach allows researchers to compare their new architectures with others, and designers will be able to explore the design space in order to select the best architecture for their application.

Section 4.1

SOC Modeling and Simulation Based on Java

Henry Fu and Richard F. Hobson
Schools of Engineering and Computing Science, Simon Fraser University

Abstract: An instruction level SOC simulator is proposed to speed up the functional verification and optimization of SOC hardware-software specification. The simulator is written in the Java programming language, and a Java hardware interface model is proposed to interface the hardware to the simulator. Clock synchronization techniques are used in the simulator to model real-time processing.

Key words : SOC, modeling, instruction-level simulation, Java

1. INTRODUCTION

System-on-chip (SOC) [2] is usually formed with reusable and programmable cores, such as embedded processors and DSPs, software programs and hardware IP attachments, such as input/output interface, codec, memory blocks, etc. SOC design is so complex that functional verification is desired to start in the early design stages through high-level simulation, while further simulations will be carried out step by step at lower design levels, such as at RTL and gate levels.

Traditional digital hardware systems are defined by using hardware description language (HDL), such as Verilog and VHDL. These two HDLs are suitable for RTL and gate level specification, but not so efficient in the higher algorithmic level specification [3]. In order to define SOC in a higher level of abstraction, newer hardware specification languages based on the C++ [7][11] and Java [5][12][8] programming lan guages have been proposed.

This paper presents a high level model of the SOC and a high level SOC simulator, targeting on reducing simulation time as required by the traditional HDL models.

For a processor, the highest software abstraction is the instruction set. Simulation of a processor at the instruction set level is apparently the fastest software approach. Our approach is to model all the SOC's programmable cores at the instruction set level, and glue the other hardware to the programmable cores as attachments, such that the attachments become parts of the programmable cores. Figure1 shows this modeling approach. It would appear that the SOC contains only a number of programmable cores. We can then apply only instruction level simulation to simulate the whole SOC.

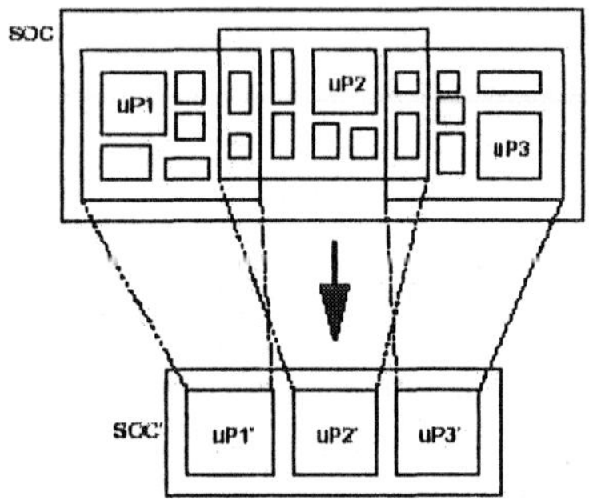

Figure1. SOC Model

Designing an instruction level simulator may take some effort at first. However, as in the case of SOC design that a programmable core can be reused for so many times, the speed advantage can easily outweigh the effort. Even with feature configuration changes on the programmable core, it is relatively easy to update changes on the instruction level simulator.

A high-level hardware description language is required to define the programmable core interface and the hardware attachments, and to interface the hardware attachments to the programmable cores. The Java programming language is chosen as the hardware description language, simply because Java fulfills the following requirements:

a) It is high-level, object-oriented [9], and is easy to use.
b) It supports multi-threading, therefore, can be used on parallel processing simulation.
c) It supports networking, such that the high-level simulator can be easily adapted to cluster computing [1].
d) It has a powerful graphical user interface library.
e) It has been proved suitable for hardware -software description [5][12][8].
f) Speed problem can be improved with the just-in-time compiler [4].

2. PREVIOUS WORK

Some previous work on using the Java programming language as a specification language for hardware-software systems are discussed below:

Helaihel and Olukotun [5] proposed the Java programming language as a single specification language for hardware-software system modeling at algorithmic or behavioral level. A technique of control and dataflow analysis was presented to extract concurrency at various levels of granularity from the design specification.

Young, et. al. [12] described another approach in using Java as a hardware-software specification language. Based on a methodology called successive, formal refinement, a specification written in unrestricted Java can be transformed gradually to a restricted and formal model of embedded system specification.

Kuhn, et. al. [8] presented an approach differing from the above with regards to overlapping support of varying abstraction levels. While the other two approaches concentrate in algorithmic level, this approach enables hardware-software specification at RTL, algorithmic and system levels.

3. CHIP ARCHITECTURE MODELING

Our approach differs from the previous work in that we emphasize the use of the instruction level simulation technique. We first write instruction level simulator for the programmable cores using Java, and interface other hardware to the simulator by instantiation of the component objects. An instruction level simulator can be optimized for speed relatively easily. It is similar to optimizing the algorithm of a Java program. For the previous work, all the programmable cores have to be described using Java hardware specification statements. It is non-straight forward and rather difficult to optimize the hardware specification of programmable cores for simulation speed.

The SOC model we proposed is based on the objected-oriented approach that enhances the reusability of the instruction level simulator and components. A hardware system can be defined as a top-level component in the hierarchy. A component can be formed with other sub-components, and sub-components can be formed with even lower-level components. Pre-defined component bodies are stored in a component library. The declaration of a new hardware system is relatively easy and fast. The architectural description consists of the component declaration and component body declaration [6] parts.

Component Body Declaration

Component body declaration specifies the behavior of a component to be modeled. Each component body contains a set of statements, which specify the functionality but not the structure of the component.

A component body is formed with body declaration, input ports, output ports, and parameter setting. By setting the parameters of a component, different properties of a component can be defined. For example, we can set the bus width or data path width of a component.

Figure 2 shows an example of component body declaration. The component is defined as a member of the component library. The parameters can be passed to a component at its instantiation. Data can be input to a component through parameter passing at a method. Output values can be obtained from method returns. For example, if we want to get the sum of integers a and b, we can call the method "sum", and pass a and b to the method through parameter passing. The result with bus width adjusted can be obtained from the method return statement.

```
package comp_library;
public class Cp {
    private int busWidth;
    public Cp(int busWidth) {           // parameter setting
        this.busWidth = busWidth;
    }
    public int sum(int a, int b) {        // input/ouput
        return widthAdjust(a + b);
    }
    private int widthAdjust(int value) { ... }
}
```

Figure2. Example of component body declaration

Component Declaration and Connection

Component declaration initiates a new component from a component library, and specifies the parameters of the component. The component library contains a large number of pre-defined component body templates. For instance, a register file component body stored in a component library does not posses the data path width and size information. A register file statement in the component declaration will import the register file component body to a design, and set up the data path width and register file size parameters. Connections between components are also described in the component declaration part.

Figure3 shows an example of component connection. Suppose E is the parent component. We have sibling components A, B, C, and D to be connected. The program defines the connections in the architectural level. We first input all the components A, B, C and D from the component library, and instantiate the components, respectively. The detailed, low-level bus and data bits information, input or output port, etc. need not be clearly specified at the connection, and components are connected through parameter passing. This is quite different from VHDL or Verilog.

```
import comp_library;
public class E {
    private A a = new A();
    private B b = new B();
    private C c = new C();
    private D d = new D();
    private boolean x0, x1, x2, y0, y1, y2, a0,
        z0, z1, z2;
    public void update(boolean x0, boolean x1,
        boolean x2, boolean y0,
        boolean y1, boolean y2) {
        this.x0 = x0; this.x1 = x1; this.x2 = x2;
        this.y0 = y0; this.y1 = y1; this.y2 = y2;
        a0 = a.getResult(y0, y1, y2);
        z0 = b.getResult(x0, a0);
        z1 = c.getResult(x1, a0);
        z2 = d.getResult(x2, a0);
    }
    public boolean getZ0() { return z0; }
    public boolean getZ1() { return z1; }
    public boolean getZ2() { return z2; }

}
```

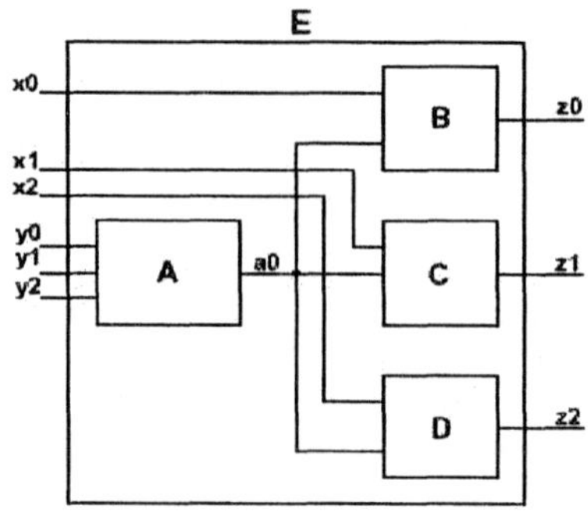

Figure3. Component connection example

In the example, the system is operated by calling the "update" method. Data x0 to y2 are input through parameter passing, and the intermediate value a0, and output data z0 to z2 are updated. Values at the output ports can be obtained by calling the get result methods, such as getZ0, getZ1 and get Z2.

As shown in Figure 4, the connection strategy is based on the following approach:

a) As least one programmable core is in the simulation system. The instruction-level simulator cannot operate if there is no programmable core.
b) Non-programmable components are connected to a programmable or non-programmable component through instantiation, parameter passing and method calling.
c) Each programmable core and its hardware attachments form a subsystem.
d) If a direct connection exists between two subsystems, a pseudo non-programmable component, such as a buffer should be added in between the connection to avoid component declaration conflict.
e) Components shared by more than one subsystem, such as processor interconnect, shared memory and communication FIFO are declared static in Java.

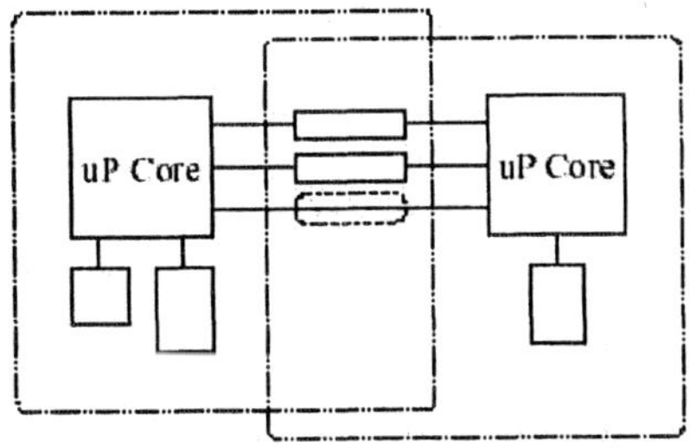

Figure 4. Hardware interfacing in SOC

4. INSTRUCTION LEVEL SIMULATOR

Instruction-level simulation is a high-level simulation technique. It takes a machine language program as its input, and runs the program according to the definition of the processor instruction set. The detailed hardware structure of the processor is not required.

Our approach is to use the instruction-level simulator for both hardware and software verifications. The instruction-level simulator can be used to fine-tune the target system.

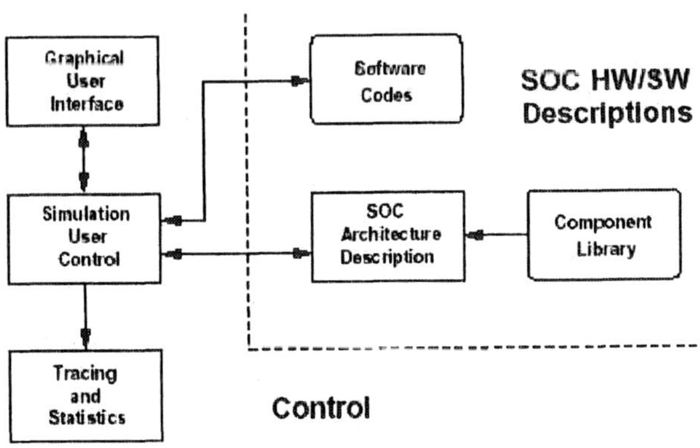

Figure 5. Structure of the instruction level simulator

Figure 5 shows the structure of our instruction-level simulator. The simulator is cycle accurate and has the hardware described by the Java HDL. The component library provides reusable components such that the time for writing the hardware description can be minimized.

Software codes are input to the simu lator as input. Users can control the simulation process interactively using the graphical user interface (GUI). Where the GUI can be decoupled from the system for speed enhancement.

Address tracing, memory and I/O tracing, and other statistics information can be recorded to data files on a hard drive. As the recording process is quite time consuming, this feature can be turned off when not required.

One very important feature of our simulator is that it supports parallel, multiple clocks and multiprocessor host simulation. It is also capable to support cluster-computing environment [1]. We can simply utilize Java threads for parallel process implementation. Inter-thread communications and thread context switching are relatively lightweight, when compa red to Unix context switching.

For the thread assignment, we can assign one thread to each programmable core, or one thread to each different clock. The latter case may require less context switching. However, for a multiprocessor host computer, or the case in computing using a cluster of computers, the former case may be more efficient, since the load can be more evenly shared.

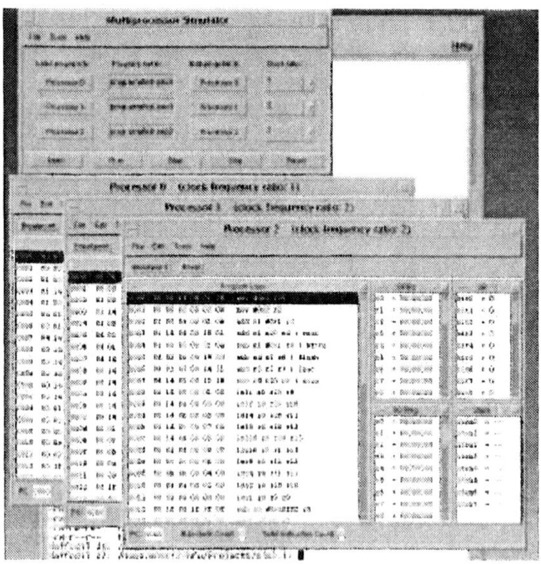

Figure 6. Graphical user interface of the simulator

Figure 6 shows a snapshot of the GUI of our simulator. There are some buttons in the top main window for interactive control, such as Open, Run, Step run, Stop and Reset. We can also click at the buttons to edit and compile a source code. Monitor windows can be opened to show the programmable cores' internal information, such as the register contents and the system stack contents. The simulator has a built-in disassembler to convert the machine code to assembly language. We can set breakpoint on a particular program line, or to start running on a selected program line. There are also some on-screen statistics such as the sub-clock tick count and instruction count. In order to simplify the simulation process, an SOC system must be divided into subsystems, and subsystems must be relatively independent and can communicate with each other in a simple way. Our approach is to divide an SOC to subsystems of programmable cores with their hardware attachments, respectively. Each subsystem is simulated using a thread, and the subsystems can communicate with each other through data passing in shared memory or FIFO, as in the common SOC design.

5. PARALLEL DISCRETE SIMULATION

The simulator has two simulation modes: synchronous and asynchronous. The synchronous mode is for debugging purposes. This is useful for step run, breakpoint or to trace the process flow. The asynchronous mode has higher efficiency as all the processes can run independently.

Synchronous Simulation

For debugging and verification with respect to real-time processing sequence, we need to use synchronous simulation to monitor the processes at each clock cycle. Some assumptions have to be made to enhance high-level simulation. First, we assume all processor (programmable core) clocks are derived from a global clock source. This eliminates multiple clock phases as required for low-level simulation. Second, we assume the processor clock frequencies can be represented in a small ratio. For example, the clocks at 150MHz and 100MHz can be represented as 3:2. Third, embedded processors are pipelined, which can fetch one instruction per clock cycle. Our simulator instruction fetch timing can then be greatly simplified. We also assume that all events are clock rising-edge triggered, this is true for most processor simulation cases.

Figure 7 shows a simple and fast timing strategy that we use for regulating the SOC parallel processing. The clock frequencies are 100MHz for clock 1, and 150MHz for clocks 2 and 3, respectively. So that the clock period ratio is 3:2:2. We will take the greatest common factor as a sub-clock tick. We advance the sub-clock tick until a clock rising-edge (next event) is met, then the processors at that clock rising-edge will wake up and execute, when finished, the processors will go to sleep. If no more processor clock rising edge is found at a sub-clock tick, the simulation clock will be advanced to the next sub-clock tick immediately without waiting.

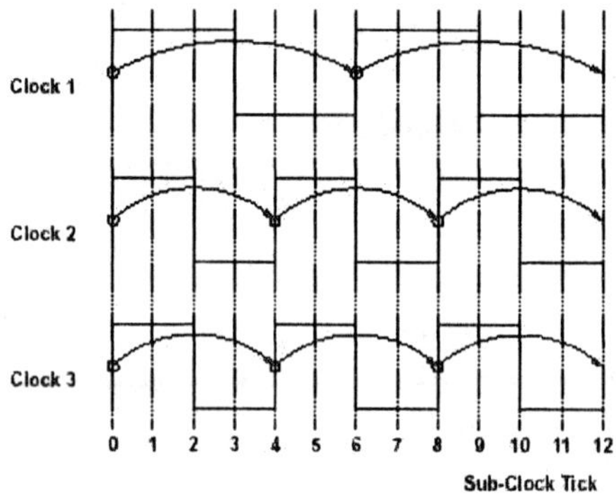

Figure 7. Synchronous clock simulation

Asynchronous Simulation

For continuous program run, asynchronous simulation is more efficient than synchronous simulation, since events can keep running until a stall signal is raised. Asynchronous simulation is also required if clock skew and clock phase is to be considered.

The proper communication time sequence is maintained by using local clocks in the subsystems. Output messages are time-stamped, and the input messages are consumed in the ascending time stamp order. Local clock time is updated according to the consumed input message time stamp and the event service time.

A subsystem thread will be put into a wait mode only if at least one input buffer is empty, and all non-empty buffers have event messages with time stamps larger than that expected in the empty buffers, and no output data to send. The thread will be activated again when a new event message arrives.

We assume that deadlock can be detected, or recovered, or prevented by using proper protocol in the input software, so that the instruction level simulator does not have to handle this situation.

6. CONCLUSION AND FUTURE WORK

We proposed a hardware specification language with Java programming language constructs, which differs from the previous work in the use of an instruction level simulator to replace the detailed hardware specification of a programmable core. The simulation speed then can be greatly increased. The simulator can be run on cross platform Java virtual machines.

Preliminary results show that our approach actually has good speed advantage over the VHDL simulation. We have performed some examples using one to several processors . A comparison with VHDL simulation in our test shows that, our instruction level simulation approach is around two to three orders of magnitude faster than the VHDL simulation on the same platform.

In the future, we will investigate the translation of Java hardware description to a general HDL, such as Verilog for synthesis. Also for synthesis purposes, the programmable cores have to be redefined using the proposed hardware specification language, since the instruction level simulator does not provide enough netlist information for synthesis.

We are planning to write protocols to coordinate a cluster of workstations and to simulate the parallel processes at even higher speed.

REFERENCES

1. R. Buyya. High Performance Cluster Computing: Programming and Applications, Volume 2, Prentice Hall PTR, 1999.
2. H. Chang, L. Cooke, M. Hunt, G. Martin, A. McNelly, and L. Todd. Surviving the SOC Revolution, Kluwer Academic Publishers, 1999.
3. S.T. Cheng. Compilation, Synthesis, and Simulation of Hardware Description Languages – The Compositional Models of HDL's, Ph.D. Dissertation, University of California, Berkeley, 1998.
4. T. Cramer, R. Friedman, T. Miller, D. Seberger, R. Wilson and M. Wolczko. Compiling Java Just In Time, IEEE Micro, pp. 36-43, May/June 1997.
5. R. Helaihel and K. Olukotun. Java as a Specification Language for Hardware-Software Systems, ICCAD, pp. 690-697, 1997.
6. R.F. Hobson, K. Balasubramanian, G. Chen, K. Cheung, A. Dyck, S. Evenson, and H. Fu. Design Considerations for Real-Time System-on-Chip Applications, Invited Session on System on Chip (SoC) for Real-Time Information Processing, the 5th World Multi-Conference on Systemics, Cybernetics and Informatics, July 22-25, Orlando, 2001.
7. http://www.systemc.org/.
8. T. Kuhn, W. Rosenstiel, and U. Kebschull. Description and Simulation of Hardware/Software Systems with Java, DAC, pp. 790-793, New Orleans, June 1999.
9. S. Kumar, J.H. Aylor, B.W. Johnson, and W.A. Wulf. Object-Oriented Techniques in Hardware Design, Computer, vol. 27, no. 6, pp. 64-70, June 1994.
10. H. Tomiyama, A. Halambi, P. Grun, N. Dutt, and A. Nicolau. Architecture Description Languages for Systems-on-Chip Design, in Proc. APCHDL, pp.109-116, Oct. 1999.
11. D. Verkest, J. Kunkel and F. Schirrmeister. System Level Design Using C++, DATE, Paris, March 2000.
12. J.S. Young, J. MacDonald, M. Shilman, A. Tabbara, P. Hilfinger, and A.R. Newton. Design and Specification of Embedded Systems in Java Using Successive, Formal Refinement, DAC, 1998.

Section 4.2

RTOS Modeling Using SystemC

Mohamed Abd El-Salam, Ashraf Salem, Gamal Aly
Mentor Graphics Egypt, Computer & Systems Dept, Ain Shams University, Cairo, Egypt.

Abstract: In this paper we present our methodology for modeling a priority-based preemptive real time operating system (RTOS) kernel in SystemC. We use the current modeling constructs of SystemC 2.0 and throughout our development of the kernel's system calls we propose new constructs that can be used in RTOS modeling. We show then the interaction of the embedded software module consisting of the RTOS kernel and the running application tasks with a hardware module representing a bus functional model (BFM) of a generic microcontroller.

Key words: SystemC, RTOS, Co-verification, Co-simulation, Virtual prototyping

1. INTRODUCTION

SystemC, a new modeling methodology based on C++, is now used to describe both hardware and software of an entire system throughout the design flow [1,2]. SystemC fits smoothly within current System on Chip (SoC) design methodologies [3,4] & provides great flexibility in modeling.

The language is entirely based on C++ and its constructs are provided as a C++ class library, for modeling at different levels of abstraction from system behavioral to Register Transfer Level (RTL). In addition, it embeds a lightweight simulation kernel for cycle-based simulation [1,5].

In SystemC 1.0, modeling constructs address RTL to play the same role as VHDL and Verilog in modeling H/W. In SystemC 2.0, new constructs were added to give the language an advantage in describing systems at higher level of abstraction. Current development is to extend SystemC capability to describe embedded S/W & model RTOS [1,6].

Topics like, Co-design [3,7] and Co-verification [8,9,10] can then be easily achieved using the same modeling language in a homogenous development environment where mature C++ development tools are used to debug various parts of design and then use a process of gradual refinement for synthesizing directly from C/C++ specification [11,12].

In this paper we show how to model an abstract RTOS kernel in SystemC describing its features, architecture, system calls and finally its interaction with a BFM of a generic microcontroller.

The rest of this paper is organized as follows. In Section 2, we present an overview on RTOS used terminology. In Section 3, we describe the kernel's model architecture. In Section 4, we show how the embedded S/W will be interacting with the H/W module. In Section 5, we present an example showing how to write a S/W application using the RTOS kernel model. Conclusion is in Section 6.

2. TERMINOLOGY & DEFINITIONS

Real time refers to any system that responds to inputs and supplies outputs fast enough to meet hardware or user requirements [13,14]. A *real time operating system (RTOS)* is a software task that's linked with application tasks; it uses processor resources to bring order & standardization to running tasks. A *kernel* is the basic part of an operating system. A *task* is a functional unit in a controller job. An *event* comes from outside a task and might cause the system to switch between tasks. A *priority* means, which task moves to the head of the line and gets done first. A *scheduler* refers to an event that can regularly wake up a task with no other intervention. *Preemption* means, that the running task has been kicked out not just moved to the head of the line. A *priority-based preemptive multitasking* means, more urgent tasks go to the head of the line or preempt the running task. A *counting semaphore* is a counter that a task waits to run until it reaches zero, more count units means greater suspension depth. A *mailbox* is a way of communication, where tasks can exchange information by passing messages to a mailbox. A *message* is an event where a task can be asleep until a message arrives.

3. RTOS KERNEL MODEL

The following features are supported for each task in our SystemC kernel model: waiting for an interval event; a feature associated with a scheduler, waiting for a timeout event, handling Interrupts, message & semaphore events.

A task waiting for any of these events will go to sleep until it's notified with the arrival of the event. Upon notification, the task goes to the READY state. We also developed a priority/preemption scheme that is used in task switching, in which a task can be in one of 3 states: RUNNING, ASLEEP or READY, as shown in Figure 1.

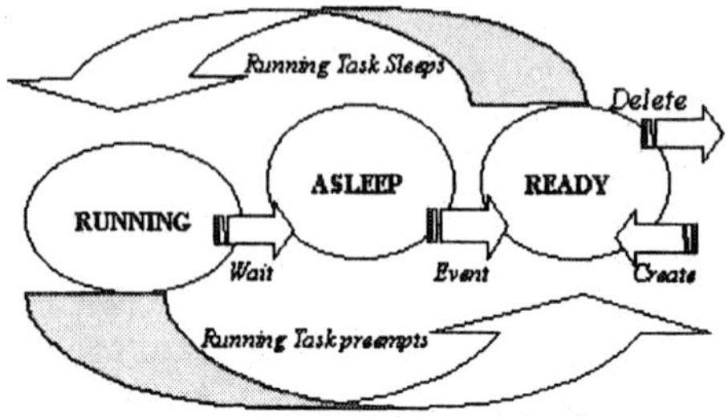

Figure 1. Task State

The RTOS Kernel involved 3 interacting modules, as shown in Figure 2. A Real Time Clock module from H/W side. Central, Table, System Calls & Tasks modules from S/W side.

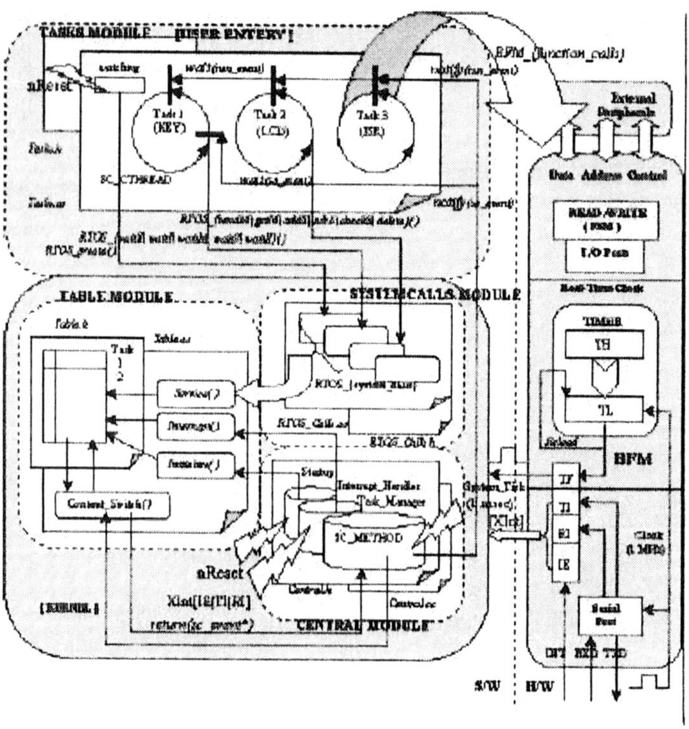

Figure 2. a) RTOS Kernel b) BFM

Real Time Clock – This is the driving force of the Central Module. We used the timing feature of the microcontroller where one timer/interrupt modeled in hardware is used to drive kernel operation with a default timing resolution 1 msec and the capability of adjusting it, as shown in Figure 3.

```
#define TH_COUNT 250   /*When Simulation Clock is 1 MHz, TH = 250 */
#define ROLL_OVER 4           /* 4 Count Rolls give 1 msec Tick */

SC_MODULE (RT_Clock){                     /* Ports */
  sc_out< bool >     System_Tick;   /* System Tick */

  sc_in_clk       Clk;              /* Simulation Clock */
  sc_in< bool >   nReset;           /* System Reset */

  void do_RT_Clock();

  unsigned int Roll_Over;                     /* Roll Over Counter */
  sc_uint<8> TH, TL;                      /* Timer Registers */
  bool TF;                          /* Timer Interrupt Flag */

  SC_CTOR (RT_Clock) {
    SC_CTHREAD (do_RT_Clock, Clk.pos ());
    TH = TH_COUNT; TL = TH_COUNT; TF = 0; Roll_Over = 0;
  }
};
```

Figure 3. Real Time Clock Module

Central Module – This module is responsible for task management, interrupt handling & kernel table initialization, as shown in Figure 4.

```
SC_MODULE(Central){                   /* Ports */
  sc_inout< bool >   System_Tick;   /* System Tick */
  sc_inout< bool >   eXInt;            /* External Interrupt  */
  sc_inout< bool >   TxInt;            /* Serial Tx Interrupt */
  sc_inout< bool >   RxInt;            /* Serial Rx Interrupt */

  sc_in_clk        Clk;           /* Simulation Clock */
  sc_in< bool >    nReset;            /* System Reset */

  void Task_Manager();      /* Task Management & Context Switching */

          /* Identifying Type of interrupt and Notifying ISR */
  void Interrupt_Handler();

  void Startup();                   /* Kernel Table initialization */

  SC_CTOR(Central) {              /* Processes & Sensitivity List */
    SC_METHOD(Task_Manager);
    sensitive_pos << System_Tick << Clk;
    SC_METHOD(Interrupt_Handler);
    sensitive_pos << RxInt << eXInt << TxInt << Clk;
    SC_METHOD(Startup);
    sensitive_pos << nReset << Clk;
  }
};
```

Figure 4. Central Module

Table Module – This is a data structure that keeps a record on every S/W task created upon startup and gets updated every system tick, interrupt and system call service handled, through a group of manipulation functions *{Context_Switch(), Interrupt(), Service() and Initialize()}*. Fields for each task record are shown in Figure 5.

```
enum State {ASLEEP, READY};
enum Int_Type {APPL_TASK, SERIAL_INT, eXt_INT};

typedef struct TASKINFO {
/* Task Id No.: To identify each Task */
char Task_IdNo;

/* Priority: Lowest No. being highest priority */
unsigned int Priority;

/* Task State: READY or ASLEEP */
State Task_S;

/* Message box Flag: To signal message arrival */
unsigned int Msgbx_F;

/* Message */
char *Msg;

/* Wake me when this time expires */
unsigned int Timeout;

/* Wake me every scheduled time */
unsigned int Interval;

/* Counting Semaphore */
unsigned int Semaphore;

/* eXt Int Flag: To signal external Interrupt */
unsigned int Interrupt_F;

/* Type of Task: Application, ISR-eXt, ISR-Serial */
Int_Type Int_T;

/* Notify Me at this Location when */
sc_event Msg_event;              /* Message arrives */
sc_event Semaphore_event;     /* Semaphore reaches 0 */
sc_event Timeout_event;          /* Timeout expires */
sc_event Interval_event;           /* Interval over */
sc_event Interrupt_event;       /* external interrupt */
sc_event Run_event;    /* If I'm still not preempted */
} Task, *TaskPtr;
```

Figure 5. Task Info

System Calls Module – This module contains all kernel related system calls. They are invoked from a running task, to perform an RTOS service. These calls then invoke the *Service()* manipulation function in the Table Module to update tasks' status based on the service required. RTOS_Calls are briefed in Table 1-a.

Notice that, an *RTOS_wait* call will use SystemC *wait(sc_event)* to cause a suspension of the task's thread of execution until its event arrives.

Tasks Module – This module represents the programmable entry for the user to implement his tasks. It consists of 2 types of threads, one that simply represents an application software task for ex, writing to an LCD or scanning a keypad, another dedicated to an interrupt service routine (ISR), handling one type of interrupts. We used events and dynamic sensitivity introduced in SystemC 2.0 to implement preemption feature in the kernel, where each thread is suspended every cycle of execution and then dynamically notified from the Central Module with every system-tick to keep on running if it's not preempted or go to the READY state. Within each task, the user is able to access functions located in System Calls Module for ex, sending a message to another task or waiting for a semaphore or timeout event, as well as, access member functions of the BFM to interact with hardware peripherals for ex, reading from an external memory location or writing to an I/O port. We also used global watching feature of SystemC to watch a system reset calling an RTOS system call to create and initialize the S/W tasks; i.e. register each task and set its priority and initial state in the Table Module.

a) RTOS Calls	b) BFM Call
RTOS_waitT – Sleep until timeout event	**BFM_WrXmem** – Write To Memory
RTOS_waitI – Sleep until interval event	**BFM_RdXmem** – Read From Mmeory
RTOS_waitM – Sleep until a Msg In mailbox	**BFM_FetchBusData** – Fetch Data from Bus
RTOS_sendM – Send a Msg. To another task	**BFM_BusBusy** – Bus Transaction Going On
RTOS_getM – Fetch a Msg when it arrives	**BFM_WrPort** – Write To I/O Port
RTOS_waitX – Sleep until external interrupt	**BFM_RdPort** – Read From I/O Port
RTOS_addS – Add #units to task's semaphore	**BFM_FetchPortData** – Fetch Data From Port
RTOS_subS – Sub. #units from task's semaphore	**BFM_PortBusy** – Port Transaction Going On
RTOS_checkS – Return semaphore count	**BFM_WrSBUF** – Write To Serial Buffer
RTOS_waitS – Sleep until semaphore counts 0	**BFM_RdSBUF** – Read From Serial Buffer
RTOS_create – Register task in kernel's Table	**BFM_WrIntReg** – Write To Interrupt Register
RTOS_delete – Un-register task from Table	**BFM_RdIntReg** – Read Interrupt Register

Table 1. a) RTOS Calls b) BFM Call

4. INTERACTION WITH BFM

In our methodology, the embedded S/W is compiled for the host other than the intended CPU. This model of execution is known as the compiled software model (CSM). As the running tasks execute, they are able to call member functions of the BFM. BFM_Calls are briefed above in Table 1-b.

The performance of this model is very suitable for Co-verification at this level of abstraction and faster than using detailed interaction of a hardware description language (HDL)-based simulation with Instruction Set Simulator (ISS) model of the targeted CPU [8,9,10,15]. We used an abstract BFM of a generic microcontroller [16] to test the functionality of the embedded S/W. Here, we present a brief description on its components, as in Figure 2-b.

Real time Clock – It's modeled as a 16-bit counter (TH/TL) working in 8-bit auto-reload mode, where TL will be loaded with the value in TH when the former overflows. An overflow flag (TF) is set by H/W when timer overflows-TF is reset automatically then by H/W signaling a Tick for kernel.

Write/Read Cycle – It's modeled as a finite state machine, where a running task can initiate a bus transaction to read or write from an external data memory or memory mapped I/O via address decoding.

Interrupt control – It's modeled as an Interrupt register that can be accessed from the Task Module to set or reset different bit locations in it. One external device is allowed to interrupt the microcontroller by setting an interrupt flag (IE).

Serial port – It's modeled as a buffer (SBUF). Writing to SBUF loads data to be transmitted and reading from SBUF accesses received data. We used a shift register mode of fixed baud rate. A Transmission interrupt flag (TI) is set at the end of character transmission to allow the S/W task to know when to load next outgoing character. TI must be cleared by S/W then. Also a Receive interrupt flag (RI) is set at the end of character reception to allow S/W task to read incoming character. RI must be also cleared by S/W.

I/O Port – It's modeled as an 8-bit register.

H/W peripherals – These are external devices such as LCD, Keypad. They are wrapped in GTK+ GUI widgets that interact with X-events to give the look and feel of a Virtual System Prototype [17,18], as in Figure 7.

5. S/W APPLICATION EXAMPLE

We implemented a simple example, showing 2 application tasks working together on the top of the RTOS kernel and interacting with the BFM and its associated H/W peripherals, as shown in Figure 6.

A High priority Task 1, acts as a scheduler that's executed every 100 msec to scan a H/W switch connected to I/O Port 1, display scanned data on a LED module connected to I/O Port 2 and send a message to Task 2.

A Low priority Task 2 is waiting for a message to arrive from Task 1 to display the data on an SSD counter connected to I/O Port 4 and save the most recent value in an external data memory at location 0x2010. Memory & I/O transactions can be traced in a VCD waveform viewer, as shown in Figure 8.

```
void Tasks::Task1() {
 sc_event Interval_event;
                                       /* S/W Reset Action */
if(nReset.read() == false)
 RTOS_create("Task_1",0,&Task1_run_event,APPL_TASK);
                                  /* A High Priority Task */
while (true){
             /* Suspend until dynamically notified */
 wait(Task1_run_event);
                               /* Scheduler: 100 msec */
 RTOS_waitI("Task_1", 100, &Interval_event);
 BFM_RdPort(0x1);                      /* Port 1 - Switch */
 while(BFM_PortBusy())
  wait();                  /* Pass Control To H/W Thread */
 Test_Data = BFM_FetchPortData();

 BFM_WrPort(0x2, Test_Data);           /* Port 2 - LED */
 while(BFM_PortBusy())
  wait();
                               /* Send a Message To Task 2 */
 RTOS_sendM("Task_2", "Count");
 }
}

void Tasks::Task2(){
 sc_event Msg_event;

if(nReset.read() == false)
 RTOS_create("Task_2",1,&Task2_run_event,APPL_TASK);
                                      /* A Low Priority Task */
while(1) {
 wait(Task2_run_event);
     /* Waiting for a "Count" Message */
 RTOS_waitM("Task_2", &Msg_event);
 if(!strcmp(RTOS_getM("Task_2"), "Count"))
  {
                                   /* Port 4 - SSD Counter */
   BFM_WrPort(0x4, Test_Data);
   while(BFM_PortBusy())
    wait();
                          /* Data Memory Starts at 0x2000 */
   BFM_WrXmem(0x2010, Test_Data);
   while(BFM_BusBusy())
    wait();
  }
}
```

Figure 6. Application Tasks example

6. CONCLUSION

We have presented our methodology for modeling an abstract RTOS Kernel in SystemC and showed its interaction with a generic micro-controller. Using SystemC to model all parts of the system showed great flexibility in design and enabled faster simulation [8].

Finally, it's obvious that by proper design of the RTOS and the BFM, performance estimates can be obtained for software execution, which can be used to drive the co-design process and speedup system integration phase early in design cycle.

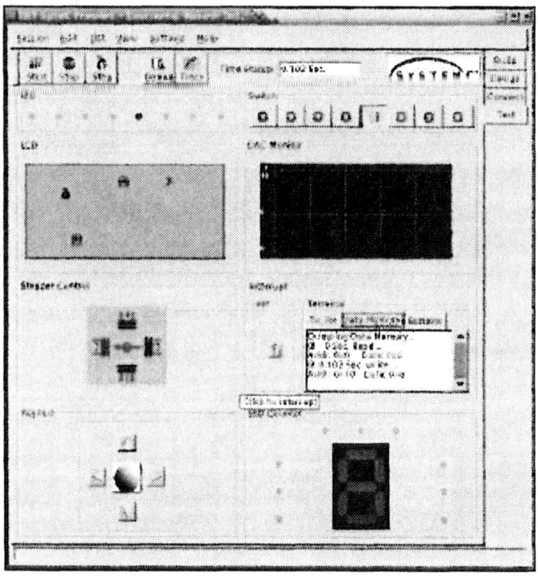

Figure 7. RTOS Based SystemC FrameWork

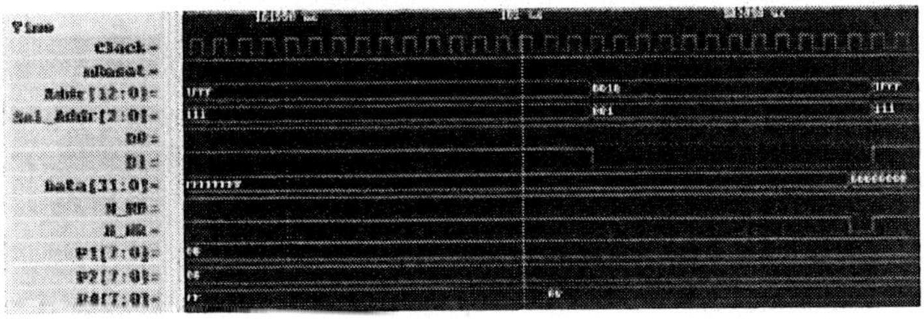

Figure 8. Simulation Wave Form in a VCD Viewer

REFERENCES

[1] SystemC User's Guide Version 1.1, Functional Specification For SystemC Version 2.0, Sept. 2001, *http://www.SystemC.org*

[2] S. Y. Liao, "Towards a new standard for system-level design," in Proc. 5th International Workshop on Hardware/Software Co-Design, May 2000.

[3] W. H. Wolf, "Hardware/Software Co-Design of Embedded Systems," in Proc. IEEE, July 1994.

[4] R. K. Gupta, G. De Micheli, "A Co Synthesis Approach to Embedded System Design Automation," in Proc. Design Automation for Embedded Systems, Jan. 1996.

[5] S. Liao, S. Tjiang, R. Gupta, "An Efficient Implementation of Reactivity for Modeling Hardware in the SCENIC Design Environment," in Proc. DAC'97, June 1997.

[6] J. Gerlach, W. Rosenstiel, "System Level Design Using the SystemC Modeling Platform," Worshop on System Design Automation, SDA 2000.

[7] D. E. Thomas, J.K. Adams, H. Schmit, "A Model and Methodology for Hardware-Software Co-Design," in Proc. IEEE Design and Test, Sept. 1993.

[8] L. Séméria, A. Ghosh, "Methodology for Hardware/Software Co- verification in C/C++," in Proc. ASP-DAC, 2000.

[9] R. Klein, "Miami, a Hardware/Software Co-verification System," in Proc. 7th IEEE Rapid Systems Prototyping Workshop, 1996.

[10] D. Harris, D. Stokes, R. Klein, "Executing an RTOS on Simulated Hardware using Co-verification," in Proc. Embedded Systems Conference, San Jose, Sept. 2000.

[11] G. De Micheli, "Hardware Synthesis from C/C++ Models," in Proc. DATE'99, March 1999.

[12] A. Ghosh, J. Kunkel, S. Liao, "Hardware Synthesis from C/C++," in Proc. DATE'99, March 1999.

[13] T. W. Schultz, "C and the 8051 Building efficient applications, Volume II", Prentice-Hall, 1999.

[14] Ward, P., S. Mellor, "Structured Development for Real-Time Systems", Prentice Hall, 1985.

[15] B. Bailey, R. Klein, S. Leef, "Hardware-Software Co-Simulation Strategies for the Future," Mentor Graphics Co., *http://www.mentor.com*

[16] T. W. Schultz, "C and the 8051 Hardware modular programming and multitasking, Volume I", Prentice-Hall, 1998.

[17] A. Ghosh et al., "A Hardware/Software Co-simulator for embedded system design and debugging," in Proc. ASP-DAC'95, 1995.

[18] C. Passerone, L. Lavagno, M. Chiodo, A. Sangiovanni-Vintentelli, "Fast Hardware/Software Co-simulation for virtual prototyping and trade-off analysis," in Proc. DAC'97, 1997.

Section 4.3

Modeling, Synthesis and Implementation of Communicating Hierarchical FSM

Valery Sklyarov
University of Aveiro, Department of Electronics and Telecommunications, IEETA, 3810-193 Aveiro Portugal

Abstract: The paper suggests a technique for the design of high-speed communicating hierarchical finite state machines (HFSM). The considered specification of a HFSM provides support for modularity, hierarchy and parallelism. It is shown that an implementation of a HFSM can be provided on the basis of a reusable circuit with reprogrammable RAM-blocks. Hierarchical mechanisms have been realized with the aid of a stack memory. The proposed synthesis method allows the given specification to be formally translated into bitstreams for the RAM-blocks.

Key words: FPGA, logic synthesis, hierarchical specification, and hierarchical finite state machine

1. INTRODUCTION

With the advent of Field Programmable Logic Devices (FPLDs) it became possible to design and implement digital systems and their components without the need for technological steps dealing with silicon. This process is very similar to the process for the development of software for general-purpose computers. Initially FPLDs were used for implementing unique, and usually relatively simple, digital circuits. The effectiveness of this approach forced enormous research efforts in this area for developing new generations of FPLDs with a huge number of available logic primitives and much more powerful capabilities. Today FPLDs are considered to be an alternative for ASICs, and they have already been very efficiently used in a very large number of practical applications, such as co-processors for general-purpose computers, problem-oriented digital systems, embedded controllers, prototyping boards, and so on.

There are many different techniques and CAD tools that can be employed for the design of digital circuits on the basis of FPLD. They combine formal synthesis methods and heuristic approaches mainly based on previous experience. As a rule a new circuit is composed of both original sub-circuits and reusable parameterizable components that have been already designed and included into the respective libraries. Formal synthesis methods permit the construction of the circuit from the given specification, which can be presented in a variety of forms, such as hardware description languages, general purpose languages reoriented to hardware design (such as SystemC), state-based languages, etc. The existing synthesis tools permit circuits to be constructed that in general cannot be reused. Any desired modifications in the behavior of the circuit require the synthesis process to be repeated from the beginning, including debugging, testing, and other steps that are considered as a part of a design process. This paper shows that for some practical applications (such as digital circuits that implement control algorithms) the problem can be significantly simplified. The technique proposed combines specification and synthesis methods. The specification method allows modular, hierarchical and parallel control algorithms to be described. The synthesis method enables us to translate the given specification into a reusable control circuit that is implemented based on commercially available FPGAs with distributed RAM blocks. Note that integrating of FPGA configurable logic blocks (CLB) with flexible distributed RAM is supported today by different FPGA families.

2. HIERARCHICAL GRAPH SCHEMES

The specification of control algorithms has been provided with the aid of graphical languages called hierarchical graph schemes (HGS) [1], which belong to a group of operator schemes of algorithms [2]. The first language of this group was proposed in [3] and it is called logical scheme of algorithms (this language was also described in [2]). Graphical specification languages (called graph-scheme of algorithms - GSA) of this group were exploited for design purposes in a number of publications that were summarized in the book [4]. A HGS is considered to be a GSA that supports hierarchical mechanisms, which can be implemented in the model of a finite state machine (FSM) that is called a hierarchical FSM (HFSM) [5]. There are many other methods for the specification of control algorithms that have similarities with this group of languages, such as algorithmic state machines [6], statecharts [7], specifications based on Petri nets [8], the PRALU language [9], etc. An extension of HGSs that permits support to be provided for parallel execution of control algorithms was proposed in [10]. Formal specification of extended HGSs, which combines the capabilities considered above and is used in this paper, was presented in [11,12].

Note that for a number of practical applications HGSs can easily be constructed from a program described in a general-purpose language such as C/C++. Let us consider some examples. Fig. 1 shows a C++ function *SolveP* that finds a minimal row cover of a Boolean matrix with using an approximation algorithm [13] (this algorithm was also implemented in the FPGA -based circuit that was described in [14]). The algorithm [13] includes such operations as: finding a column of the Boolean matrix that has a minimal number of values "1" (we call it *min_col*); finding a row that has the value "1" in *min_col* and has maximal number of values "1" (we call it *max_row*), setting mask registers that enable us to exclude rows and columns that have already been included into the row covering and to consider just the remainder of the matrix (we call it *set_mask*). We assume that *SolveP* is a member function of a class *covering* that is responsible for the considered problem and variables *Rg_const*, *h_mask*, *v_mask*, and *result* are data-members of this class. The sequence of control operations described in the function *SolveP* can be converted to a HGS that is also depicted in fig. 1. Rectangular nodes of the HGS contain either *microinstructions* or *macroinstructions*. Any *microinstruction* Y_j, includes a subset of *microoperations* from the set $Y=\{y_1,...,y_N\}$. A *microoperation* is an output signal, which causes a simple action in the datapath. Any *macroinstruction* incorporates a subset of *macrooperations* from the set $Z=\{z_1,...,z_Q\}$. Each *macrooperation* is described by another HGS of a lower level and can activate a new macrooperation. For example, z_1 and z_2 invoke a macrooperation z_4 that counts the number of values "1" in a Boolean vector (we call it *number1s*). Each rhomboidal node contains one element from the set X, where $X=\{x_1,...,x_L\}$ is the set of *logic conditions* (i.e. input signals, which communicate the result of a test).

Fig. 2 shows the code for z_4 and the respective HGS that describes the sequence of control operations [15]. We assume that variables *vector*, *counter*, and *ccv* shown in fig. 2 are also data-members of the class *covering*. As we can see from fig. 2 the HGS does not have macrooperations and that is why this is an ordinary GSA [2]. State transition table in fig. 2 will be used in section 5. Each line of the table describes transition from the state a_{from} to the state a_{to} and $X(a_{from},a_{to})$ is a product of input variables, which causes this transition, $Y(a_{from})$ is a set of microoperations in the state a_{from}.

In the example the class *covering* models a datapath for the problem. It includes data (such as *h_mask*, *v_mask*, *vector*, etc.) associated with pre-defined components for a register-transfer level. The required sequence of elementary operations in the datapath can be specified by the control algorithm (or by the corresponding HGS).

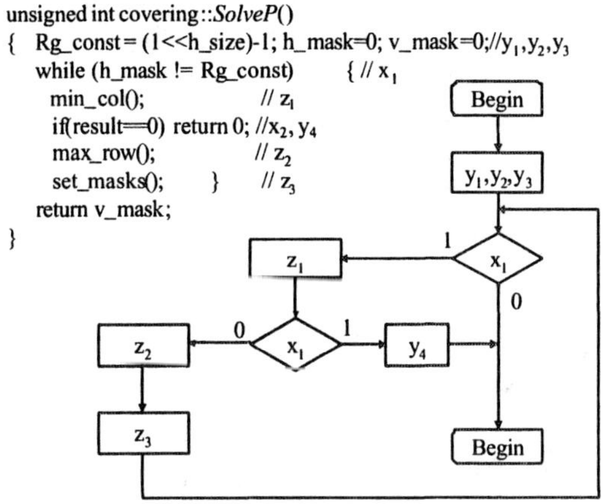

Figure 1. HGS that describes the function *SolveP*

```
short int covering::number1s()
{       counter=0;
        for(ccv=0; ccv<size; ccv++)
        {       if (vector & 1) counter++;
                *this >>= 1;                    }
        return counter;
}
```

a_{from}	$Y(a_{from})$	$X(a_{from},a_{to})$	a_{to}
a_0 (000)		1	a_2 (010)
a_2 (010)	y_1,y_2	$\bar{x}_1$	a_1 (001)
		x_1	a_5' (101)
a_3 (011)	y_3	1	a_4 (100)
a_4 (100)	y_4,y_5	$\bar{x}_1$	a_1 (001)
		x_1	a_5' (101)
a_5' (101)		$\bar{x}_2$	a_4 (100)
		x_2	a_3 (011)

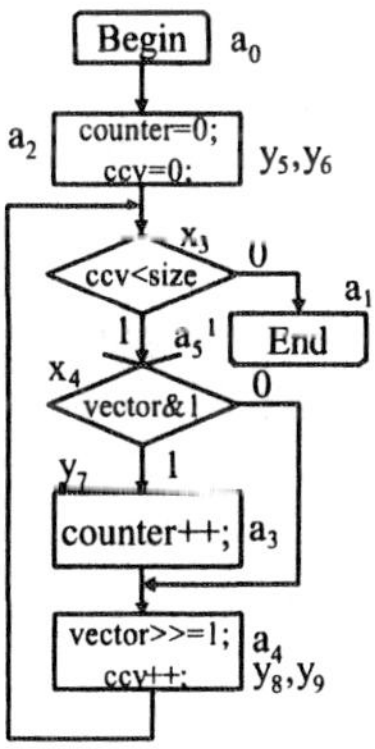

Figure2. HGS that describes the function *number1s*

Suppose we select a group of problems that employ a similar computational model, which permit an initial architecture of a datapath to be selected and this architecture might need to be extended in future. For example we can consider computations over Boolean and ternary matrices. As a result the datapath is constructed so that basic data is stored and primary operations over matrices can be carried out. It might include such components as storage for a matrix, mask registers for rows and columns, etc. Suppose that reusability of the selected architecture is provided in two possible ways. The architecture can be extended as much as we like since we can do this through inheritance in object-oriented programming. Thus the initial architecture can be considered as a base class and the desired additional functionality can be implemented in a derived class that models an extended datapath (this technique was considered in [15]). On the other hand the desired changes in the functionality might be achieved through modifications of the control algorithms (i.e. HGS) that describe operations in the reusable datapath. Thus we have to be able to construct a control circuit that permits us to utilize the same hardware (i.e. the same datapath) for implementing different algorithms.

3 COMMUNICATING HIERARCHICAL FSM

It is known that a HGS can be formally translated to the respective control circuits modeled by a HFSM [11]. A HFSM can support parallel execution of macrooperations but such an implementation is very resource consuming [10]. The considered approach is based on the following assumptions:

1. Parallelism is supported only at the top-level, i.e. in the main HGS;

2. Suppose that we have the following set $Z = Z_s \cup Z_p$, $Z_s \cap Z_p = \varnothing$, where Z_p contains macrooperations that can be executed in parallel with other macrooperations, Z_s is composed of macrooperations that cannot be executed in parallel with other macrooperations. The considered technique assumes that all macrooperations from the set Z_p must be implemented through communicating modules and each of them can be a HFSM. Macrooperations from the set Z_p can be implemented using any mechanism (mainly through a HFSM considered in [11]).

3. Communication mechanisms are provided with the aid of a special synchronization method [12], i.e. they can be embedded into the respective HGSs and described within the same specification.

Thus if we want to modify any parallel algorithm, we can change the respective embedded fragments of HGSs. In other words HGSs themselves permit to change communication mechanisms. As a result, in order to provide reusability of control circuits and alterability of control algorithms it is necessary to construct a HFSM with modifiable functionality.

4. RAM-BASED FSM

Reprogrammable control units can be modeled by FSMs with modifiable behavior. Fig3, a depicts a feasible hardware implementation of such an FSM. It is composed of RAM-based blocks (level RAM - LRAM) depicted in Fig3, b and a register (Rg). The programmable multiplexer (PM) whose structure is shown in Fig.3, c permits any input $x_q \in X$ to be selected in such a way that $p=x_q$ and the value 1 can be specified by the multiplexer RAM (MRAM). State transition RAM (STRAM) enables us to generate codes for the next states and outputs. By modifying the contents of MRAM and STRAM we can implement any FSM behavior we want (of course within the scope of the predefined constraints, such as the size of Rg - R, the number of inputs - L, the number of outputs – N, and the number of levels G).

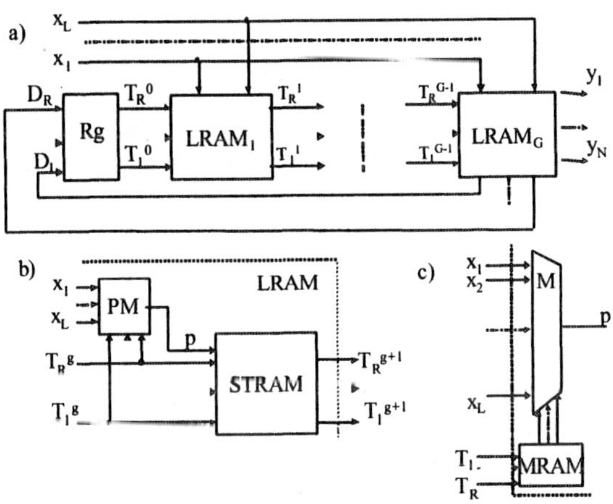

Figure3. FSM with modifiable behavior (a), the structure of LRAM (b), the structure of programmable multiplexer (c)

To implement a control circuit on the basis of the considered model an arbitrary state transition graph of a FSM has to be altered in such a way that allows any state transition to be effected by a single input variable. It might be done by inserting dummy states [5,15] as shown in fig. 4.

Now the LRAM shown in fig. 3,a allows us to implement any state transition caused by any single input and the next state can be either a real state that is required by the behavior or a dummy state. The latter is used just for providing reprogrammability and it is not stored in the FSM register. Dummy states can only appear between levels (i.e. LRAMs) during propagation of the respective signals in the combinational circuit. Thus they practically do not reduce the speed of state transitions and any state transition can be performed during one clock. Obviously any subset of the output signals (microoperations) $y_1,...,y_N$ can easily be generated either as for a Moore FSM or as for a Mealy FSM (the last case is shown in fig. 3, a).

A simple project for the Xilinx Foundation Software (version 3.1) with an implementation of the model shown in fig. 3,a for FPGA XC4010XL is given in [16].

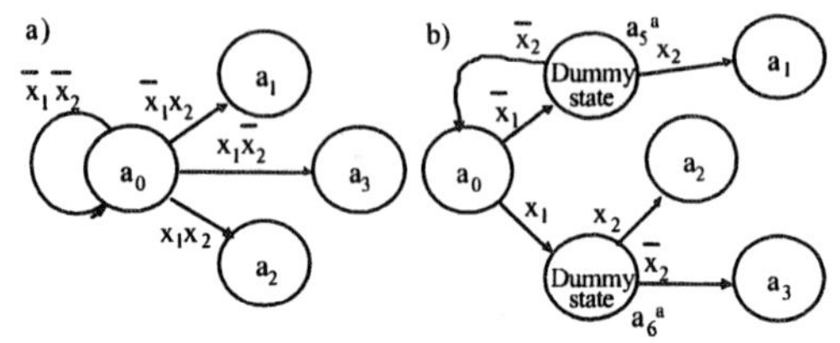

Figure 4. A fragment of a FSM state transition graph (a) and splitting transitions (b)

Changes in the behavior of a FSM can be provided either by reloading or by re-switching of RAM-based (ROM-based) blocks [15]. The last case is demonstrated in fig. 5 [15].

5. SYNTHESIS AND IMPLEMENTATION OF A HFSM

HFSMs can be implemented on the base of the model shown in fig. 3,a. However in this case the register (Rg) has to be replaced with a stack memory [11]. Fig. 6 [15] demonstrates the use of this memory for HGS shown in fig. 1.

Suppose we have a set of individual HGSs, and these can be used for describing control sequences in the functions *min_col*, *max_row*, *set_masks* and *number1s* considered in section 2. The graph in fig. 6 demonstrates hierarchical invocations of these functions for solving the covering problem [13]. Let us consider how we can synthesize a HFSM that implements HGSs for all the functions including the function *SolveP* (see section 2) on the basis of a reusable circuit. Suppose the circuit with the structure shown in fig. 3,a has already been constructed and implemented in an FPGA. This circuit can be programmed for realizing different control algorithms within the pre-defined constraints identified in section 4. Let us assume that our HGSs satisfy these constraints. In order to synthesize the entire control circuit we can construct bitstreams that have to be loaded in reprogrammable blocks (such as MRAM and STRAM) for all individual HGSs. The required hierarchical calls (see fig. 1) can be implemented through re-switching segments of memory (see fig. 5) as shown in fig. 6. Initially the HGS for the function *SolveP* will be executed at level 0 of the stack. This HGS can call HGSs for the functions *min_col*, *max_row* and *set_masks* (see fig. 1), which will be executed at level 1 of the stack (see fig. 6). HGSs for the functions *min_col* and *max_row* call HGS for the function *number1s*, which will be run at level 2 of the stack. Changing levels is povided with the aid of the method [11]. In order to activate a HFSM for any function it is necessary to indicate the respective block of memory (see fig. 5, 6) and an initial state for the HFSM. The latter may be the same for all HGSs z_1-z_4 (see fig. 1, 2).

Figure 5. Changing blocks (segments) of memory

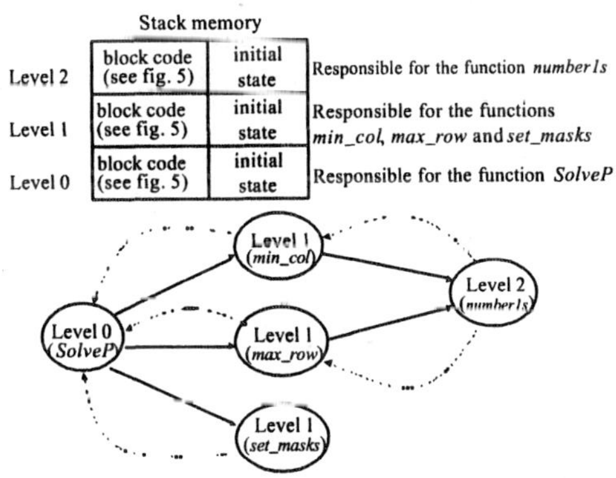

Figure 6. Using stack memory for an activation of different macrooperations in fig. 1

Synthesis of a HFSM for any individual HGS includes the following steps that are based on the results of [11]: 1) marking HGSs with labels $a_1,\dots,a_M$ that will be considered as HFSM states (see labels $a_0,\dots,a_4$ in fig. 2); 2) adding dummy states [5] if required (see the state a_5^1 in fig. 2); transforming the HGSs to a state transition table [11] (see fig. 2); 4) state encoding (in the trivial case the codes can be the binary values of the indexes as shown in fig. 2); 5) programming of MRAMs and STRAMs. Since the components in fig. 3,a have predefined connections the steps considered above make it possible to implement the circuit with the required functionality just by reprogramming memory blocks.

For our example in fig. 2 these blocks have to be programmed as follows: $LRAM_l$ - $MRAM_l$: 0:0, 1:0, 2:1, 3:0, 4:1, 5:2, 6:0, 7:0; $LRAM_l$ - $STRAM_l$: 0:2, 1:2,

2:1, 3:1, 4:1, 5:5, 6:4, 7:4, 8:1, 9:5, a:0, b:0, c:0, d:0, e:0, f:0; $LRAM_2$ - $MRAM_2$: 0:0, 1:0, 2:1, 3:0, 4:1, 5:2, 6:0, 7:0; $LRAM_2$ - $STRAM_2$: 0:0, 1:0, 2:1, 3:1, 4:2, 5:2, 6:3, 7:3, 8:4, 9:4, a:4, b:3, c:0, d:0, e:0, f:0. This allows all state transitions to be realized.

Another example, which demonstrates a simplified practical specification of a system that controls individual operations in an assembly line together with implementations of RAM-based FSM can be found in [16].

6. EXPERIMENTS

All the experiments were performed with XS40 boards (XStend V1.2 of XESS) that contain the Xilinx XC4010XL FPGA and with the Xilinx Development System (series 3.1). STRAM and MRAM blocks were constructed from FPGA CLBs configured as either ROM or RAM. All these blocks were created by the Xilinx LogiBLOX Module Generator. In the trivial case specifying the desired behavior has been provided by preparing *.*mem* files for MRAMs and STRAMs. For dynamic modifications we used dual port RAM library components. Finally, HFSMs with dynamically alterable behavior were implemented on statically configured FPGAs. It is important to appreciate that the proposed technique makes it possible to realize any modification in the behavior within pre-defined constraints.

7. CONCLUSION

We have attempted to develop an approach to the specification and synthesis of fast control systems implemented on an FPGA chip. The specification permits a net of communicating HFSMs to be described using a formal graphical language. The proposed synthesis steps translate a given specification into the bitstreams for the memory blocks of a reusable control circuit that can be preliminarily implemented in hardware. Using RAM-blocks with dual port capabilities allows run-time modifications of the hierarchical control algorithms, making these algorithms flexible and extensible.

ACKNOWLEDGEMENTS

This work was supported by the Portuguese Foundation of Science and Technology under grant No. POSI/43140/CHS/2001.

REFERENCES

[1] V. Sklyarov, Hierarchical Graph-Schemes, Latvian Academy of Science, Automatics and Computers, Riga, no 2, pp. 82-87, 1984.
[2] S.Baranov, Logic Synthesis for Control Automata, Kluwer Academic Publishers, 1994.
[3] Y.Yanov About Logic Schemes of Algorithms. Problems of Cybernetics, N 1, 1958, pp. 75-127.
[4] S.Baranov, Synthesis of Microprogramming Automata, Leningrad, Energia, 1974.
[5] V.Sklyarov, Synthesis of Finite State Machines Based on Matrix LSI. Minsk, Science and Techniques, 1984.
[6] Christopher R. Clare, Designing Logic Systems Using State Machines, McGraw-Hill, Inc, 1973.
[7] D.Harel Statecharts: A Visual Formalism for Complex Systems, Science of Computer Programming, 1987, N8, pp. 231-271.
[8] R. J.Machado, J.M.Fernandes, A.J.Esteves, H.D.Santos, An Evolutionary Approach to the Use of Petri Net based Models: From Parallel Controllers to HS/SW Co-Design. Kluwer Academic Publishers, 2000, pp. 205-222.
[9] A.Zakrevskij, V.Sklyarov, The Specification and Design of Parallel Logical Control Devices. Proceedings of PDPTA'2000, June, Las Vegas, USA, 2000, pp. 1635-1641.
[10] V.Sklyarov, Synthesis of Control Automata Described by Parallel Hierarchical Graph-Schemes. Automatics and Computers, Riga, no 5, pp. 68-76, 1987.
[11] V.Sklyarov, Hierarchical Finite-State Machines and Their Use for Digital Control, IEEE Transactions on VLSI Systems, 1999, Vol. 7, No 2, pp. 222-228.
[12] V.Sklyarov, Graphical Description and Hardware Implementation of Parallel Control Algorithms. Proceedings of PDPTA'99, June, Las Vegas, USA, 1999, pp. 1390-1396.
[13] A.Zakrevskij, Logical Synthesis of Cascade Networks, Moscow, Nauka, 1981.
[14] I.Skliarova, A.B.Ferrari, Design and Implementation of Reconfigurable Processor for Problems of Combinatorial Computations, Proceedings of the Euromicro Symposium on Digital System Design – DSD'2001, Warsaw, Poland, September 2001, pp.112-119.
[15] V.Sklyarov, Modelação em C++, Síntese e Implementação de Circuitos Digitais com base em FPGA, Electrónica e Telecomunicações, January, 2002.
[16] http://webct.ua.pt/ Configurable computing.

Section 4.4

A Modeling Method for Reconfigurable Architectures

Lilian Bossuet, Guy Gogniat, Jean-Philippe Diguet, Jean-Luc Philippe
University of South Brittany, Lorient, France– Research Center-BP 92116
56 231 Lorient Cedex

Abstract: Field Programmable Gate Arrays (FPGAs) are now integrated in complex electronic systems and are not only used during the prototyping phases. Moreover, their application space is getting larger thanks to their flexibility, which is obtained through the reconfiguration. For these reasons component industries are proposing more and more FPGAs with different architectures, sizes, capabilities. Designers are then faced to a larger choice in the design space but they do not have any tool to efficiently compare different reconfigurable architectures. In this paper a reconfigurable architecture zodeling method is presented. The proposed model is applied to a domain representative component in order to show its ability to describe a complex architecture. The modeling method, which is based on a functional view, is able to describe complex, heterogeneous and multi-granularity architectures. In future works the modeling method will be combined to a generic FPGA architecture estimation tool.

Key words: FPGA, Modeling, design environment

1. INTRODUCTION

Constant improvement of semiconductor technologies conducts to a transistor size, delay and consumption reduction and to an increase in packaging density. In parallel, production costs of Application Specific Integrated Circuits (ASICs) are continuously rising. Consequently, ASIC founders accept to run an ASIC development cycle only in case of mass products. Hence, lot of electronic designers turn towards Field Programmable Gate Array (FPGA) technology which tends to be a promising alternative.

FPGAs are now integrated in complex electronic systems, and are not only used during the prototyping phases. Moreover, FPGA's dynamic reconfigurable properties begin to be widely used, like in Run Time Reconfigurable systems [1] or in Software Radio systems. Consequently, application space of FPGAs increases and spreads on ASICs domain. Hence, silicon vendors propose a large spectrum of FPGA devices with several levels of complexity (till ten million of gates in Xilinx VIRTEX II) and several levels of granularity (LUTs, registers, multipliers, RAMs, processor core, etc.)

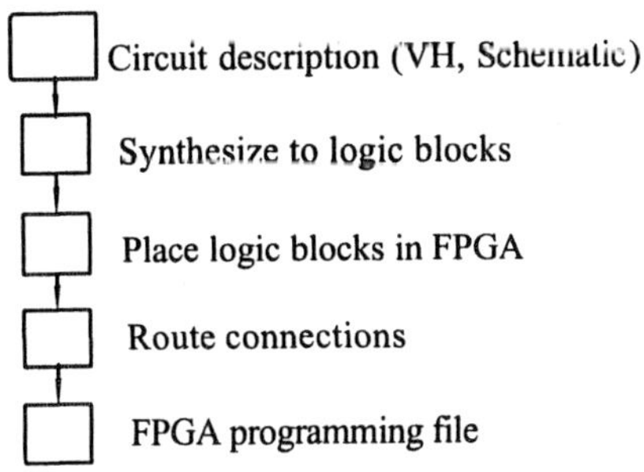

Figure 1. FPGA Design Flow [2]

Designers are then faced to the difficult choice of the target FPGA device that is critic since it can strongly affect the final system's performances. To help them in that task, it is necessary to develop tools that compare FPGA's architectures. With such tools designers could improve adequacy between application and architecture and then choose the best one for their application. Today, a way to compare two architectures is to measure application's performances on different hardware targets. In that case the comparison is made at the end of the design flow, which is shown in Figure1. However, the time consuming place and route step is a major inconvenience when the designer needs to compare numerous devices. Moreover, making the comparison at the end of the design flow means to compare all the flow's steps (in particular the place and route algorithms) and not only the FPGA architecture's performances. Since, the steps of the design flow are very different between two manufacturers, the architecture's performances comparison is only possible between different devices of the same manufacturer that conduct to a limited design space exploration!

A solution to the problem of architecture comparison is to define a retargetable FPGA design environment. Figure 2 shows the differences between actual dedicated FPGA design environments (a) and the retargetable FPGA design environment that we propose (b). For dedicated environment, each architecture (expect for the same manufacturer's architectures) is linked to a specific design tool. For retargetable environment, a unique design tool is requested for all architectures. Hence, architectural comparison is possible since synthesis algorithms are common.

However, such generic environment is possible only if a reconfigurable architecture modeling method is defined. This method must permit to describe efficiently the architectures to compare. In this paper we focus on that point which is a key aspect in the definition of the environment. We can notice that a similar wok has been carried out, at the Berkeley University, on a retargetable scheme for Digital Signal Processor (DSP) architecture selection [3].

In our approach, from an architecture instance (based on the modeling method presented in the paper), a performance estimation method is performed in order to compare different architectural solutions. The modeling method aims to characterize a large panel of reconfigurable architectures. Its objective is to support the description of commercial FPGA's architectures but also no commercial FPGA's architectures emerging from research works. Thus, the proposed modeling method will support the evolution of future FPGA architectures.

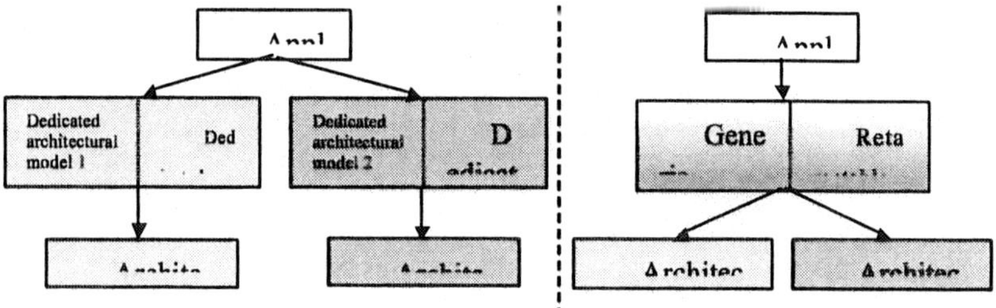

a) Dedicated FPGA design b) Retargetable FPGA design

Figure 2. FPGA design environment a) dedicated b) retargetable

This paper is organized as follows, in section two we present a survey dealing with reconfigurable architectures and synthesis tools which are based on a reconfigurable architecture modeling. Section three, explains our reconfigurable architecture modeling method. Then, in section four, a Xilinx Virtex II device is modeled to illustrate our approach. Finally, section five gives us the opportunity to conclude and to propose some extensions to this work.

2. SURVEY

Reconfigurable FPGA architectures

Current FPGAs are composed of configurable logic elements, which are connected through routing resources. Some FPGAs have also embedded RAM memories, processor core (such ARM 9 for Altera Excalibur device [4]), phase locked loops and inputs/outputs configurable blocks. As regards the logic blocks, they are usually build with LUTs and registers. Nevertheless, for some devices (like

Xilinx Virtex device [5]), the logic blocks are build with coarse-grained operator like multiplier or adder. The different elements of a FPGA are arranged in several architectural forms. Two architectural styles are predominant; the island style (the greater part of the Xilinx's architectures) and the hierarchical style (the greater part of the Altera's architectures), figure 4 shows the island style architecture of a Xilinx Virtex II device. These architectural styles have mainly an influence on the routing resources layout. Numerous research projects focus on new reconfigurable architectures definition. Such new architectures innovate the logic blocks size, form and granularity level. For example, the project carried out at the Twente University (The Netherlands) defines an original architecture called Field Programmable Function Array (FPFA), [6]. FPFA is composed of a matrix of processor tile, each processor tile contains multiple reconfigurable ALUs, local memories, control unit and communication unit.

FPGA architecture modeling

Among works dealing with reconfigurable architecture modeling, two are especially significant. Both correspond to place and route tools. Even if the study of place and route algorithms is not the issue of this paper, these tools are relevant to our work since they exploit a reconfigurable architecture modeling method in order to be technological independent. In this paper we only focus on the reconfigurable architecture modeling part of the tools. The first one is called Versatile Place and Route (VPR). It was developed by Vaughn Betz and Jonathan Rose at the Toronto University (Canada) [2]. The second one was developed by Loï c Lagadec and Bernard Pottier at the UBO University (France) [7, 8].

In the VPR tool, the modeling method is structural and is strongly oriented to island style architectures. A reconfigurable architecture is described in a structural way: number of LUTs in CLB, width of routing channels, connection styles, etc... This type of modeling allows to quickly modify a parameter of the architecture (i.e. the number of LUT inputs) in order to evaluate the impact on the device performances. However, this modeling type is tightly specialized for island style architectures with a fine level of granularity (LUTs, registers). Hence, its extension to a more complex architecture with other patterns of computation and connection is difficult.

In the UBO tool, the modeling method is functional. Each element of the architecture is described by the function it is able to perform (for example, a four inputs LUT can execute a logical function with four inputs and one output). The functional modeling enables to describe a large panel of architectures (hierarchical or island style) and the description is technological independent. The designer can define new types of element since the functionality is known. This modeling is attractive, however it deals with a low-level granularity and each element is characterized by only one function. This is due to the underlying place and route algorithms. In case of complex architecture composed of memories, coarse-grained and/or multi-function operators these features can restrict the modeling of the architecture. Main characteristics of both reconfigurable architecture modeling are summarized in table 1. If the designer wants to describe architecture composed of numerous distinct elements in a technological independent way, the functional modeling approach is more adapted than the structural modeling.

Name	VPR	UBO's tool
Modeling type	Structural	Functional
Island style architecture	Yes	Yes
Hierarchical architecture	No	Yes
No-homogeneous architecture	No	Yes
Routing modeling	Yes	Yes
Logic blocks modeling	Partially	Yes
Embedded memories modeling	No	No
Arithmetical elements modeling	No	No
Input/output blocks modeling	Yes	No

Table1 Main characteristics of both reconfigurable architecture modeling types

To conclude this survey, the following remarks can be done: the modeling must permit to describe functionally architecture and to represent hierarchical or island styles. Several levels of granularity, numerous types of elements and heterogeneous architectures must be considered. Finally, the modeling must be able to take into account the future architectural evolutions.

3. MODELING PRESENTATION

In our approach, the architecture's elements are functionally described and the modeling allows representing different architectural styles and different architecture's elements. However, the routing resources are not described since the aim of this modeling is to support a performance estimation tool and not a place and route tool. Nevertheless, the routing resources must be taken into account in the estimation tool as they have significant effects on device performances, they can have a power consumption upper than the half of the device power consumption [9] and they can slow down signals in the device [2]. Thus, connection resources are taken into account by connection costs in the proposed modeling.

There are two types of element in the modeling; the hierarchical element and the functional element.

The hierarchical element is used to describe the architecture's hierarchy. It can be composed of functional elements and other hierarchical elements. It is defined by five characteristics:

- its name, it is a label,
- the number of such elements in the current hierarchy's level,
- its reference, that is composed of two numbers, the first one for the current hierarchy's level (0 is the uppermost level), the second one for the number of that element in the current hierarchy's level,
- the list of (hierarchical or functional) elements that are embedded in this hierarchical element,
- the connection costs between embedded elements (routing resources consideration).

The functional element is used to describe the architecture's resources. For classical FPGA architecture, it corresponds to logic element, input/output block, embedded memory and arithmetic element (i.e. adder or multiplier) But a functional element can describe various types of resources due to the functional approach. It is defined by seven characteristics:

- its name, it is a label, as for a hierarchical element,
- the number of such elements in the hierarchical container element,
- its reference, that is composed of two numbers, the first one for the current hierarchy's level (0 is the uppermost level), the second one for the number of that element in the current hierarchy's level,
- its maximum use, actually routing a FPGA becomes difficult when the architecture's elements are overused, so for example only 99% of them can be really used,
- its utilization costs, they correspond to latency and power consumption,
- the list of functions that this element can realize,
- the function utilization rules, these rules are used to specify the possibility of parallelism or exclusiveness between functions.

Figure 3 illustrates the modeling; different levels of hierarchy are represented. This feature is important to model hierarchical architecture style.

In this modeling, there are two types of costs, the connection costs associated to the hierarchical elements and the utilization costs associated to the functional elements. The utilization costs correspond to delay and power consumption costs, which represent the latency and the power consumption of a functional element. The connection costs model the delay and the power consumption of routing resources between two elements in a same hierarchical element. These costs are requested to take into account the impact of routing during the estimation process.

To efficiently estimate an application it is necessary to have information about the resources that can support the architecture and the related performances (speed, consumption). In our modeling method all these information (through the functional modeling and the modeling costs) are precisely defined which guaranty that the estimation process can be performed efficiently.

The modeling cost can be defined with two approaches. If the designer wants to rapidly evaluate the performances of several architectures, he can define relative costs since, in that case, the aim is not to obtain accurate estimations but to identify the best architecture in term of relative performances. If the designer wants to compare precisely several architectures, the costs must be more precise to take into account the whole characteristics of each architecture. In that case, the designer can settle the different costs through delay and consumption measures. However this task can be very tedious in order to obtain precise characterisations. In our work, costs are defined relatively since we want to explore rapidly several architectures.

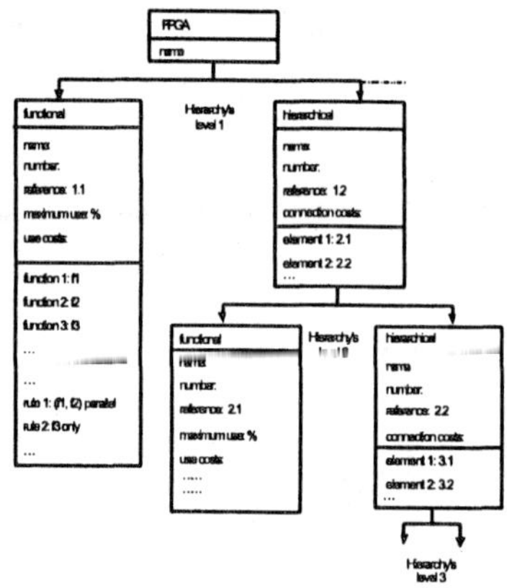

Figure 3. Modeling illustration

4. XILINX VIRTEX II MODELING

Several experiments on the following devices, Xilinx XC4000, Xilinx Virtex, Xilinx Virtex II and Altera Apex, have been carried out at the laboratory in order to validate the proposed modeling. In this section the results with a Xilinx Virtex II XC2V1000 (package FF896) device are presented. This device's architecture is illustrated in figure 4.

FPGA: Virtex II XC2V1000 (package FF896)

ELEMENT 0
Type: hierarchical
Name: Virtex II XC2V1000
Number: 1
Reference: 0.1
Element1: 1.1
Element2: 1.2
Element3: 1.3
Element4: 1.4
Connection costs
1.1-1.1: 2ns
1.1-1.2: 2ns
1.1-1.3: 2ns
1.1-1.4: 2ns
1.2-1.2: 1ns
1.2-1.3: 1.5ns
1.2-1.4: 1.5ns
1.3-1.3: 1ns
1.3-1.4: 1ns
1.4-1.4: 1ns
END ELEMENT 0

ELEMENT 1
Type: functional
Name: IOB
Number: 432
Reference: 1.1
Utilization max: 100%
Latency: 10ns
Power consumption: 2 mW +(0.4mW/1MHz)*f
Function 1 (f1): input
Function 2 (f2): output
Function 3 (f3): input/output
Rule 1: (f1, f2, f3) exclusive
END ELEMENT 1

ELEMENT 2
Type: hierarchical
Name: CLB
Number: 1280
Reference: 1.2
Element1: 2.1
Connection costs
2.1-2.1: 0.2ns
END ELEMENT 2

ELEMENT 3
Type: functional

Name: SLICE
Number: 4
Reference: 2.1
Maximum use: 98%
Latency: 4ns
Power consumption: 1 mW +(0.4mW/1MHz)*f
Function 1 (f1): logic (1, 2, 3 or 4 inputs) whit asynchronous output
Function 2 (f2): logic (1, 2, 3 or 4 inputs) with synchronous output
Function 3 (f3): fast adder 1 bit with asynchronous output
Function 4 (f4): fast adder 1 bit with synchronous output
Function 5 (f5): fast adder 2 bit with asynchronous output
Function 6 (f6): fast adder 2 bit with synchronous output
Function 7 (f7): 16 bit shift register
Function 8 (f8): ROM 16*1 bit
Function 9 (f9): ROM 16*2 bit
Function 10 (f10): ROM 32*1 bit
Function 11 (f11): RAM 16*1 bit
Function 12 (f12): RAM 16*2 bit
Function 13 (f13): RAM 32*1 bit
Rule 1: (f1, f2, f3, f4, f5, f6, f7, f8, f11) parallel two by two
Rule 2: (f9, f10, f12, f13) exclusive
END ELEMENT 4

ELEMENT 5
Type: functional
Name: SelectRAM
Number: 40
Reference: 1.3
Maximum use: 100%
Latency: 15ns
Power consumption: 1 mW +(0.4mW/1MHz)*f
Function 1 (f1): Single-port RAM 16384*1 bit
Function 2 (f2): Single-port RAM 8192*2 bit
Function 3 (f3): Single-port RAM 4096*4 bit
Function 4 (f4): Single-port RAM 2048*9 bit
Function 5 (f5): Single-port RAM 1024*18 bit
Function 6 (f6): Single-port RAM 512*36 bit
Function 7 (f7): Dual -port RAM 16384*1 bit
Function 8 (f8): Dual -port RAM 8192*2 bit
Function 9 (f9): Dual -port RAM 4096*4 bit
Function 10 (f10): Dual -port RAM 2048*9 bit
Function 11 (f11): Dual -port RAM 1024*18 bit
Function 12 (f12): Dual-port RAM 512*36 bit
Rule 1: (f1, f2, f3, f4, f5, f6) exclusive
Rule 2: (f7, f8, f9, f10,f11,f12) parallel two by two
END ELEMENT 5

ELEMENT 6
Type: functional
Name: Multiplier
Number: 40
Reference: 1.4
Maximum use: 100%
Latency: 15ns
Power consumption: 1 mW +(0.4mW/1MHz)*f
Function 1 (f1): MUL 18*18 unsigned
Function 2 (f2): MUL 17*17 unsigned
Function 2 (f2): MUL 8*8 unsigned
Function 4 (f4): MUL 7*7 unsigned
Function 5 (f5): MUL 6*6 unsigned
Function 6 (f6): MUL 5*5 unsigned
Function 7 (f7): MUL 4*4 unsigned
Function 8 (f8): MUL 8*8 signed
Function 9 (f9): MUL 6*6 signed
Function 10 (f10): MUL 5*5 signed
Function 11 (f11): MUL 4*4 signed
Function 12 (f12): MUL 3*3 signed
Function 13 (f13): absolute value 2*18 bit
Function 14 (f14): two's complementer 2*18 bit
Function 15 (f15): two's complementer 2*9 bit
Rule 1: (f1, f2, f3, f8, f13, f14, f15) exclusive
Rule 2: (f9, f6) parallel
Rule 3: (f10, f5) parallel
Rule 4: (f6, f6) parallel
Rule 5: (f11, f4) parallel
Rule 6: (f11, f12) parallel
END ELEMENT 6

Table 2. Virtex II XC2V1000 (package FF896) device mod .

The Virtex II has an island style architecture. Nevertheless, there are two hierarchy's levels for this architecture. At the upper hierarchy's level there are the Input Output Blocks (IOBs), the Configurable Logic Blocks (CLBs), the embedded RAM memories (SelectRAM) and the multiplier blocks. The CLBs contain four logic slices. Each slice is realized with two four inputs LUTs, two registers and two fasts carry chains

Hence, in this architecture's model there are six elements; two hierarchical elements and four functional elements. The hierarchical elements are; the FPGA composed of hierarchical elements called CLB and three functional elements called IOB, Multiplier and SelectRAM, the CLB are composed of functional element called Slice. The Virtex II XC2V1000 (package FF896) model is shown in table 2.

Each functional element is characterized by its utilization costs (latency and power consumption), its list of functions and its function utilization rules. The number of functions depends of the functional element, for example a Slice can realize nine exclusive functions and four two by two parallel functions.

The modeling of the Virtex II presents the use of the modeling and its ability to describe complex architecture with several types of elements (memories, arithmetic elements, logic elements). Moreover, once a device in

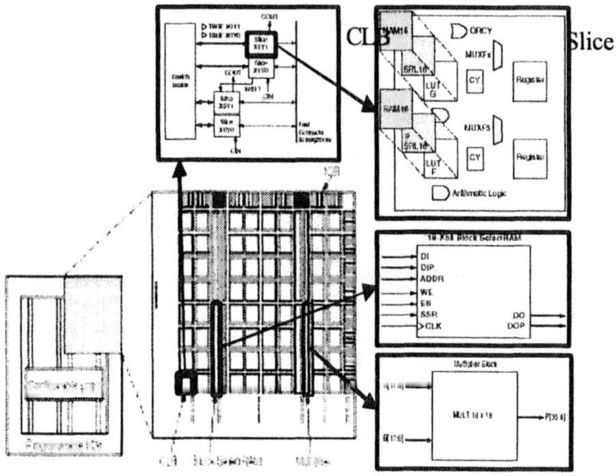

Figure 4. Virtex II architecture

a family is modeled, it is very easy to generalize the model for all the device of that family since the architecture is the same, just the number of elements changes (like the number of logic blocks, the number of input/output blocks, or the number of memory bits, etc.)

4 CONCLUSION

The definition of an open design environment for reconfigurable architectures imposes to define a reconfigurable architecture modeling method. This paper presents an original modeling method based on a functional approach. The proposed modeling allows describing a large architecture range, including heterogeneity, hierarchy and multi-granularity levels. It results from the study of actual and emerging architectures as well as the study of reconfigurable architecture modeling methods. Table 3 summarized the main characteristics of the proposed method in comparison with the VPR and UBO approaches (Table 1). This modeling is natural and can easily evolve in order to take into account new features of future architectures.

The aim of our work is to propose a modeling, which could federate the reconfigurable domain researchers. The use of a common modeling will permit to fill up the important lake of reconfigurable architectures comparison tools. The interest of our approach is double: first with an open design environment the researchers will be able to compare their new architectures with others, and second the designers will be able to explore the design space in order to select the best architecture for their application.

Our actual works focus on a performance retargetable estimation method for reconfigurable architectures that is based on the proposed modeling method. This estimation tool will constitute one of the tiles of the open design environment.

Name	**Proposed modeling**
Modeling type	Functional
Island style architecture	Yes
Hierarchical architecture	Yes
No-homogeneous architecture	Yes
Routing modeling	No
Routing impact modeling	Yes
Logic blocks modeling	Yes
Embedded memories modeling	Yes
Arithmetical elements modeling	Yes
Input/output blocks modeling	Yes

Table 3: Results with proposed modeling

REFERENCES

[1] Katherine Compton. *Architectures For Run-Time Reconfigurable Systems*. Master's Thesis, Dept of ECE, Northwestern University, Evaston, IL USA. December 1999.

[2] Vaughn Betz, Jonathan Rose and Alexander Marquardt. *Architecture and CAD for Deep Submicron FPGAs*. Klumer Academic Publishers, 1999.

[3] Naji Ghazal, Richard Newton, Jan M. Rabaey. *Retargetable Estimation Scheme for DSP Architecture Selection*. ASP-DAC 2000, Yokohama, Japan, Jan 2000, pp. 377-380.

[4] *Technical report*, 2001. http://www.altera.com

[5] *Product Specification*, 2000. http://www.xilinx.com

[6] Paul M. Heysters, Japp Smit, Gerard J.M. Smit and Paul J.M. Havinga. *Mapping of DSP Algorithms on Field Programmable Function Arrays*. FPL'2000, August 2000.

[7] Loï c Lagadec. *Abstraction, modélisation et outils de CAO pour les circuits intégrés reconfigurables*. PhD Thesis, Université de Rennes1, 2000.

[8] Loï c Lagadec, Bernard Pottier. *Object oriented meta-tools for reconfigurable architectures*. In Conference on Reconfigurable Technology FPGAs and reconfigurable Processors for Computing and Application, 2000.

[9] Varghese George, Hui Zhang, Jan Rabaey. *The Design of a Low Energy FPGA*. In Proc. Int. Symp. On Low Power Electronics and Design (ISLPED'99), pages 188-193, 1999.

5

DESIGN TECHNIQUES

This chapter looks at four different aspects of developing design techniques for SOC integrated circuits.

In the first section, "The Syslib-Picasso Methodology for the Co-Design Specification Capture Phase", by Luc Filion, Jérome Chevalier, Guy Bois and El Mostapha Aboulhamid, of École Polytechnique de Montréal and Université de Montréal, Canada, SYSLIB, a high-level C++-based library for designing embedded systems, is introduced. The authors use a common language from the highest to the lowest level of development to alleviate the refinement tasks. SYSLIB can be seen as an intermediate level to decompose a purely C++ high-level specification into several modules suitable for partitioning. SYSLIB also brings a step-by-step refinement procedure to design specifications in approved implementation languages, e.g. C++ and Cynlib.

In section 5.2, "Automatic Porting of Binary File Descriptor Library", by Maghsoud Abbaspour and Jianwen Zhu of the University of Toronto, Canada, the authors present a technique that can automatically retarget the GNU BFD library, the foundation library for a suite of binary tools. In addition to having all the advantages enjoyed by open-source software, by aligning to a de facto standard, their technique is also systematic by using a formal model of an abstract binary interface (ABI) as a new element of an architectural model. The technique is also simple, as a result of leveraging free software to the largest extent. The tool that has been developed using this technique is extremely easy to use, and the porting of the GNU linker takes only tens of lines of ABI specification and a few seconds of computation time.

In the third section, "Code Compression on Transport Triggered Architectures", by Jari Heikkinen, Jarmo Takala, and Jaakko Sertamo of Tampere University of Technology in Finland, the authors examine Entropy coding to improve code density of processor architectures. Entropy coding is used to improve code density of application-specific programmable processors based on a transport triggering paradigm. Three DSP applications are used as benchmark applications and Huffman coding is applied to compress the instructions of a processor architecture tailored for each benchmark. The results show that entropy coding provides a notable improvement in code density.

The final section, "An Approach To Flexible Multi-Level Network Design", by V. Cheushev, J. Kolodziejczyk, T. Luba,C. Moraga, P. Sapiecha, V. Shmerko and S. Yanushkevich, of the University of Calgary, deals with the design of multi-level logic networks using evolutionary network synthesis. A variety of techniques are available to the designer using a flexible approach. The main contribution of this paper is an extension and improvement of the state-of-the-art of gate-level evolutionary networks synthesis, allowi ng flexible scanning of the subspaces of possible network solutions. A B-decomposition of the network search space is more preferable, because it does not require any multiplexers to merge a network from sub networks. This decomposition provides additional and unique possibilities for massive parallel processing.

Section 5.1

The Syslib-Picasso Methodology for the Co-Design Specification Capture Phase

Luc Filion [1], Jérome Chevalier [1], Guy Bois [1] and El Mostapha Aboulhamid [2]

[1] *Electrical Engineering Departme nt, École Polytechnique de Montréal, P.O. Box 6079, Succ. Centre-Ville, Montréal, Québec, Canada, H3C 3A7.* [2] *Department of Computer Science and Operational Research, Université de Montréal, C.P. 6128, Succ. Centre-ville, Montréal, Québec, Canada, QC H3C 3J7.*

Abstract: This paper presents Syslib, a high-level C++-based library for designing embedded systems. Syslib splits the specification capture phase (of the co-design flow) into three distinct levels and provides services to aid a designer to develop systems. A designer captures a specification respecting pre-defined constructs representing the system semantics, like modules, ports and channels and refines his descriptions to reach the implementation languages. The functional level allows the definition of the main algorithm of an application at an untimed high level of abstraction. The behavioural level introduces the notion of time and the scheduling of the algorithm's modules. This phase allows the insertion of hints to determine the partitioning. Finally, in the third level, Syslib helps a designer to map to an architecture by providing a complete API to link software modules with an RTOS (real-time operating system) and to support hardware blocks with module intercommunications. Validation of specifications at a high level decreases simulation time. Syslib takes place in a complete co-design methodology called Picasso. Results to be expected from the methodology are a shorter development time period by efficient refinement steps with less translation errors.

Key words: co-design, system specifications, system-level language, semantics, Syslib

1. INTRODUCTION

As years go by, embedded system complexity is increasing exponentially. Transistor and component density per chip grow incredibly and will soon make the work of designing such systems unmanageable. More than ever, we must encourage the introduction of principles, standards and rules, such as system specification abstraction or reuse. Many workgroups are heading in this direction, namely SystemC [1], SpecC [2] and OCAPI [3], which are certainly well known at the moment. These organisations all suggest new methods for hardware (and sometime software) modelling based on C++ and object oriented paradigm. SystemC still is the most popular effort on the market. This approach consists of a progressive refinement of a user's specification, starting from an abstracted high-level representation to a bus -cycle accurate model. From the SystemC design flow in current version 2.0, it is not clear about how to fully supp ort software modules with an RTOS and how to map them to an architecture, unless perhaps a platform is completely written in SystemC [4] which may not be always possible. SpecC, on the other hand, gives examples of architectures and could represent another solution, but still, designers hesitate to switch to non-standard languages.

This work propose Syslib, a new C++ system library that covers all the specification capture phase of co-design. Syslib does not claim to compete with existing languages like SystemC, for two reasons. First, Syslib is intended to be a complementary language to Cynlib from Forte Design Systems [5] in order to work at the system level. Second, the integration of Syslib in a complete function-architecture co-design methodology strengthens its efficiency and interest. One of the goals of this work is to provide a good source of inspiration for people working on C/C++ standardization efforts at system level.

We characterise Syslib as a **system-level language**, as a way to specify a system without implementation details (hardware or software). At a high level of abstraction, we must remain generic. We qualify Syslib as a hybrid solution to describe both hardware and software. However, a Syslib specification has to be refined to be eventually implemented on a real SoC. We identified the common elements (semantics) to be reused in different levels of abstraction, starting from the high-level algorithm validation, passing to the partitioning and finishing with standardised implementation languages. Doing so, we maximize the reuse of Syslib constructs at every level of the specification part.

The paper is organized as follows. Section II of this article discusses about the semantics of a system, a topic that must be explained. In section III, we describe Syslib and the methodology in which it figures. In section IV, we describe an example about how to use the Syslib library and finally, the simulation engine supporting the execution is given in Section V.

2. SEMANTICS

A new movement is on the rise in the co-design community. Since nobody wanted to agree on a common language to "co-design", some had the idea that it would be more efficient to focus on semantics rather than on languages [6].

We analysed the semantics generally found in a system in the specification perspective. Some definitions have been taken from the specialized language (for telecoms) SDL [7].

We first analysed what a system is composed of, and noticed the following. A system is generally built up with several **modules**: defined by **behaviours** or algorithms. A module itself could be built up with many other modules, implying **hierarchy**. Those modules communicate together via **ports**. Ports collect input **variables** for the module's processing and output results also as variables. For module synchronisation or message passing, we noticed the use of **events**. Since variable transfers could be accompanied of events, we defined **protocols** working over **channels**.

All of the definitions above could be defined as abstract syntax, letting the concrete syntax be the language that will implement all this functionality. In order to avoid confusion concerning the meaning of various terms, it is important to properly define and document the semantics associated with the vocabulary used.

3. THE SYSLIB-PICASSO METHODOLOGY

Once the fundamental concepts defined, we may apply them to an embedded system design methodology. We integrate the Syslib library in a complete co-design methodology. We describe three distinct levels of specification and validation in a general co-design flow. These levels are all defined with a Syslib API. A GUI tool, called **Picasso**, supports the methodology we propose. This tool, developed by our team, supports the concepts above and provides services to graphically enter a system specifications. Figure 1 supports the description of our methodology.

3.1.1 Functional level or Syslib^{FL}

In this first level, we validate the general idea of the algorithm or the application to be implemented on a SoC, without bothering about any architectural detail or about partitioning. Software programmers commonly perform this step with the help of an OO language. At this level, we shall define neither timing considerations nor synchronisation issues.

Syslib^{FL} is a C++-based library to specify embedded systems at afunctional level of abstraction. It uses C++-constructs to support the semantics explained in section 2. A designer declares a module with the use of a *class*. In this class are instantiated different *class member* variables defining different abstract data types as well as event definitions. At this level, these types respond to the general needs of a system. They will be changed to more appropriate types at lower refinement steps. Other members of a module's class are ports for communication between modules.

These ports are defined differently to handle data communication or event signalling for synchronisation. Other members include what we call behaviour definitions. These are the heart of every module. A behaviour is simply a procedure associated with a sensitive port (carrying an event). It is executed when an event change is detected. In the behaviour's algorithm, we interface the communications with methods for data read/write and event peeking, consumption and notification.

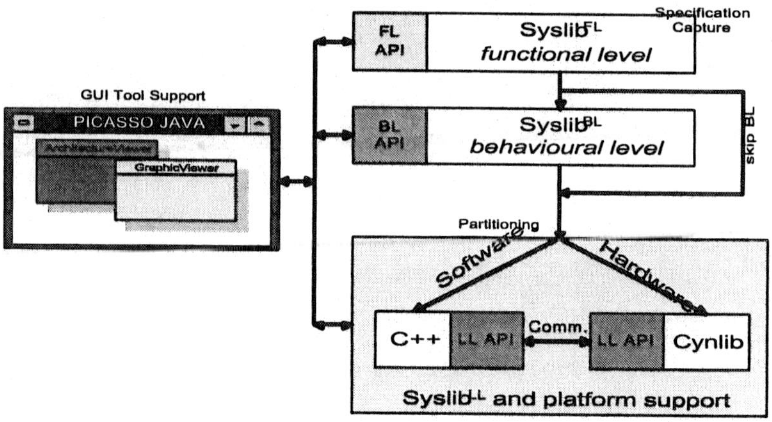

Figure 1: The complete Syslib-Picasso methodology. Starting at a functional level, we refine to the behavioural level or we pass directly to the architectural level, where we partition and link the design to architectural details. These operations are performed with Picasso, a GUI tool.

When a module is completed and ready to be used, it is then time to *instantiate* it as a C++ object. Connections between modules are created via channels. A channel can be built of a value and a FIFO for temporary storage before the targeted modules can read their messages. A channel can be seen as a *socket*, which lets two modules of any kind exchange data, and is given restricted features. When instantiating modules and channels, a designer creates a hierarchy of modules. This hierarchy, linked with ports can be instantiated as well in other modules.

Syslib^{FL} does not hide the C++ constructs and syntax from the designer, but keeps them simple and easy to understand. In fact, Syslib is based on Cynlib [8]. Cynlib is a C++-based open source hardware modelling language. It brings similar functionality to that of VHDL or Verilog, but it uses the C++ syntax and OO concepts.

3.1.2 Behavioural level or Syslib^{BL}

Having passed the first level, a designer may want to enrich his design with timing considerations, making a refinement procedure appear. Syslib^{BL} is an additional function set and library that is used to emulate timing operations at a somewhat high level of abstraction. This level is intended to make decisions before the partitioning phase. Within this step, it is possible to either make some modules wait for specific values (coming in from ports) or for a predefined period of time, with knowledge that a value may be delayed. In this level, we may introduce module execution priorities, timing constraints or module preemption, to better model the real behaviour of a system. Also, a designer will profile the execution of modules and take performance measurements (brought through the Syslib ^{BL} API) to know if a (a group) of module should be implemented into hardware or software.

Syslib^{BL} is currently under development and will not be discussed in this paper.

3.1.3 Architectural level, low level or Syslib^{LL}

Finally, when functional and behavioural analyses have been accomplished, we may think of implementing the modules into their effective domain, which are hardware and software. From this point, some modules have to be coded into a hardware modelling language for the subsequent hardware-domain steps (e.g. synthesis). The same apply for software modules, which need to be matched to a real-time operating system for proper handling and management of modules (threads). Syslib ^{LL} is the lowest level for the specification capture phase of the co-design flow and acts with the partitioning phase. After partitioning decisions, the designer progressively refines his specification to what we could call a final solution. If he chooses to implement a module in hardware, he will adapt his code to Cynlib. If he decides to implement a module in software, he will refine it to C++ and match the low-level RTOS layer Syslib^{LL} and its API. Finally, he will bind to each module channels for hardware/software communications. These channels imply a shared-memory pattern communication.

The main interest of Syslib^{LL} is to support the C++ software modules by linking the designer's modules to the microprocessor's RTOS features. At this level, software modules are now purely C++ and should interact with the RTOS API, which supports the architectural services: interruption handlers, memory allocation and so on. For our methodology, we have chosen the µC/OS-II micro-kernel [9], a simple and efficient RTOS with enough features to satisfy any designers. The goal here is to correctly link the API's methods to the interface we previously introduced in Syslib^{FL} and Syslib^{BL}, to minimise the refinement step.

In addition, we must integrate the Cynlib hardware modules to some architecture interfacing to ease the interaction between hardware modules and other (either hardware or software) modules. An additional package to interface Cynlib modules and the architecture itself must be supplied. This package redefines for example the *read/write* methods used at high-level with adapted methods for memory accesses.

Once the capture finished, the system is ready to be validated with commercial tools, like MTI ModelSim and Mentor Graphics' Seamless CVE. Cynlib modules are simulated within ModelSim using the FLI technology [10]. If modifications need to be applied, it is always possible to roll back to the specification capture step and change the code, at high or low level and simulate again.

```
#include "syslib.h"// SyslibFL services
#include "gcdfunc.h"
class GCD: SysModule {
private :// hidden from other modules
   // behaviour procedure prototypes
   proc();
public :// visible from other modules
   GCD();// for module instantiation
   // data ports
   SysPortIn<SysInt>pA;
   SysPortIn<SysInt> pB;
   SysPortOut<SysInt>pC
   // event ports
   SysPortIn<SysEvent> pGo;
   SysPortIn<SysEvent> pGrant;
   SysPortOut<SysEvent> pFinish;
};
// constructor
GCD::GCD() {
   SysBehaviour(&GCD::proc, pGo, pGrant, 0);
}
GCD::proc() {
// to proceed, both pGrant and pGo must be set
if (pGrant.peek() && pGo.peek()) // peek events
  {
    // consume the received event
    pGo.consume(); pGrant.consume();

    // read A and B from channel
    pA.read();
    pB.read();
    // get GCD from function
    pC = findGCD(pA, pB);
    //propagate value to channel
    pC.write();
    pFinish.notify();
  }
else
  return;// do not block and simply return.
}
```

Figure 2: A Syslib^{FL} module example of a greatest common divisor algorithm. Ports and constructor are public members: open for communication with other modules. The behaviour definition lies in constructor. Ports values are manipulated like variables. Header and source could also have been in separated files.

This methodology is thus dedicated to the specification phase of the co-design flow, with the Syslib library as a strong branch of it. The Picasso tool interacts in the three levels of specification. The methodology proposed by Picasso is similar to the VCC tool from Cadence: it loses flexibility (due to an almost-fixed architecture) but gains in development time (simpler and more intuitive).

4. EXAMPLE

In this section, we present a small example of a Syslib[FL] module. This serves to introduce the reader to the Syslib constructs and its similarities with Cynlib. The example given in figure 2 is a simple Greatest Common Divisor module. The algorithm has been removed (and replaced with a function call) to only keep the structure of the module. The module (class) is defined in lines 1 to 17, in a *header*, where a designer declares the module's constructor and ports as public, and the rest as private members. Lines 18 to 40 contain the module *source*. In the constructor (lines 19-21), we define the behaviour of the module, i.e. we provide the parameters that determine the module's execution. The module's procedure `proc` will be called when a change occurs either on the `pGo` port or the `pGrant` port. This completes the module definition. The remaining function represents the behaviour function and could be seen as a *callback function*. In this procedure, we first look for all the conditions to be respected: the `pGrant` and `pGo` events must be activated. We only pook their respective values and consume them only if both are activated, otherwise the module returns and wait for a rescheduling. Data (`pA` and `pB`) are consumed from their respective channels. Finally, the module performs the GCD and returns the value into the `pC` port, which can be used as a member variable. We propagate the value to the channel using the write function and finally, we activate the `pFinish` event.

The module can now be connected to a testing environment (e.g. a testbench), compiled and tested. In figure 3, we see the connections between modules with the use of channels. The channel definition (lines 11-13) is passed a data type to define the channel width and an integer to define a FIFO depth (no depth means no FIFO), by the use of *templates*. Event channels cannot stack information. Since a module will be scheduled when its ports (connected to a channel) are activated, it is not necessary to stack those values. An event does not contain any information but a state: active or not. We see the bind operations in lines 21-23. Channels are defined with an output port of a module and an input port of another (different) module.

```
#include "syslib.h"
#include "GCD.cpp"
#include "TestBench.h"

class Global : public SysModule {
private:
   // modules
   GCD mGCD1;
   TestBenchModule mTB1;
   // channels (only 3 shown)
   SysChannel<SysInt,4> cA;
   SysChannel<SysInt,4> cC;
   SysChannel<SysEvent> cGo;
public:
   Global();
};

Global() : SysModule(), mGCD(), mTB1()
{
   // ports-channels binding
   cA.bind(mTestBench.pFirstValue, mGCD.pA);
   cC.bind(mGCD.pC, mTestBench.pResultValue);
   cGo.bind(mTestBench.pStart, mGCD.pGo);
}

void main(void)
{
   // instantiate all modules
   Global g();
   // let engine begin
   SysOSStart();
}
```

Figure 3: A Syslib^{FL} module example of a greatest common divisor. This figures shows the instantiation of modules, definition of channels and binding operations between ports and channels. The main function instantiate starts the simulation.

After validation at functional level, the module can be refined to behavioural level, where it could be expanded with *wait* calls when one of the two events (figure 2, line 24) is not activated. Execution time could be evaluated through Syslib^{BL} API calls.

At lower level, the specification is almost kept untouched when partitioned into software. Some changes to take advantage of the µCOS-II features are suggested, like adjusting the thread priority, setting up mailboxes or critical sections for shared resources, and so on. However, if a module is partitioned in hardware, it must be ported to Cynlib. Syntax similarities between Cynlib and Syslib make this transition a relatively easy task. Support from a translation tool could be certainly used.

5. SIMULATION ENGINE

The Syslib simulator is written in C++. It provides a stable environment to develop modules written with the use of the Syslib library, at all levels. The simulation engine is inspired from Cynlib and µC/OS-II. It makes use of the ANSI/ISO C++ Standard Template Library (STL) for stability and quick development turnaround. STL's specialized algorithms for container manipulations and search algorithms are extremely useful. A detailed API supports the exchanges between the modules and the simulator. The simulator work is explained in this section. The reader can look at figure 4, for an example of the work of the simulator.

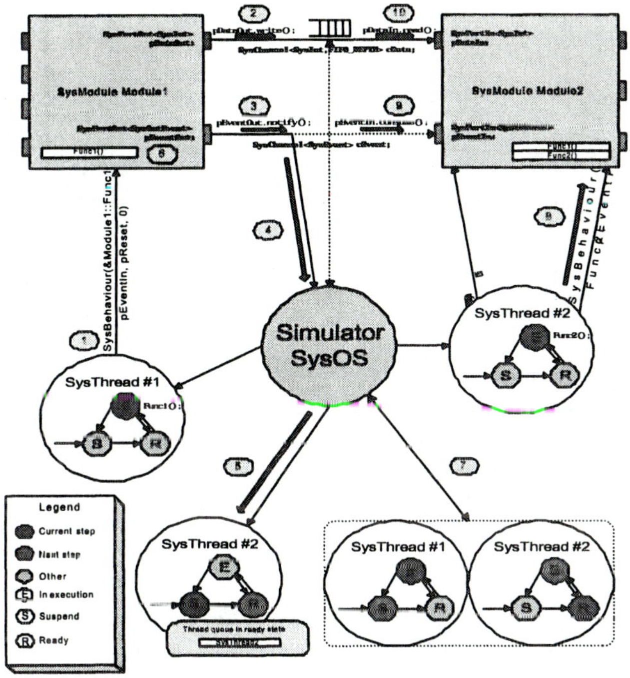

Figure 4: An example of the simulation engine with two modules. SysThread #1 executes (1) the behaviour Func1() of SysModule #1. The module first writes (2) data to its port and next (3) notifies an event. The event is trapped (4) by the simulator engine and the SysThread2 (5) toggles to the R state. The SysModule #1 completes its execution (6). The simulator toggles (7) the SysThread #1 to the S state and the SysThread #2 to E state. The SysThread #2 executes (8) the module's behaviour sensitive to the event changes. The event is consumed (9) and the data value is read from the port (10).

6. CONCLUSION

We proposed, in this paper, a new system-level library to design embedded systems. We believe it is important to base our work on the popular streams emerging from the general R&D results in the domain of embedded systems, SoC and modeling. Thus, object-oriented system modeling is a visionary concept to increase system design productivity, reuse and simulation performance. In addition, the use of a common language from the highest to the lowest level of development alleviates the refinement tasks. Syslib could be seen as an intermediate level to decompose a purely C++ high-level specification into several modules susceptible for partitioning. Syslib brings a step-by-step refinement procedure to lead the specification to approved implementation languages, e.g. C++ and Cynlib.

ACKNOWLEWDGEMENTS

We would like to thank Mentor Graphics, Micronet and GRIAO for their financial support. We would also like to thank Marc Bertola, from the CIRCUS group, who has worked on this project in various ways.

REFERENCEs

[1] S. SWAN, *An Introduction to System Level Modeling in SystemC 2.0.* Open SystemC Initiative (OSCI) Cadence Design Systems, Inc. May 2001.

[2] GAJSKI, ZHU, DOMER, GERTSLAUER, ZHAO. *SpecC Design Methodology.* Kluwer Academic Publishers, 2000.

[3] SCHAUMONT P., CMAR R., VERNALDE S., ENGELS M., BOLSENS I., *Hardware Reuse at the Behavioral Level.* Design Automation Conf. '99, New Orleans, June 1999

[4] PANDA P.R. *SystemC – A modeling platform supporting multiple design abstractions.* 14th Intern. Symp. on System Synthesis. 2001. Montreal, Canada. October 2001.

[5] Forte Design Systems. *Cynlib Language Reference Manual Release 1.4.0.* http://www.cynapps.com/CynAppsNew/3sp-cynlib.html. Last visit October 2001.

[6] Accellera Webpage. http://www.accellera.org/index.html. Last Visited November 2001.

[7] F. BELINA, D. HOGREFE, *The CCITT-Speficiation and Description Language SDL.* Elsevier Science Publishers B.V. North-Holland. 1989.

[8] L. Filion, G. Bois, E.M. Aboulhamid, *Syslib: A System-Level Library Extended from Cynlib for SoC,* HDLCON 2002, SanJose, March 2002.

[9] LABROSSE J. J. *MicroC/Os-II. The Real-Time Kernel.* R&D Books. Lawrence, KS, USA. 1999.

[10] HÉNEAULT Y., FILION L., BOIS G., ABOULHAMID E.M., *A Fast Hardware Co-Specification and Co-Simulation Methodology Integrated in a H/S Co-Design Platform.* 8th Intern. Conf. on Microelectronic. Rabat Morocco. November 2001.

Section 5.2

AUTOMATIC PORTING OF BINARY FILE DESCRIPTOR LIBRARY

Maghsoud Abbaspour, Jianwen Zhu
Electrical and Computer Engineering
University of Toronto

Abstract Since software is playing an increasingly important role in system-on-chip, retargetable compilation has been an active research area in the last few years. However, the retargetting of equally important downstream system tools, such as assemblers, linkers and debuggers, has either been ignored, or falls short of production quality due to the complexity involved in these tools. In this paper, we present a technique that can automatically retarget the GNU BFD library, the foundation library for a suite of binary tools. Other than having all the advantages enjoyed by open-source software by aligning to a de facto standard, our technique is systematic, as a result of using a formal model of abstract binary interface (ABI) as a new element of architectural model; and simple, as a result of leveraging free software to the largest extent.

Keywords: Hardware/software codesign, retargetable compilation, embedded software, ASIP.

1. INTRODUCTION

New products in consumer electronics and telecommunications are characterized by increasing functional complexity and shorter design cycle. It is generally conceived that the complexity problem can be best solved by the use of system-on-chip (SOC) technology. And the design cycle problem can be best solved by pushing functionality as much as possible to software. However, the conventional wisdom here that "software is cheaper than hardware", is not necessarily true unless the software development platform, typically includes operating system, compiler, assembler, linker, and debugger are readily available. Unfortunately, all these tools depend intrinsically on the processor architecture, which in an SOC context is usually designed to adapt to the application. The development platform has to be retargeted to the new processor architecture and this task is by no means trivial.

The field of retargetable compilation has evolved to the point where an architecture description language (ADL) can be used to model a processor microarchitecture, and a compiler can be generated automatically from such a architecture specification. While research in compiler generation is becoming mature, few efforts address the automatic generation of other tools. To the best of our knowledge, only the automatic generation of assembler and simulator has been published. This is partly due to the fact that these "downstream" tasks are perceived to be trivial compared to optimizing compiler. While this perception has been persistent enough to be reflected in all computer science curriculums, it is no longer valid. Take the linker as an example, while the traditional linker does nothing but threading the object file together, the modern linker has to handle features such as shared libraries and dynamic linking as a result of modern operating system, static constructors and templates as a result of modern programming languages such as C++ and Java, and even inter-procedural optimization as a result of modern compiler theory. The most recent Vernon of Free Software Foundation's `binutils` package, which delivers exactly the downstream development tool suite, has a daunting size of 250k lines of C code.

Neither the manual development nor the automatic generation of software with such complexity is reasonable to fit into an SOC development cycle. Most companies chose to port, or reuse the majority of an existing package, while manually rewriting the architecture dependent part of it. The de facto standard of such a package is the GNU `binutils` package, partly due to the fact that it is designed to be "portable", partly due to the fact that it is free software and accessible to everyone in the world. While this package has been ported to virtually every known processor in the world, it still has to be manually ported to every new processor ever created. Unfortunately, the skill required to port this package is arguably limited only to the group of people worldwide which can probably fit into one room

A tool that can automatically port this mature, robust and standard package then seems both ideal and feasible: the architecture-dependent part of the package is relatively small after all. It is not trivial however, since the interface between the architecture dependent and independent part is neither cleanly de- fined nor well documented.

In this paper, we focus on the automatic porting of the GNU Binary File Descriptor (BFD) library, which implements an API that manipulates object files and forms the foundation of all GNU's downstream development tools. The contribution of this work is two fold: First, we present a model of abstract binary interface (ABI) as a new element of architecture model. While ABI is one of the essential information for retargetting, it hasn't be a subject of architecture specification in previous work. Second, to strike a balance between elegance and standard compatibility, we have developed an automatic technique that can generate an implementation of obscure, poorly documented interface from a cleanly defined ABI model.

In the sequel, we will first review the related work in Section 2 and give a brief tutorial of the BFD library in Section 3 from the perspective of BFD user. We will then present the relevant ABI model in Sectio n 4. In Section 5, we describe BFD's architecture-dependent interface in detail and illustrate how C code can be automatically generated from the ABI specification to implement this interface. Finally, we describe our tool implementation and show experimental result in Section 6.

2. RELATED WORK

The first step towards retargetable compilation is the establishment of architectural model and the definition of corresponding architectural description languages (ADLs). The earliest forms of ADLs are various code generator generators (CGGs). The CGGs typically use tree pattern specifications to drive the generation of the instruction selector. However, such specification is often tied to a particular compiler implementation, for example, a particular intermediate representation.

A more recent effort is the the set of computer system description languages in the Zephyr project, where [Ramsey and Fernandez, 1995] is devoted to the description of binary encoding of instruction set, [Ramsey and Davidson, 1998] focuses on the semantics of instruction sets, and [Bailey and Davidson, 1994], [Cifuentes and Simon, 1999] describes the calling conventions. However, the models are not integrated and different aspects of the same architecture are scattered in different specifications in different languages. In addition, there is no explicit support of instruction level parallelism (ILP). The machine description language MDES [Gyllenhaal, 1994] of the impact compiler, seems to be the most sophisticated in describing ILP. In MDES, the architecture model is accurate enough to describe the superscalar, VLIW, as well as new architecture features such as predicated and speculative execution.

Architecture descriptions for embedded processors, for example, the DSP processors and application-specific instruction set processors (ASIPs), have also received intense interest in the recent years. MIMOLA [Marwedel, 1984] used a hardware description language to describe the structural model of the processor, code selection and register allocation is then performed by pattern matching. Since ILP can be concisely represented using a structural model, some recent work adopted a similar approach. For example, CHESS [Prate et al., 2001] uses a graph-based model, while EXPRESSION [Halambi et al., 1999] uses a network of abstract components including those that capture memory hierarchy.

Among all above-mentioned work, few addressed the issue of retargeting downstream tools. The most sophisticated seems to be the NewJersey Machine-Code Toolkit [Ramsey and Fernandez, 1995], which can, for example, automatically generate instruction encoding and decoding routines from an abstract ADL specification. However, its ADL does not allow a complete specification of ABI, and leave important issues such as the organization of relocation closure to the application. Furthermore, the tools generated have by far less capability than standard tools such as GNU `binutils`, and therefore is not yet practical for use in a production environment.

3. BINARY FILE DESCRIPTOR LIBRARY (BFD)

GNU's BFD library is a package that contains a set of common routines to manipulate object files [S.Chamberlain, 1991] There are three types of object files.

- relocatable object files, which hold code and data suitable to be linked with other object files to create an executable or shared object file, or another relocatable object;
- executable files, which hold a program ready to execute;
- shared object files, which hold code and data suitable to be linked in two contexts: first, the link-editor can process them with other relocatable and shared object files to create other object files; second, the runtime linker can combine them with a dynamic executable file and other shared objects to create a process image.

Due to historical reasons, object files present with different formats, called the binary file format (BFF). The most commonly used BFFs are `a.out`, `COFF` and `ELF`. The general structure of an BFF contains four major parts: a file header containing general information as well as pointers to other parts of the file, a number of sections holding code and raw data, relocation tables and symbol table information.

One difficulty in using the BFD library is its complexity. The library itself is very large, the number of functions offered in the front end are exceptionally many. The BFD front end was designed in mind to allow the programmer to be able to retrieve all type of information about any BFF, at least the existing ones. The BFD library can be integrated with disassemblers, decompilers, debuggers, etc. Due to this generality and hence its bulkiness, it is difficult to use it without spending a great deal of time learning how to use it. Perhaps because it is too general, it often contains information that is not needed for some applications.

4. ABI MODELING

An ABI defines a binary interface for application programs that are compiled and packaged for a specific OS running on specific hardware architecture. An ABI is a protocol between different software development tools, so that software created in different languages and compiled by different compilers can still be linked and interoperate with each other. An ABI is also a protocol between the application and the OS, so that the OS loader can create the correct process image from an executable file and possibly many shared object files. Rather than presenting a complete ABI model, in this paper we focus only on part of the ABI that is relevant for BFD porting and describe the modeling of relocation and procedure linkage table (PLT). An architecture description can be specified using this model and serves as the input to our automatic porting tool.

Since software programs are compiled separately into object files, each object file may contain data or instructions that reference symbols defined elsewhere. Even for the reference of local symbols, the actual address of these local symbols cannot be resolved at compile time since the enclosing sections can be moved to arbitrary locations at link or load time. The process of calculating the correct values of these external or local symbol references, called the *relocated values*, and adjusting the bits within the corresponding instructions or data, called the *relocation field*, is called relocation. Typically, an object file contains an array of *relocation entries* in a special relocation section, each of which points to the instruction or datum to be relocated. Depending on different BFF used, the relocation entry may contain a *relocation type*, which designates the calculation method of relocated values, and a *relocation addend*, which is an integer-sized storage that can help store useful information for the calculation in case the relocation field of the instruction or datum to be relocated is not large enough to hold the information. While the exact source of this information may vary, it is always an integer value that will be added to relocated value, and hence the name.

To support shared object and dynamic linking, position-independent code (PIC) whose instructions need no relocation should be supported. The linker usually creates a *global offset table* (GOT) that contains pointer to all the global data that the executable file addresses. GOT entries can be considered as global data themselves and therefore any reference to them needs relocation. Similarly, the linker may also create a procedure linkage table (*PLT* for procedure symbols. Due to the need for lazy evaluation, that is, not providing procedure addresses until they are called for the first time, each PLT entry contain a series of architecture-dependent instructions that calls routines defined in the dynamic linker.

The calculation of the relocated value involves the following parameters, which are either information kept in the relocation field, or relocation entry, or values maintained by applications such as linker or loader:

- A: the addend, which can either be stored in the relocation field of the instruction or datum to be relocated, or the relocation entry;
- B: the base address at which a shared object is loaded into the memory during execution;
- GOT: the address of the global offset table;
- G: the offset into the global offset table at which the address of the referenced symbol resides during execution;
- L: the place (section offset or address) of the procedure linkage table entry for the reference procedure symbol;
- P: the place (section offset or address) of the instruction or datum to be relocated;
- S: the value of the referenced symbol.

The relocation model for ABI specification can thus be characterized by the calculation method as well as the identification of relocation field.

Definition 1 gives our simple model of relocation type, which is on the other hand a complete one due to the specialty of the relocation calculation expression.

```
Reloc = tuple {
    id          : int;
    expCode     : byte;
    rightshift  : int;
    bitsize     : int;
    bitpos      : int;
    complain    : {IGNORE, BIT, SIGN, UNSIGNED};
    }
```

Definition 1 A **relocation type** is a member of

Where id is an unique integer identifier, expCode = $C_7, C_6, \ldots, C_0$ encodes the expression $\Sigma_i C_i P_i$, with $P_7, \ldots P_0$ being – GOT, A, B, G, ,GOT, L,– P, S respectively; rightshift represents the number of bits at the right side of the calculated $\Sigma_i C_i P_i$ that should be dropped; bit pos and bitsize represents the bit position as well as the size of the relocation field within the instruction or datum to be relocated; complain encodes the action to take when specific type of overflow occurs.

-GOT	+A	+B	+G	+GOT	+L	-P	+S	
0	1	0	0	0	0	1	1	=0x43

==> S+A-P

Figure 1.1 Relocation calculation expression.

Example 1 shows the specification of Intel 386 relocation types.

Example 1 Intel 386 family relocation description:
R_I386_NONE= 0 0x00 0 0 0 IGNORE
R_I386_32= 1 0x41 0 32 0 *BIT*
R_I386_PC32= 2 0x43 0 32 0 *BIT*
R_I386_GOT32= 3 0x52 0 32 0 *BIT*
R_I386_PLT32= 4 0x46 0 32 0 *BIT*
R_I386_COPY= 5 0x00 0 0 0 *BIT*
R_I386_GLOB DAT= 6 0x01 0 0 0 BIT
R_I386_JUMP SLOT=7 0x01 0 0 0 BIT
R_I386_RELATIVE= 8 0x60 0 32 0 BIT
R_I386_GOTOFF= 9 0x61 0 32 0 BIT
R I386 GOTPC= 10 0x4a 0 32 0 *BIT*

Each PLT entry contains a sequence of instructions that are executed when the corresponding procedure symbol is first referenced. The instruction usually jumps to the stub code of the dynamic linker which in turn load the corresponding shared object files. While the exact way of implementing this mechanism is outside the scope of this paper, it is suffice to model the PLT entry as a sequence of binary words that represents these instructions. Note that PLT is specific to ELF format. Definition 2 gives our PLT model.

```
plt = tuple {
    size        : int;
    instrns     : [illegible];
    }
```

Definition 2 A PLT entry is a member of
Where size is the number of words (4 byte) for each PLT entry and instrns is a sequence of binary words.

Example 2 shows one PLT description for ELF object format of SPARC.

Example 2 PLT Description for SPARC ELF32 Format:
pltn=3, 0x03000000, 0x30800000, 0x01000000,

5. RETARGETTING BFD

To port the BFD library to a new processor, the architecture-dependent part of the BFD implementation needs to be generated. Unfortunately, this part does not have a clean interface due to historical reasons, and is winded together with BFF-specific code and scattered in many different C files.

The architecture -dependent interface of BFD contains type declarations, data and functions, detailed as follows.

- Types: It includes the definition of an enumeration type that defines the relocation type identifiers.
- Data: It contains some general information about the target processor, such as word size, address size and name. It includes the definition of a relocation howto table, an internal representation hat characterizes the relocation calculation methods of each relocation type. It also contains an internal representation of the PLT entries to be generated

- Functions: it contains the functions for checking relocations as well as generating dynamic sectbns. The dynamic sections, such as .dynamic, .hash, .got and .plt, are used by dynamic linker for creating process image. They are created by the linker to hold various data, symbol table, global offset table and procedure linkage table respectively. It also contains the function to relocate all relocation entries of a section.

The architecture -dependent implementation of BFD for a specific processor can be generated from the architecture specification, which includes the ABI model that we have described in Section 4.

6. IMPLEMENTATION AND EXPERIMENTS

We have implemented a retargetting tool called `bfdgen`. The tool takes two inputs: one is the architecture description from the user. This description is compiled by the ADL compiler and converted into an internal representation. The architecture description contains not only the instruction set that can be used to retarget compiler and assembler, but also the ABI information of thc processor. The other input is a set of template files provided by our tool suite for the BFF/OS configuration of interest. The template files are essentially C headers, files, as well as configuration scripts with special placeholders for architecture-dependent code. Our code generator generates code at these placeholders according to th e compiled architectural model. The generated files can be merged with GNU BFD source tree, and configured and compiled by the normal BFD building process.

To exercise the tool, we have specified the processor model for both SPARC and Intel 386, as they are representative to the RISC and CISC architectures.

Processor	ABI complexity (#line)
sparc	26
i386	13

Table 1.1. Complexity of ABI specification.

Table 1.1 shows the complexity of the ABI specification that is relevant to BFD porting.

We fed the template files for ELF/Solaris and the SPARC architecture description to bfdgen and copied the generated files to the binutils source tree. The files can be used to build both the BFD library and the GNU link editor, ld. Table 1.2 shows the generated files. It generated 13,247 lines of code in total. To verify the ported BFD and linker, we built all SPEC2000 integer benchmark including gzip, mcf, eon, vortex, vpr, crafty, perlbmk, bzip2, gcc, parser, gap and twolf using the generated linker. They linked and ran successfully.

Generated Files	#line (generated)
/bfd/config/mysparc-elf.mt	3
/bfd/archures.c	1483
/bfd/configure.host	112
/bfd/configure.in	286
/bfd/config.bfd	166
/bfd/elf32-mysparc.c	1482
/include/elf/common.h	229
config.sub	1014
/bfd/target.c	785
/bfd/cpu-mysparc.c	44
/bfd/elfcode.h	6582
/ld/configure.in	183
/ld/emulparam/ elf32_mysparc.sh	9
/ld/Makefile.in	868
/ld/config/mysparc-elf.mt	1

Table 1.2. Generated and changed files.

7. CONCLUSION

In conclusion, we have argued that the complexity involved in modern downstream tools such as assemblers and linkers have made development or automatic generation of these tools from scratch an impractical task. This has led to our effort which seeks to automatically port an existing tools that is powerful, robust and freely available. By augmenting the specification of instruction set information of a processor with ABI information, we are able to automatically port the heart of GNU's binutils package, the BFD library. Our experiment shows that this approach is both feasible and practical. The tool that we developed using this technique is extremely easy to use: the porting of GNU linker takes only tens of lines of ABI specification and seconds of running time of our tool, `bfdgen`.

REFERENCES

[1]Bailey and Davidson, 1994] Bailey, M. and Davidson, J. (1994). A formal model and specification language for procedure calling conventions. Technical Report CS-94-39, Computer Science, University of Virginia.

[2]Cifuentes and Simon, 1999] Cifuentes, C. and Simon, D. (September 1999). Procedural abstraction recovery from binary. Technical Report 448, Department of Computer Science and Electrical Engineering, University of Queensland,Brisbane QLD 4072,Australia.

[3]Gyllenhaal, 1994] Gyllenhaal, J. (1994). A machine description language for compilation. Technical report, Department of Electrical and Computer Engineering, University of Illinois, Urbana IL.

[4]Halambi et al., 1999] Halambi, A., Grun, P., Ganesh, V., Khare, A., Dutt, N., and Nicolau, A. (1999). Expression: A language for architecture exploration through compiler/simulator retargetability. In *Proceedings of the Design Automation and Test Conference in Europe*.

[5]Marwedel, 1984] Marwedel, P. (1984). The MIMOLA design system: Tools for the design of digital processors. In *Proceeding of the 21st Design Automation Conference*, pages 587–593.

[6]Prate et al., 2001] Prate, J. V.,W.Geurts, D. L., and Goossens, G. (July 2001). Processor modeling and code selection for retargetable compilation. *ACM Transaction on Design Automation of Electronic Systems*, 6(3).

[7]Ramsey and Davidson, 1998] Ramsey, N. and Davidson, J. (1998). Specifying instructions' semantics using CSDL. Technical report, Department of Computer Science, University of Virginia.

[8]Ramsey and Fernandez, 1995] Ramsey, N. and Fernandez, M. (1995). The New Jersey machine-code toolkit. In *Proceedings of the 1995 USENIX Technical Conference*, pages 289–302.

[9]S.Chamberlain, 1991] S.Chamberlain (April 1991). *libbfd: the Binary File Descriptor library.* Cygnus Support, Free Software Foundation, Inc.

Section 5.3

Code Compression on Transport Triggered Architectures

Jari Heikkinen, Jarmo Takala, and Jaakko Sertamo
Institute of Digital and Computer Systems
Tampere University of Technology
P.O.B. 553, 33101 Tampere, Finland

Abstract: Entropy coding is an efficient method for compressing information and it can be used to improve code density of processor architectures. However, optimal compression requires information of the probability of instructions, which, in general, is not available when processor is designed. However, when designing customizable processors, which are tailored for a given application, such information can be obtained at design time. In this paper, entropy coding is used to improve code density of application-specific programmable processors based on transport triggering paradigm. Three DSP applications are used as benchmark applications and Huffman coding is applied to compress he instructions of a processor architecture tailored for each benchmark.

Key words: code compression, Huffman coding, customizable processor architecture, transport triggered architecture, discrete cosine transform, Viterbi decoding

1. INTRODUCTION

The current trend in software development for embedded systems in the field of digital signal processing (DSP) is to move towards high-level language (HLL) programming and customizable architectures [1]. The reason behind this is the gap between the productivity of designers and increased complexity of DSP applications. A popular means to improve performance in DSP applications is to exploit the instruction-level parallelism and especially very long instruction word (VLIW) architectures have currently gained considerable popularity. In such architectures, a long instruction word is used to control the operation of parallel function units (FU).

The instruction word consists of dedicated fields for each of the Fus specifying the operations and operands for the FUs.

Due to the several control fields, the code density of VLIW architectures has been low [2]. However, the code density can by improved by utilizing code compression methods to the instructions. Programs are stored into the memory in compressed format, then decompressed in the instruction decoding phase, and finally executed in the processor. Such systems are called compressed-code architectures [3].

In [4], Huffman coding is used to compress general-purpose programs taken from SPEC benchmark suite for six different architectures including MIPS R4000, Sun 68020, and Sun SPARC [5]. Results indicate ompression ratios between 0.77 and 0.87 for these architectures. In the previous approach, the coding parameters are constant for each application, thus optimal compression rate cannot be obtained. Better compression ratios can be achieved when the coding is optimized for each application, which implies that an application-specific instruction decoding hardware is eeded. Such an approach can be applied when designing customizable processor architectures, where the hardware resources are tailored according to the application. In such a case, the probability of each instruction can be measured during the architecture design and optimal entropy coding can be applied resulting in more efficient compression rate.

In this paper, entropy coding is used to improve the code density of customizable processor architectures. Three DSP applications, for which tailored processor configurations are designed, are used as benchmark applications. Huffman coding is then applied to compress the instructions of these three processor designs to improve code density.

2. TRANSPORT TRIGGERED ARCHITECTURE

VLIW architectures lend themselves for computationally intensive DSP applications. They are modular; the number of function units (FU) can be easily increased. There are even VLIW architectures that support customized, application-specific function units. This may, however, restrict the flexibility, e.g., in TriMedia [6], support for multi-operand instructions, i.e., multi-operand FU, reserves several instruction fields from the VLIW instruction. VLIW architectures have been criticized for their requirements on read/write ports in the register file [7]. In order to alleviate this problem, clustered approach has been suggested, i.e., the register file is partitioned as illustrated in Fig. 1. In addition, the complexity of the bypassing network and the register file is high since the bypassing network must support bypassing of operands from all the FU outputs to all the FU inputs.

The bypass complexity of VLIW architecture can be reduced by making the bypass visible at the architectural level. Furthermore, the bypass complexity can be reduced by reducing the number of read and write connections and by reducing the number of bypass buses. This implies that besides the operations also the operand transfers (transports) need to be scheduled at compile-time. The bypass transports become visible at the architectural level implying that operations can be hidden. In this model, the data transports trigger the operations implicitly, thus the traditional operation triggered programming paradigm is mirrored, hence the name transport triggered architecture (TTA) [8]. In TTA programming model, program specifies only the data transports to be performed by the interconnection network. Therefore, only one type of operation is supported: move operation, which performs a data transport from a source to a destination. The number of move operations per instruction is equal to the number of simultaneous transports supported by the interconnection network.

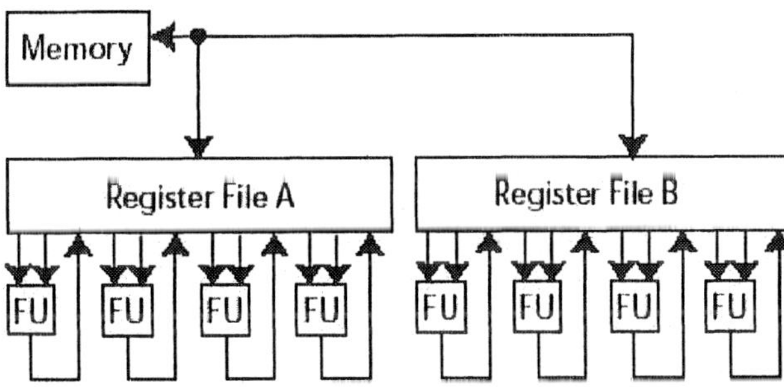

Figure 1. Principal block diagram of clustered VLIW. FU: Function unit.

A TTA processor consists of a set of FUs and register files containing general-purpose registers connected by interconnection buses as illustrated in Fig. 2. Connections to buses are established through sockets; an input socket feeds operands from the buses into the FUs and an output socket places the FU results into the correct bus. The TTA concept provides flexibility in form of modularity; functional units with standard interface are used as basic building blocks.

MOVE framework is a design environment containing a set of software tools for designing application-specific instruction set processors using the TTA paradigm [9]. The design flow consists of three principal components as illustrated in Fig. 3. The design space explorer searches for a processor configuration, which yields the best cost/performance ratio for a given application. The explorer optimizes both the resources and connectivity. The hardware subsystem generates structural hardware description of the selected processor configuration and produces statistics of timing, area, and power consumption. The software subsystem generates instruction-level parallel code for the selected processor configuration. It provides statistics, e.g., on cycle count, instruction count, and hardware resource utilization. Processors designed with the MOVE tools are called MOVE processors.

MOVE design environment supports all the trends in DSP processors mentioned previously: HLL programming, customization, and instructionlevel parallelism. Designer can tailor and optimize the resources of the programmable architecture and the HLL compiler adapts to these modifications. This approach allows processor core to be customized according to the requirements of the application in question.

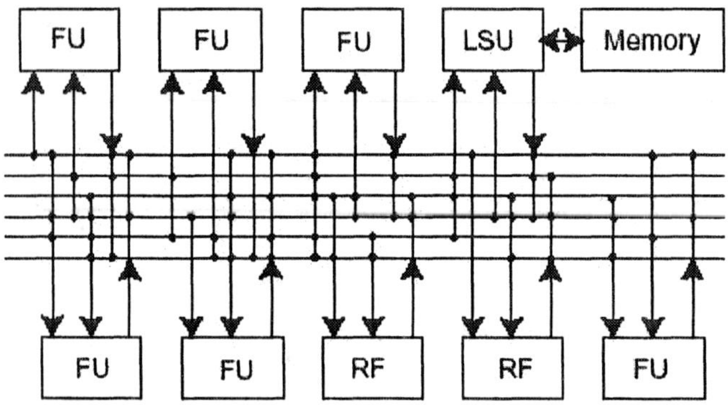

Figure 2. Principal block diagram of TTA. FU: function unit. RF: Register file. LSU: Load-store unit. Dots represent socket connections.

One drawback of MOVE architecture is its poor code density. A long instruction word, containing a dedicated field for each bus, together with a large number of instructions results in a large program code. However, MOVE architecture offers a possibility for efficient code compression. Since the processor is tailored according to the requirements of the application, application-specific compression methods can be utilized.

3. MOVE INSTRUCTION FORMAT

The instruction format of a MOVE processor consists of specific fields, slots, for all the data transports, namely moves. Since one bus can perform a single move per cycle, each instruction contains as many move slots as there are buses. The instruction shown in Fig. 4 contains two move slots and a reserved field for an optional long immediate.

Each move slot contains identification code (ID) for a guard, destination, and source identification. The guard ID is used by a guard unit to control the execution or to squash a move operation. This provides means to realize conditional execution. The source and destination IDs are used to select the source and destination sockets of the data transport, respectively. Both the destination and source IDs contain a socket address and an optional operation code, opcode. This is used to select the operation to be performed at the FU. When the opcode is not used, destination and source IDs contain only the socket addresses.

MOVE processors support also immediate extensions. Long immediate operands can be specified with dedicated fields at the end of instruction word. Short immediates are supported by storing the data into source ID field. The most significant bit of the source ID field is used as a flag to determine whether the bits of the source ID field are used to specify a source address and an opcode or a short immediate value

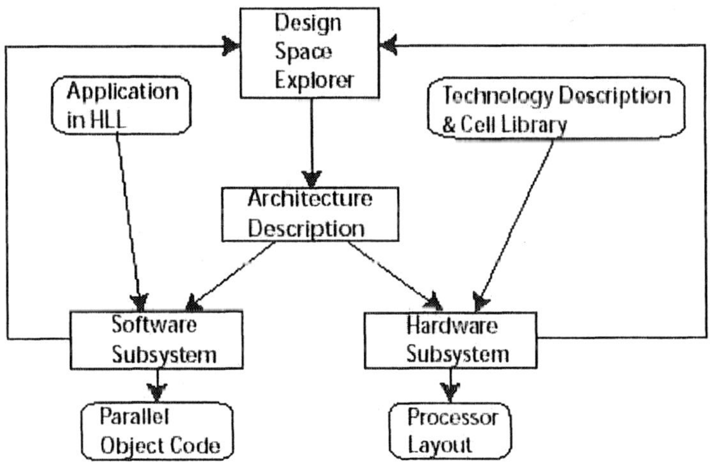

Figure 3. Principal design flow in MOVE framework.

In general, 20 bits are allocated for each move slot. The width of the guard ID field is determined by the number of Boolean registers and compare units, since the outputs of compare units with Boolean registers can be used as guard definitions. This allows compare operations to be used immediately and there is no need to transfer them to Boolean register file. The width of guard ID field is typically less than four bits. The width of the destination ID field is determined by the number of destination connections on the bus. In general, destination ID field is up to seven bits. The width of the source ID field is determined by the number of source connections or the size of the short immediate. In most cases, the size of the short immediate, typically eight bits, determines the source ID width. Taking the immediate flag bit into account, the width of the source ID field is in general nine bits.

The width of the entire MOVE instruction depends on the number of move slots and dedicated fields for long immediates. In general, MOVE processors contain five to 10 buses and one or two long immediate fields. Therefore, the instruction width varies between 130 and 260 bits.

4. ENTROPY CODING ON MOVE INSTRUCTIONS

Entropy coding is based on the fact that if *b*-bit symbols used to describe information are not uniformly distributed, there exists a code, which uses less than *b* bits per symbol. According to this principle, compression can be achieved when small-length codes are assigned to the most probable symbols. This implies that the coding results in a variable-length code.

One of the most popular entropy coding techniques is Huffman coding where symbols are arranged in decreasing order according to their probabilities. After the symbols are arranged, two nodes with smallest probabilities are merged forming a new node with a probability of the sum of the merged nodes. '0' and '1' are assigned to the branches merging into these two nodes. This process is repeated as long as there is only one node left. The compressed code for each symbol is then obtained by reading from the root node to each leaf node and obtaining the binary code from the branches along the path.[10]

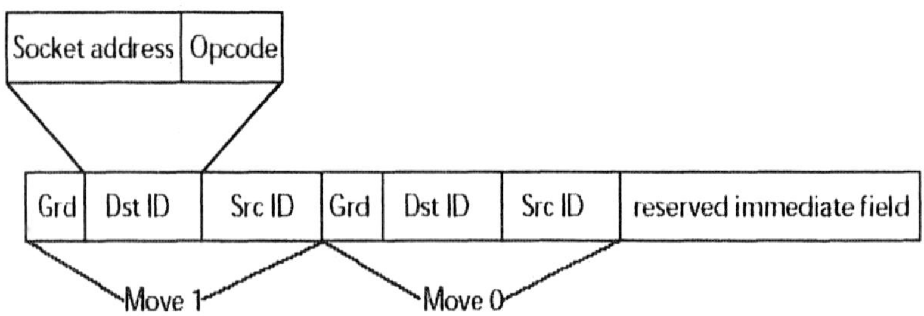

Figure 4. Principal organization of MOVE instruction.

In order to improve code density of MOVE architectures, entropy coding can be applied by interpreting the move slots in the instruction word as symbols. However, this would result in symbol explosion and, therefore, we have considered the fields inside move slot as symbols. Source and destination IDs are coded separately since, on each bus, data can be moved from each source to all connected destinations. The guard ID is not coded since, in most cases, its width is small. The guard belongs also to the critical path since guard value determines whether the operand transfer on bus should be squashed or not. Therefore, the guard must be detected early in the instruction cycle. Coding of the guard ID field would require decoding of the field, which would lengthen the critical path even more resulting in lower clock frequency.

The probabilities of each source and destination on each bus are obtained from the parallel code, which is generated by the MOVE compiler. Huffman coding is then applied and new variable-length codes for each source ID and destination ID symbol on each bus are obtained.

5. BENCHMARKS

In order to evaluate the code density, we have used two applications as examples: discrete cosine transform (DCT) and Viterbi algorithm. DCT is widely used in video, image, and audio coding. Viterbi algorithm is widely used in many decoding and estimation applications in communications and signal processing since it is computationally efficient.

The first benchmark is an 8x8 DCT realized with row-column approach, i.e., the entire two-dimensional (2-D) transform is computed with the aid of 1-D transforms. Here the constant geometry algorithm proposed in [11] has been used. Constant geometry algorithms being regular and modular allow better exploitation of the inherent parallelism.

The second benchmark is a 32-point DCT, where DCT algorithm described in [12] is used. The created C-code contains five functions, one for each processing column of the signal flow graph of the algorithm. Each processing column is written totally unrolled, i.e., no iterations are used. This way the MOVE compiler was able to detect and exploit the inherent parallelism of the algorithm. On the other hand, this sort of code results in larger program code.

Both the DCT applications were described in C-language using fractional data representation, i.e., fixed-point representation where the number range is normalized. Fractional representation is often used in DSP realization and it represents challenges for C compilers because ANSI C does not contain predefined data type for fractional representation.

The third benchmark task was Viterbi decoding, which, as the DCT application, was written in C-language. This algorithm contains more complex control flow; conditional statements are also needed.

These three benchmarks represent different kinds of DSP applications. The 32-point DCT code has been tuned for fast execution and, therefore, it requires large program code, while the 2-D DCT code is a compromise between performance and code size. Viterbi application is control dominated, more complex application resulting in a long execution time and large program code.

6. RESULTS

The design space explorer of the MOVE framework was first used to find an area-efficient architecture for each of the three applications. This was done by first performing resource optimization for an oversized and fully connected architecture configuration to find optimal resources for each of the applications. Next, the unnecessary con nections of the obtained architecture configurations were removed by running connectivity optimization. Optimized architecture configuration for 32-point DCT and Viterbi algorithm are shown in Fig. 5. After obtaining the customized architecture configurations, benchmark applications were compiled and the symbol probabilities were calculated. Finally, Huffman coding was applied and variable-length codes were assigned for each symbol.

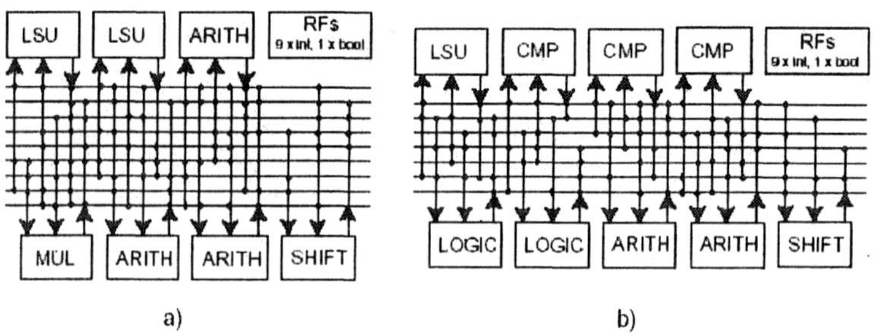

Figure 5. Optimized architectures for a) 32-point DCT and b) Viterbi decoding. LSU: load-store unit. ARITH: arithmetic unit. MUL: multiplier. CMP: compare unit. SHIFT: shifter. LOGIC: logic unit. RFs: register files.

Uncompressed code sizes were calculated by multiplying the number of instructions with the instruction size, which are presented in Table 1. Compressed code size was obtained by first determining the number of bits each new source ID and destination ID code saves compared to the original fixed-length code. Then the total number of bits saved was obtained by multiply ing the number of bits saved on each source and destination Ids with the number of their occurrences. Finally, the compressed code size was obtained by subtracting the number of bits saved from the uncompressed code size.

By utilizing Huffman coding on MOVE instructions, notable reduction in code size is obtained in all benchmark tasks as can be seen in Fig. 6. Compression ratios for the three benchmarks are given in Table 2. Compression ratios of the two DCT tasks are the same. The compression ratio of Viterbi decoding is slightly worse. This was caused by the problems in MOVE compiler, since it was not able to use the least connected architecture configuration obtained from the design space explorer. Thus, an architecture configuration with more connections had to be chosen. This resulted in larger number of symbols to be coded and thus lower code compression ratio. If the MOVE compiler had been able to use the most optimal architecture configuration for Viterbi, the compression ratio would have been the same as for both DCT benchmarks. Compared to the compression ratios of fixed-resource processor architectures in [5], results indicate a notable improvement in code density.

Table 1. Instruction widths for benchmark applications.

Application	Number of instructions	Instruction size [bits]
8x8 2-D DCT	237	155
32-point DCT	552	181
Viterbi	1076	172

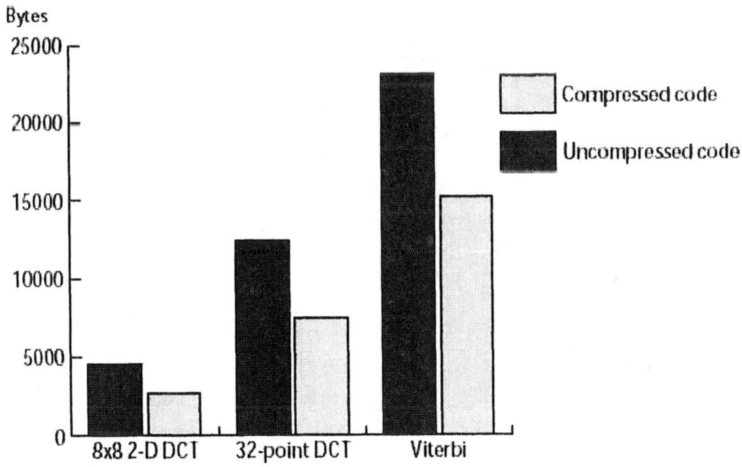

Figure 6. Uncompressed and compressed code sizes.

MOVE instructions could be further compressed by compressing the dedicated long immediate fields. When the field does not contain a long immediate value, the bits dedicated for it are wasted. By compressing the long immediate field in the case it is not used, further improvement in code density could be obtained. Furthermore, the guard ID field could be coded, when its width is larger than two bits.

Application	Compression ratio
8x8 2-D DCT	0.600
32-point DCT	0.599
Viterbi	0.658

Table 2. Compression ratios for three benchmark applications.

7. CONCLUSIONS

In this paper, entropy coding has been applied to improve the code density of application-specific processors based on transport triggering paradigm. Huffman coding was used as entropy coding method to compress the instructions. It was noted that entropy coding provides a notable improvement in code density. The code density could even be further improved by compressing also the dedicated long immediate fields when they do not specify a long immediate value.

In the future research, long immediate fields will also be compressed. Implementation of the decompression circuitry will also be examined to provide fast enough instruction decoding for the real-time requirements of DSP applications.

REFERENCES

[1] J. M. Rabaey, W. Gass, R. Brodersen, T. Nishitani, and T. Chen, "VLSI design and implementation fuels the signal-processing revolution," *IEEE Signal Processing Mag*, vol. 15, no. 1, pp. 22–37, Jan. 1998.

[2] H. Suzuki, H. Makino, and Y Matsuda, "Novel VLIW code compaction method for 3D geometry processor," in Proc. *IEEE Custom Integrated Circuits Conf.*, Orlando, FL, U.S.A., May 21-24 2000, pp. 555–558.

[3] A. Wolfe and C. Chanin, "Executing compressed programs on an embedded RISC processor," in *Proc. 25th Annual Int. Symp. Microarchitecture*, Portland, OR, U.S.A., Dec. 1–4 1992, pp. 81–91.

[4] D. A. Huffman, "A method for the construction of minimum redundancy codes," *Proc. IRE*, vol. 40, no. 9, pp. 1098–1101, Sept. 1952.

[5] M. Kozuch and A. Wolfe, "Compression of embedded system programs," in Proc. *IEEE Int. Conf. Computer Design*, Cambridge, MA, U.S.A., Oct. 10–12 1994, pp. 270–277.

[6] J. Hoogerbrugge and L. Augusteijn, "Instruction scheduling for TriMedia," *Journal on Instruction-Level Parallelism*, vol. 1, pp. 1–21, Feb. 1998.

[7] R. P. Colwell, R. P. Nix, J. J. O'Connel, D. B. Papworth, and P. K. Rodman, "A VLIW architecture for a trace scheduling compiler," *IEEE Trans. Comput.*, vol. 37, no. 8, pp. 967–679, Aug. 1988.

[8] H. Corporaal, *Microprocessor Architectures: From VLIW to TTA*, John Wiley & Sons, Chichester, UK, 1997.

[9] H. Corporaal and M. Arnold, "Using transport triggered architectures for embedded processor design," *Integrated Computer-Aided Engineering*, vol. 5, no. 1, pp. 19–38, 1998.

[10] A. K. Jain, *Fundamentals of Digital Image Processing*, Prentice Hall Int., Inc., Englewood Cliffs, NJ, U.S.A., 1989.

[11] J. Kwak and J. You, "One - and two -dimensional constant geometry fast cosine transform algorithms and architectures," *IEEE Trans. Signal Processing*, vol. 47, no. 7, pp. 2023–2034, July 1999.

[12] J. Takala, D. Akopian, J. Astola, and J. Saarinen, "Constant geometry algorithm for discrete cosine transform," *IEEE Trans. Signal Processing*, vol. 48, no. 6, pp. 1840–1843, June 2000.

Section 5.4

AN APPROACH TO FLEXIBLE MULTI-LEVEL NETWORK DESIGN

V. Cheushev, J. Kolodziejczyk, T. Luba,C. Moraga, P. Sapiecha, V. Shmerko and S. Yanushkevich

Abstract: The paper is focused on integrating the flexibility in the design of logic networks. We achieve the flexibility by several ways: evolving the design and applying the decomposition technique. To perform the circuit evolution, we use one of more powerful techniques borrowed from information theory and decision making, to evaluate the design. This technique can be used along with any multi-level networks design optimisation strategy over a fixed library of cells. As an example, we use the evolutionary network synthesis and our previously introduced concept of a target design style. Also, in order to alleviate the problem of large-size network search space, we use the partition of the space by the state-of-the-art decomposition technique.

Key words: Multi-level synthesis, evolutionary design, parallelism, and information theory approach

1. INTRODUCTION

Our study is relevant to the problem of design networks basedon FPGA. We utilize the paradigm of flexible network synthesis [10] in order to generate networks over a given design style. The methodology we use is refereed to as artificial evolution and machine learning methods of the circuitry design. Our final goal is to improve recently obtained results in evolutionary network synthesis based on information theory [9] and state-of-the-art methods of digital circuit design such as balanced decomposition [2]. Evolutionary network design approach operates with very small circuits only (2 - 6 bit multiplexers, 2 - 5bit adders, 2-bit decoder, 4 - 8-state machine, 3 -bit counter) and it is none of practical implementation (sometimes run time for the evolved simple network requires dozens hours) [1], [4], [11], [5], [12].

In this paper we justify that based on concepts of scanning window and target design style in the evolutionary network synthesis, we are able to apply decomposition technique and extremely improve recently obtained results (include functions when traditional evolutionary algorithms was failed). This is because the network search space is partition able and, therefore, parallel processing of subspaces is possible. Moreover, because genetic algorithms [15] are very good candidate for parallelization, we can achieve the massive parallel search. We are benefit from utilization of new computing resources for another purposes. In this paper we use these resources for manipulation with design styles (any limitations such as type of gates, possible connections and so on are extremely increase the time of searching process).

2. OUR STRATEGY OF THE EVOLUTIONARY NETWORK DESIGN

Adaptive search in the space of possible network solutions is the basic idea of the evolutionary network design. We interpret the search process as a scanning of a network space by a window and consider more complex problem, namely the evolutionary multi-level network design under the given design style and scenario, and continue our study reported first in [7], [10], [9]. The window is a tool that allows us 'to scan' the network search space in order to implement a desired logic function by a multi-level network. Scanning window is codified by a stream of integer numbers that is the chromosome. Each gene of the chromosome corresponds to a primitive gate assigned with an integer number. So, first, one has to choose gates and codify them, along with interconnections between gates, to produce the gene encoding.

We denote a scanning window as *GM£N* that specifies the maximal number of levels *M* and maximal number of gates *N* in each level of the network realization over a fixed cell library *L*. We consider the networks where the gate outputs in a given level are not connected to an input of another gate from this level.

The size of a scanning window *GM.N* is dynamically changed through evolutionary process in accordance with the equation $G_{mi}.N_j \subseteq GM.N$ for integer positive numbers $M_i < M$, $N_j <= \cdot N (M,N >= 2)$.

DEFINITION 1 The level back $L_{back}(w)$ with respect to $w \in \{IN;OUT\}$ species one of three scenarios: (i) inputs *IN* can be connected to the first $a < M$ levels of the created circuit, $L_{back}(IN) = a$, or (ii) outputs *OUT* can be Connected to the last $b < M$ levels of the circuit, $L_{back}(OUT) = b$, or (iii) connections of the depth a, $Lback(IN) = a$, and b, $L_{back}(OUT) = b$, with respect to inputs *IN* and outputs *OUT* accordingly are allowed.

EXAMPLE 1 The simplest case, $L_{back}(IN) = 1$ and $L_{back}(OUT) = 1$, means that connections of inputs *IN* and outputs *OUT* with the first and the last levels of the created network accordingly are allowed. If $L_{back} IN) = 2$ ^3 $L_{back}(OUT) = 3$, then the depth of the connections with respect to *IN* and *OUT* is two and three accordingly, i.e. inputs *IN* can be connected with two_rst levels, and outputs *OUT* can be connected with three last levels of the network.

In accordance to our design style, we specify the types of logic cells from the library *L*, architecture of network (permissible interconnections between cells, levels and inputs and outputs of the network), the maximal number *M* of levels, the maximal number *N* of gates in every level, and types of gates in every level of the evolved network.

EXAMPLE 2$G_{4.5}$ window over the library of cells $L = \{AND;OR\}$ can be considered as "a guide" to design *M*-level, $M \in \{2; 3; 4; 5g;$ networks under different scenarios. For the *G5£4* window, GA can create different networks (from 2-level to 5-level architecture) via dynamical reconfiguration of its size, i.e. $G_{m.n} \subseteq G_{5.4}$ for $M <= 5, N <= 4$.

In this respect, effectiveness of the window based scanning process in a given Target Design Style (TDS) and evolutionary design strategy, in general, can be estimated as an ability to utilize the freedom of the choice that is given by the network space corresponded to completely and incompletely specified initial function.

The upper bound of the number of gates in 2-level, *n* input single-output network created by a scanning window *G2.N* in *AND-OR* TDS is

$2^{n-1} + 1$ gates. This follows from the well-known fact that, in order to realize an arbitrary function of *n* variables in an *AND-OR* 2-level network, one needs not more than $2^{n-1} + 1$ AND and OR gates. Note that we consider correct (not optimal) networks only. Utilizing the scanning window *G3.N* in *OR- AND – OR* TDS, provides creation of a 3-level *n*-input *m*-output network with at most $m^{1/2} \cdot 2^{(n/2)+1} + m$ gates.

LEMMA 1 The number of different 2-level, *n*-input, single-output *AN-OR* networks with $V = m^{1/2} 2^{(n/2)+1} \cdot (1-\xi)$ or fewer gates created by *G2.N*, is at most $3nV$.

3. INFORMATION THEORY APPROACH FOR ESTIMATION OF EVOLVING NETWORK PROCESS

Here we offer information estimations to measure the efficiency of the evolution search [10], [9]. Justification of our approach is grounded on that information measuring matches, more than any other one, to the stochastic nature of evolutionary process.

We consider the evolutionary network design as a process along with the conditional entropy of the function f with respect to the current solution Net_t H(f/ Net_t) is reduced from H(f) to entropy of the correct network solution estimated by H(f/ N_{et}) = 0 ;i.e.

$$H(f) \geq H(f|Net_0) \geq H(f.Net_0|Net_1) \geq \cdots \quad (1.1)$$
$$\geq H(f.Net_0, Net_1, \ldots, Net_{m-2}|Net_{m-1}). \quad (1.2)$$

This equation follows from the well-known results on information theory if one assumes that f and N_{ett} are independent, then H (f) >= H(f/ N_{et}). A generalization of this equation toward f and N_{ett}, t=0,m=1, implies (1.1). Equation (1.1) is true in more simple form H (f) >= H(f/ N_{et})>=...>=
H (f/ N_{etm-1}) too.

Definition 2 *The mutual information $I(f;Net)$ and the conditional entropy $H(f|Net)$ are related to the entropy $H(f)$ as follows: $I(f;Net) = H(f) - H(f|Net)$. This is an amount of the logical work required to design the network Net.*

Verification of the evolved network consists in testing whether this network archives the target (desired) design style and the functionality of the given switching function. One can evaluate the efficiency of the searching process via measuring of the fitness function for example, as the percentage of the correct network (with desired functionality and style) evolved by the GA. In our study we use the mentioned traditional measure, as well as and information theory, in particular, entropy based fitness function.

4 PARTITION OF A NETWORK SEARCH SPACE

Most CAD problem are hard optimization problems. To handle the comp lexity, a technique of partitioning a problem into a number of sub-problems is widely used in CAD. We realize our idea of parallel scanning of a network search space based on it partition into a number of subspaces.

Definition 3 *An entropy based fitness function is the conditional entropy of a target function f given current function Net:*

$$H(f|Net) = -\sum_{i=0}^{1}\sum_{j=0}^{1}\frac{k(i,j)}{k}\log\frac{k(i,j)}{k(j)} \qquad (1.3)$$

where p_{ij} is the probability that f and Net take values i and j accordingly, and p_j means the probability that the output of Net takes value j. Note, that formula (1.3) follows from the definition of conditional entropy.

Example 3 *Consider the computation of the entropy-based fitness function for 2-bit multiplier* $(x_1x_0y_1y_0)$ *that returns a 4-bit number. Here is how the most significant bit* $f = |0...01|$ *has been changes during computation. The fitness function is calculated as follows:* $H(f|Net_0) = -{}^{15}/_{16}\log{}^{15}/_{16} - {}^{1}/_{16}\log{}^{1}/_{16}$ = ***0.337*** **bit** ***for*** $Net_0 = |1...1|$, $H(f|Net_{16}) = -{}^{12}/_{16}\log{}^{12}/_{12} - {}^{3}/_{16}\log{}^{3}/_{4} + {}^{1}/_{16}\log{}^{1}/_{4}$ = ***0.203*** **bit** ***for*** $Net_{16} = |0000010100000101|$, $H(f|Net_{2503}) = -{}^{14}/_{16}\log{}^{14}/_{14} - {}^{1}/_{16}\log{}^{1}/_{2} - {}^{1}/_{16}\log{}^{1}/_{2}$ =***0.125*** **bit** ***for*** $Net_{16} = |0..0101|$. $H(f|Net_{2506}) = -({}^{15}/_{16}\log{}^{15}/_{15} - {}^{1}/_{16}\log{}^{1}/_{1}$ =***0*** **bit.** *It follows from the above that* Net_0, Net_{16} *and* Net_{2503} *are invalid networks.*

We consider any space C of possible network solution can be partitioned into R>=2 subspaces.

In our assumption we state that a network space corresponds to the given switching function that can be decomposed into sub functions, each of these sub functions corresponds to a sub network space. Each sub space R of the network solution can be searched independently and simultaneously with a different windows $G_{mi.ni}$ and TDSs. Let us denote this evolutionary process as $G_{R(Mi.Ni)}$ There are two different ways to combine obtained sub-networks -

$SubNet_0, \ldots, SubNet_{R-1}$.

The first approach is based on implementation of multiplexes MUXs i.e resulting network created as

$$Net = \{SubNet_t, MUX\}, \qquad t = 0, \ldots, R-1.$$

The second approach does not require additional gates to create the final network.

A set of partitions of a network search space is generated through decomposition of an initial function. We explore the principle of decomposition: a problem of large dimension is decomposed into several problems of smaller dimensions, which can be solved separately and with smaller of effort. In our study we have used two types of partition of the network search space: Shannon (S) decomposition and its information theoretical notation-entropy based S-decomposition, a well as so-called balanced decomposition [2].

In our study we explore the property of conditional entropy to choose a variation with respect of an initial function (minimal entropy) and entropy based S-decomposition. In logic design, S-decomposition of a switching function is defined as $f = \bar{x}f|_{x=0} \vee xf|_{x=1}$. Entropy based S-decomposition of a function f with respect to variable x, is expressed by the relation below[14]

Example 4 *The conditional entropy of the function* $f = [10111110]$ *with respect to variable* x_1 *is* $H(f|x_1) = -{}^1/_8 \log {}^1/_4 - {}^1/_8 \log {}^1/_4 - {}^3/_8 \log {}^3/_4 - {}^3/_8 \log {}^3/_4$ *=0.8113* bit. *By analogy,* $H(f|x_2) = -{}^1/_8 \log {}^1/_4 - {}^1/_8 \log {}^1/_4 - {}^3/_8 \log {}^3/_4 - {}^3/_8 \log {}^3/_4$ *= 0.8113* bit, *and* $H(f|x_3) = -0 \cdot \log 0 - {}^1/_4 \log {}^1/_2$

$-{}^1/_2 \log 1 - {}^1/_4 \log {}^1/_2$ *= 0.5* bit. *So, variable* x_3 *conveys more information than variables* x_1 *and* x_2.

We showed in [8] that the final network *Net* created from sub networks *Sub Nett* obtained via S-decomposition, includes $R = 2d - 1$ multiplexers where d is the number of decomposed variables.

S-decomposition has some obstacles that do not allow, in some cases, to realize partitioning of a network space efficiently. We use also another method to solve the partitioning problem, B-decomposition, that relies on decomposition of a switching function with either parallel or serial decomposition applied a teach phase of the synthesis process [2].In the parallel decomposition, the set of output variables Y of a multi-output function f is partitioned into subsets, $Y0$ and $Y1$; and the corresponding functions, $f0$ and $f1$; are derived so that, for either of these two functions, the input support contains fewer variables than the set of input variables X of the original function f. An objective of the parallel decomposition is to minimize the input support of $f0$ and $f1$: In the serial decomposition, the set of input variables X is partitioned into subsets, U and V; and functions, A and B are derived so that the set of input variables of A is $V \cup W$, where W is a subset of U; the set of input variables of B is $U \cup Z$, where Z is the set of output variables of A; and B has fewer input variables than the original function f; i.e. $f = B(U, A(V, W))$.The B-decomposition is an iterative process, at each step, either parallel or serial decomposition of a selected component is performed. The process is carried out until all resulting sub functions are small enough to _t blocks with a given number of input variables.

The idea of intertwining parallel and serial decomposition has been effectively exploited in the system for logic synthesis, *EMAIN*, developed at the Institute of Telecommunications, Warsaw University of Technology.

5. EXPERIMENTS

We have implemented the algorithm as EvoDesign program in C++ for OS LINUX on computer with Pentium II 300MHz processor. Our experiments have included two phases. At the first phase, the search network space was divided into subspaces via a decomposition of an initial function. At the second phase, GA was applied, given population size 60 ¡ 70, crossover *Pcros* = 0.7; the number of experiments evaluated to 50 and maximal generation number 10^4. Here we verify our approach on parallel evolving of networks using the entropy based S-decomposition and Bdecomposition for several benchmarks. In our program *EvoDesign,* we combined GA, the modules to control TDS, the tools for partitioning a network search space, including program *DEMAIN* for B-decomposition and package *Info* [14] for entropy based S-decomposition. We introduce our results in Table 1.1, where

$$<*>\ \mathcal{G}_{1(2\times3)},\ \mathcal{G}_{2(3\times4)},\ \mathcal{G}_{6(4\times5)};$$

<**> $\mathcal{G}_{2(2\times3)}$, $\mathcal{G}_{1(3\times4)}$, $\mathcal{G}_{5(4\times5)}$; <***> $\mathcal{G}_{1(3\times4)}$, $\mathcal{G}_{4(4\times5)}$. The parameters for S-decomposition were 1) $L_{back} = 3$, $T_{select} = 2$, $T_{discrim} = 90\%$, $P_{mutat} = 0.014$ for 5xp1 was 1.2%, for con1, rd84, alu2, sao; $P_{mut} = 0.02$ for inc, and $P_{mut} = 0.022$ for rd53, rd73, misex1, 9sym; the number of runs was 50. For B-decomposition, $L_{back} = 4$, $T_{select} = 2$, $T_{discrim} = 90\%$, $P_{mutat} = 0.014$, $P_{cross} = 0.7$ and the number of runs was 100.

For example, for rd53, the network space was divided into four subspaces based on S-decomposition with respect to two variables, and the solutions were found for each of the subspaces in 4-level *G4£4* TDS. In particular, the following result was obtained: 21 gates and three multiplexers *MUX*. It is a 4-parallel and independent search process. Let us compare these results with the networks evolved via B-decomposition of the initial function. Note, that the number of sub networks *R*¤ was calculated by the program *DEMAIN* (meanwhile, R= *2d* for S-decomposition). Two subspaces only were exploited. For 5-level *G* 5.5TDS window, we obtained a network with 16 gates (without multiplexers) .One can observe that both types of decomposition give us extremely better results in gates and levels against the SIS system [13], a state-of-the art synthesis tool: a 6-level network with 34 gates, but in 0.1 *sec.*

We compared our algorithm EvoDesign with exact algorithm for finding a minimal *NAND* networks (pruning technique) [3] and SIS [13] using script rugged. In Table 1.2 we give experimental results. As can be seen, the evolutionary strategy still requires on the average more gates and time than Pruning, but at the same time, in most test the evolutionary strategy leads to circuits with less gates than with SIS. Moreover, there are cases (decom ,l, m and decom ,p ,q) where EvoDesign obtains sub-optimal solutions, but more than ten times faster than by Pruning. Even though the sample is relatively small, it supports the evidence that the evolutionary circuit design is reaching a level of competitiveness with respect to well established methods used in industry. Further improvements are presently under development.

CONCLUDING REMARKS

We showed that the idea of evolutionary network design can be more captivating and attractive, if the concept of a given design style is realized. In this case, the designer really becomes an expert, i.e. designer needs no special software to implement a design style, the style is provided by evolutionary search for the best solution. The main contribution of this paper is an extension and improvement of the state-of-the-art of gate-level evolutionary networks synthesis, namely, concept of a TDS allows us to achieve flexible scanning of the subspaces of possible network solutions. We have found that, in general, B-decomposition of the network search space is more preferable, because it does not require any multiplexers to merge a network from sub networks. In any case, the decomposition gives additional and unique possibilities for massive parallel processing. In our current work we focus on different practically preferable styles including AND/EXOR, Kronecker, ESOP.

Table 1.1. Efficiency of the network search space partition via entropy based S- and B-decomposition

Test	*I/O*	*Entropy based S-decomposition*[1]				*B-decomposition*[2] [2]		
		$\mathcal{G}_{R(M\times N)}$	*G+MUX*	*L*	*t*	$\mathcal{G}_{R^*(M\times N)}$	*G*	*t*
rd53	5/3	4(4 × 4)	21+3	6	8	2(5 × 5)	16	57
con1	7/2	16(5 × 5)	13+15	9	0.25	4(5 × 5)	22	61
rd73	7/3	16(4 × 4)	83+15	8	28	3(5 × 5)	28	40
inc	7/9	16(3 × 6)	10+15	7	0.3	9(5 × 5)	82	220
5xp1	7/10	16(4 × 8)	121+15	8	1.52	10(5 × 5)	61	84
rd84	8/4	32(5 × 5)	191+31	10	40	6(5 × 5)	32	92
misex1	8/7	16(4 × 4)	1+15	8	0.5	9(5 × 5)	61	49
life	9/1	32(4 × 5)	169+31	9	24	5(4 × 5)	34	41
9sym	9/1	32(4 × 4)	144+31	9	376	6(5 × 5)	37	39
misex20	10/1	64(4 × 5)	18+63	10	0	< * >	45	38
newpla2	10/4	64(4 × 4)	217+63	10	7	< ** >	26	5
sao2	10/4	64(5 × 4)	120+63	11	209	20(5 × 5)	165	120
alu2	10/6	64(4 × 6)	510+63	10	4448	12(5 × 5)	71	106
misex46	11/1	128(4 × 5)	24+127	11	0	6(4 × 5)	36	22
misex47	11/1	128(4 × 5)	12+127	11	0	< *** >	26	11
Total	**129/57**		**1654+677**	**137**	**5292**		**742**	**1049**

Test	*I/O*	*SIS* [13]		
		G	*L*	*t*
rd53	5/3	34	6	0.1
con1	7/2	30	8	0.1
rd73	7/3	118	11	0.2
inc	7/9	171	10	0.3
5xp1	7/10	94	10	0.3
rd84	8/4	182	13	0.48
misex1	8/7	155	7	0.4
life	9/1	138	16	0.2
9sym	9/1	132	14	0.3
misex20	10/1	56	11	0.1
newtpla2	10/4	74	8	0.2
sao2	10/4	208	16	0.15
alu2	10/6	604	19	2.9
misex46	11/1	52	10	0.1
misex47	11/1	34	9	0.1
Total	**129/57**	**2317**	**182**	**2.8**

Table 1.2. The comparison of SIS approach, pruning technique and evolutionary strategy

Test	*SIS*		*Pruning*		*EvoDesign*	
	I/O	*Gate*	*Gate*	$t(sec)$	*Gate*	$t(sec)$
c17	5/2	9	6	0.03	15	528.58
cm42a,e,f	4/2	14	9	19.74	11	36.79
cm42a,g,h	4/2	13	8	9.43	10	50.87
cm42a,i,j	4/2	13	8	9.47	10	43.47
cm42a,k,l	4/2	12	7	1.39	10	40.39
cm42a,m,n	4/2	13	8	9.33	10	40.26
decod.f,g	5/2	17	10	68.61	17	1369.43
decod.h,i	5/2	18	11	1603.07	17	1769.8
decod.j,k	5/2	18	11	1592.00	17	1377.88
decod.l,m	5/2	19	12	27818.76	16	1246.88
decod.n,o	5/2	18	11	1620.73	17	1397.72
decod.p,q	5/2	19	12	26921.29	17	1205.56
decod.r,s	5/2	19	12	26445.97	17	2246.47
majority	5/1	14	9	37.43	12	475.38
x2,k,l	3/2	11	8	8.11	10	26.64
x2,m,o	4/2	11	10	119.66	11	254.54
z4ml.27	3/1	10	8	0.6	14	1359.7

REFERENCES:

[1] T .Aoki, N.Homma and T.Higuchi. Evolutionary design of arithmetic circuit. IEICE Transactions on Fundamentals E-82-A(5):798-806,1999.

[2] J.A.Brzozowski and T.Luba Decomposition of Boolean Function specified by Cubes, Part 1:Theory of serial decomposition using blankets. Part 2(with M.Nowicka):Practical Results. n research Report CS-97-01, Dept of CompSci., University of Waterloo,Canada,1997,Revised in October 1998.

[3]R.Drechsler and W.Gunther. Exact circuit synthesis. In International Owrkshop on Logic Synthesis, Lake Tahoe,1998

[4]D.Quaglirella et al.(Eds) Genetic Algorithm and Evolution Strategies in Engineering and Computer Science. Recent Advances and Industrial Aplications. John Wiley and Sons Ltd. 1998.

[5]M.Siper et al(Eds)Evolvable Systems: From Biology to hardware lecture notes in compuer science, vol 1478,Springer 1998.

[6] L. J_ozwiak and A. Chojnacki. Functional decomposition based on information relationship measures extremely effective for symmetric functions. In Proc. EUROMICRoConf., pages 150.159, 1999.

[7] T. €uba, C. Moraga, S. Yanushkevich, M. Opoka, andV. Shmerko. Evolutionary multi-level network synthesisin given design style. In Proc. IEEE Int. Symposium onMultiple-Valued Logic, 2000.

8] T. €uba, C. Moraga, S. Yanushkevich, V. Shmerko, andJ. Kolodziejczyk. Application of Design Style in Evolu-tionary Multi-Level Networks Synthesis, In Proc. IEEEInt. Symposium on Digital System Design (DSD'2000),Netherlands, pages 156.163, 2000.

[9] V. Cheushev, S. Yanushkevich, C. Moraga, V. Shmerko, and J. Kolodziejczyk. Remarks on circuit verification through the evolutionary circuit design, In Proc. IEEE31-th Int. Symp . on Multiple-Valued Logic, pages 201.206, 2001.

[10] V. Cheushev, S. Yanushkevich, C. Moraga, V. Shmerko .Research Report, Flexibility in Logic Design. An Approach Based on Information Theory Methods, Forschungsbericht 741, University of Dortmund, Germany.

[11] C. Moraga and W. Wang. Evolutionary methods in the design of quaternary digital circuits. In Proc. IEEE 28thInt. Symposium on Multiple-Valued Logic, pages 89.94, 1998.

[12] Proc. NASA/DoD Workshop on Evolvable Hardware, Pasadena, California. 1999.

[13] SIS Release 1.2. In UC Berkeley Soft. Distr., July, 1994.

[14] V. P. Shmerko, D. V. Popel, R. S. Stankovic, V. A. Cheushev, and S. N. Yanushkevich. Information theoretical approach to minimization of AND/EXOR expressions of switching functions. In Proc. IEEE Int. Conf. on Telecommunications in Modern Satellite, Cable and Broadcasting Services, pages 444.451, 1999.

[15] K. S. Tang, K.F. Man, and S. Kwong. Parallel genetic algorithms: Implemental hardware solutions. Circuit and Systems, 10(2):3,8.11, 1999.

6

MEMORY

It is projected that memory blocks will consume over 90% of the area in most next generation SOC integrated circuits. It is therefore of some importance to examine memory blocks from the point of view of the technology, the design and also specific applications. In this chapter we present five sections that cover these various aspects of embedded memory in SOC level designs.

Embeddable nonvolatile memory (NVM) will undoubtedly play an important role in the success of System-on-Chip (SoC) technology, and in section 6.1, "Survey of Emerging Nonvolatile Embedded Memory Technologies" by Bruce F. Cockburn of the University of Alberta, Canada, three promising emerging nonvolatile memory technologies are reviewed. Flash memory, which is presently the dominant NVM type, is available as an option for standard CMOS processes; however, flash memory has undesirable properties for potential SoC application including a relatively slow write time, limited write cycle endurance, and the requirement for high-voltage supplies and transistors. This section examines the three emerging NVM technologies—ferroelectric memory (FeRAM), magnetoresistive memory (MRAM), and ovonic unified memory (OUM)—that could potentially replace flash memory, and possibly also much of the volatile memory, in future SoC designs.

Typical streaming multimedia and real-time applications are increasingly suffering from the memory access latency and the effects of sub-optimal cache refreshes. In section 6.2, "Configurable Parallel Memory Implementation For System-on-Chip Designs", by Jarno Vanne, Eero Aho, Kimmo Kuusilinna, and Timo Hämäläinen of Tampere University of Technology, Finland, the implementation of a novel Configurable Parallel Memory Architecture (CPMA) that offers many design and run-time configuration possibilities is presented. CPMA is shown to considerably decrease the processor-memory bottleneck by widening the memory bandwidth, decreasing the number of memory accesses, and diminishing the significance of memory latency. The authors also show that these implementations need not to be excessively expensive in terms of silicon area.

In section 6.3, the same authors present "XOR-scheme Implementations In Configurable Parallel Memory". This work addresses a primary problem of continuously feeding data to processing nodes in parallel data processing SOC systems. In this work, a general XOR-scheme implementation for the CPMA (described in section 6.2) is presented for the distribution of data to the memory modules. The implemented schemes allow several additional conflict free access formats to be used. Parameters for constructing a XOR-scheme can be flexibly changed at run-time without the need to redesign the complete hardware system. Results show significant reduction in the overall memory accesses, which can be viewed as one of the key approaches to solve the processor-memory performance bottleneck.

In section 6.4, Mohammed Sayed and Wael Badawy of the University of Calgary, Canada, present "An Embedded Memory Architecture for MPEG-4 Applications with Mobile Devices". This novel embedded memory architecture for MPEG-4 synthetic video object decoding is targeted for use in a new chipset for third generation mobile phones. The embedded memory architecture stores video frame and motion vectors, then it computes, in memory, the predicted frame; this is used to generate the MPEG-4 video sequence. The embedded memory is optimised with parallel reading and writing operations. The prototype of the proposed architecture is designed using Verilog HDL then synthesized using TSMC standard cell libraries in a 0.35μm CMOS technology.

In the final section, "Assessment of MPEG-4 VTC and JPEG2000 Dynamic Memory Requirements" by Gauthier Lafruit, and Jan Bormans of IMEC in Leuven, Belgium, the authors consider the dynamic memory requirements for ASAP schedules of MPEG-4 and JPEG2000 wavelet codecs, using a queueing model. The usual assumption that the tight data dependencies of the MPEG-4 parent-children tree coding induce higher memory requirements than the independent block coding approach of JPEG2000, is demonstrated to be incorrect. In particular, an MPEG-4 wavelet decoder always requires less memory than a JPEG2000 decoder. This work shows that the memory requirements for an MPEG-4 wavelet encoder are often lower than those of its JPEG2000 counterpart and the authors also show that no strict symmetry exists in the memory requirements of the encoder and decoder.

Section 6.1

Survey of Emerging Nonvolatile Embedded Memory Technologies

The Challenge

Bruce F. Cockburn

Dept. of Electrical and Computer Engineering, University of Alberta, Edmonton, Canada

Abstract: Embeddable nonvolatile memory (NVM) could play an important part in the success of System-on-Chip (SoC) technology. Flash memory, which is presently the dominant NVM type, is available as an option for standard CMOS processes; however, flash memory has undesirable properties for potential SoC application including a relatively slow write time, limited write cycle endurance, and the requirement for high-voltage supplies and transistors. This article briefly surveys three emerging NVM technologies —ferroelectric memory (FeRAM), magnetoresistive memory (MRAM), and ovonic unified memory (OUM)—that could potentially replace flash memory, and possibly also much of the volatile memory, in future SoC designs.

Key words: nonvolatile memory, ferroelectric memory, magnetoresistive memory, ovonic unified memory, System-on-Chip

1. INTRODUCTION

At present, system-on-chip (SoC) designers are confronted with a variety of alternative embedded memory technologies including static random-access memory (SRAM), dynamic RAM (DRAM), and flash memory. Each of these memory options has both strengths and weaknesses. SRAM offers ready compatibility with logic, and fast read and write speeds at the cost of large 4- or 6-transistor storage cells and relatively high dynamic and stand-by power consumption. DRAM offers high storage density, but embedding DRAM in an SoC design requires expensive modifications to standard CMOS logic processes or the use of special merged logic - DRAM processes, and usually entails compromises to the performance of the resulting logic and/or memory [11]. Neither SRAM nor DRAM offers nonvolatility, which is essential for some applications and would be attractive for many others. Flash memory and electrically-eraseable programmable read-only memory (EEPROM) macros are both available in standard foundry processes, but the minimum available feature sizes are typically one to two generations behind those of state-of-the-art processes. Flash memory has endurance limited to about 10^5 write cycles, so it cannot be viewed as a potential replacement for conventional volatile memories in SoC designs. Erase times in flash memory are very slow (e.g. 100 ms or more) and programming times are still relatively slow (e.g. 1 to 10 μs). Finally, provisions must be made for providing high erase and programming voltages, either from off-chip sources or from an on-chip charge pump [2].

Several competing nonvolatile memory (NVM) technologies are under active development that could potentially replace flash memory. Three in particular are touted as having the potential to become a universal memory for SoC designs, combining the nonvolatility of flash memory with access times approaching those of SRAM and cell densities approaching those of DRAM. Ferroelectric memory (FeRAM or FRAM) is under development by numerous companies, including Ramtron, Fujitsu, Intel, Samsung, and collaboratively by Infineon and Toshiba [5]. Magnetoresistive memory (MRAM) is being developed at Motorola [7] and collaboratively by IBM and Infineon Technologies [6]. Motorola believes that MRAM has the potential to have wide application in portable computers, consumer electronics, and wireless devices. A third candidate NVM technology, called ovonics unified memory (OUM), is being investigated by STMicroelectronics, British Aerospace, and collaboratively by Intel, Azalea Microelectronics and Ovonyx [10]. Intel believes that OUM may be easier to integrate with standard CMOS processes than MRAM, but the industry's largest player appears to be backing research efforts in both OUM and FeRAM. All three of these emerging NVM technologies show promise for integration in CMOS processes, although it is unclear how soon they will become available at leading edge line widths.

The next three sections review the characteristics of FeRAM, MRAM, and OUM, respectively. Basic cell operation is described and the major technical challenges are mentioned. A final section concludes the review.

2. FERROELECTRIC MEMORY

Among the three NVM technologies reviewed in this article, ferroelectric memory is furthest along he path of commercialization and mass production. Ramtron (Colorado Springs, CO) and Fujitsu are presently manufacturing stand-alone FeRAMs of up to 256-kbit capacity for use in niche applications such as game computers and portable data-collection instruments. Ramtron and Fujitsu are also collaborating in the development of embedded FeRAM for SoC applications.

Considerable research and development activity is also occurring in many other companies and research centers (e.g. Samsung, Hyundai, Hitachi, NEC, Toshiba, Infineon, IMEC, University of Toronto). There is already quite an extensive literature describing FeRAM technology (e.g. [13, 17, 18, 19, 21]). Reference [17] is a textbook devoted entirely to FeRAM, while [18] is an excellent survey of FeRAM technology and circuit-level techniques.

FeRAM exploits an inherent electric dipole that is present in the domain structure of certain materials called dielectrics. The two most common dielectrics used in FeRAMs are the perovskite PZT, with chemical comp osition $Pb(Zr,Ti)O_3$, and the layered peroskite SBT, with composition $SbBi_2Ta_2O_9$. In its virgin/annealed state, a volume of ferroelectric is partioned into numerous randomly-oriented polarization domains and thus the bulk material will not have a net measurable electric polarization (i.e. electric dipole). However, when a suitably strong external electric field is applied, a new pattern of domains is established in the ferroelectric in which there is a net polarization that is aligned with the applied field. As illustrated in Fig. 1, the relationship between the applied electric field and the resulting polarization can be described by a hysteresis loop that looks very similar to the hysteresis loop that relates the applied magnetic field and the resulting magnetic field for a ferromagnetic material. When the applied field is removed, a net remnant polarization, +Pr or -Pr, is retained in the ferroelectric. This polarization can be retained for 10 years or more under reasonable operating conditions. Once a residual field is set up in the material, the field will be retained until a sufficiently strong external field is applied that reorients the remnant field. In both ferroelectric and ferromagnetic materials, the remnant field is lost if the material is heated up beyond a critical Curie temperature T_c.

An FeRAM cell is formed by creating a transistor-capacitor structure very similar to that of a DRAM cell. In the FeRAM cell, however, a film of erroelectric occupies at least part of the capacitor dielectric. Write operations are achieved by applying an electric field across the capacitor plates. The voltage and capacitor dimensions must be selected so that the ferroelectric is well polarized in either the

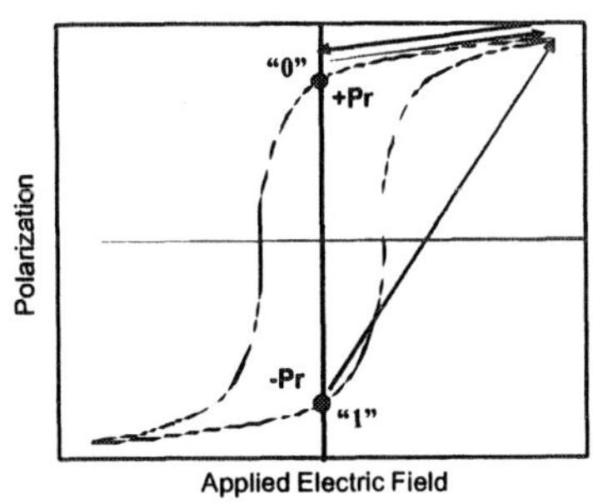

Figure 1 Ferroelectric hystersis loop

top-to-down or down-to-top directions to properly record logic 0s and 1s. In an FeRAM cell, unlike in a DRAM cell, the voltage of the capacitor plate terminal is usually pulsed in the opposite direction to the bitline signal to maximize the applied electric field strength experienced by the ferroelectric. Sensing is achieved by means of a destructive read, which involves measuring the current flowing out of the capacitor top plate when a known polarization is written into the cell. A write operation that changes the polarization state in a cell will cause a larger current pulse on the bitline than a write operation that re-writes the same state as before. The sense-amplifier that judges whether a cell signal represents a state reversal or not must compare the cell signal with a reference signal that lies midway between the two expected distributions of cell signals. There are many possible ways of generating the required reference signal [18]. Once the state of a cell has been determined, the original state of the cell must be restored in a write-back operation to avoid data loss.

There is a helpful analogy to be made between DRAM and FeRAM. Both memories can be manufactured using single-transistor, single-capacitor (1T-1C) cells, or using double-transistor, double-capacitor (2T-2C) cells. However, stand-alone DRAMs all use the 1T-1C design to maximize the storage density, whereas it appears that most production FeRAMs use less dense 2T-2C cells to provide simpler and more robust sensing. If one considers only 1T-1C cells, then both DRAM and FeRAM cells can be arranged according to either the open or folded bitline architectures. In both memory types, a write operation to a cell requires the assertion of a wordline and the driving of an appropriate voltage, either high or low, on the corresponding bitline that is connected electrically to the top plate of the cell capacitor. However, in the case of DRAM, the bottom capacitor plate is usually held at a fixed voltage (a mid-voltage in stand-alone DRAMs), while the bottom plate is pulsed high and/or low in voltage in conjunction with the bitline. The read operations in both DRAM and FeRAM are destructive.

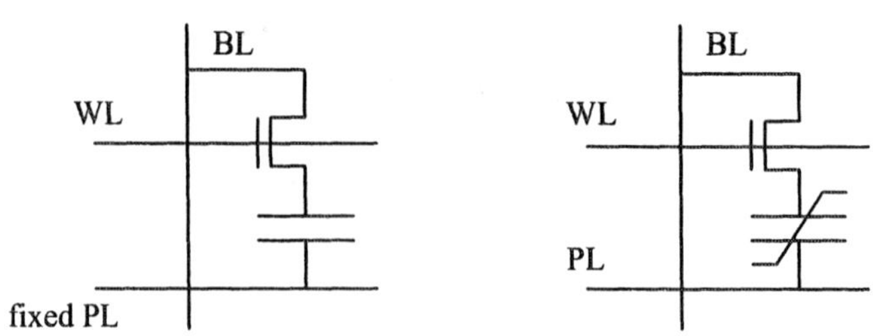

Figure 2. Ferroelectric (left) and DRAM (right) memory cells

There are two undesirable behaviors that have proved to be serious technical challenges to achieving successful FeRAM. Fatigue is the phenomenon where the height of the ferroelectric hysteresis loop collapses over time, with the strength of the remnant polarization decreasing as the number of write cycles increases. Although FeRAM has for many years had write-cycle endurance that is much greater than that of flash memory (e.g. $\sim 10^{12}$ cycles vs. $\sim 10^{5}$ cycles), it has not been considered sufficient to allow FeRAM to be considered a full replacement for volatile memories. Interestingly, Ramtron recently announced improvements to its technology that extend endurance up to $\sim 10^{16}$ cycles [1], creating what it has dubbed "unlimited" endurance FeRAM [1]. Imprint is a second phenomenon in FeRAMs in which the hysteresis loop shifts horizontally over time if the dielectric material is left in the same storage state. Thus, after a "0" has been stored for a long period of time, the strength of the remnant polarization of a newly written "1" will be less than the polarization for a re-written "0". Evolutionary advances to ferroelectric and electrode materials have increasingly controlled both the fatigue and imprint problems.

FeRAM has so far achieved only limited commercial success in competition against the dominant NVM technology, flash memory. Stand-alone flash memory densities are now comparable to stand-alone DRAMs. System and chip designers have learned to accommodate the major drawbacks of flash memory, such as limited endurance, the requirement for high voltages for erasure and programming, and the relatively slow erase and programming times. The capacities of stand-alone FeRAMs lag those of stand-alone flash memories by two to three orders of magnitude, although planned FeRAM storage capacities are increasing at a faster rate than either DRAM or flash memory. The market's verdict so far seems to be that there are relatively few present applications that are willing to pay a premium for the improved performance characteristics of stand-alone FeRAM. The prospects for FeRAM may improve over the coming years, however, with the growing use of portable devices, such as personal data assistants (PDAs) and so-called 3G and 4G cellular telephones, that can be expected to have demanding NVM requirements. FeRAM may begin to shine as an embedded memory technology for SoCs. If the processing and economic challenges of FeRAM manufacture can be addressed, then FeRAM may have a reasonable chance in competition with embedded flash memory, which is not available in state-of-the-art CMOS technologies.

3. MAGNETORESISTIVE MEMORY

Over the past two decades, several research efforts have sought to develop a practical semiconductor memory based on magnetic phenomena. In 1989, a group of employees left Honeywell to found a spin-off company, NonVolatile Electronics

Inc. (Eden Prairie, MN), focussed on developing and commercializing practical MRAM technology. The resulting technology is reportedly being licensed to Honeywell, Motorola and Union Semiconductor Technology Corp. (Taipei, R.O.C.). Another line of MRAM research has resulted from a long-term program in applied magnetics at IBM and a more recent collaborative alliance between IBM and Infineon. Since the 1980s, the U.S. Defense Advanced Research Projects Agency (DARPA) has actively funded MRAM research because of the potentially attractive properties of MRAM in military application including fast access times, operation over a wide temperature range, and radiation hardness. Several Japanese companies (e.g. Fujistu, Toshiba, Hitachi, NEC, Mitsubishi, and NTT) were reported to be closely following the development of MRAM in the late 1990s, although today the world-wide effort appears to be centred primarily in North America and Europe (e.g., Motorola, IBM, Infineon, Siemens, Philips, IMEC, and several European universities) [3].

Giant magnetoresistive (GMR) heads were a significant factor behind major increases in magnetic disk areal storage densities since the mid-1990s, and it was once thought that GMR could also form the basis for a semiconductor MRAM. Based on a magnetoresistive effect discovered in 1988, GMR devices are made using thin-film lithographic techniques. Their operation involves sensing the presence of a magnetic field by its effect on the horizontal resistance of a thin sandwich of materials. In some implementations, the sandwich consists of a very thin layer of copper (e.g. ~10 atoms thick) deposited between two layers of cobalt alloy. The extreme sensitivity of GMR heads permitted increased areal storage densities in magnetic disks, and it was hoped that the same GMR effect could be exploited to sense the state of very small semiconductor MRAM storage cells. Honeywell's research program into GMR MRAM has led to a 1-Mb prototype. In 2000 the company was aiming to achieve write speeds of 20 ns and read speeds of 10 ns using its GMR technology.

A more recent alternative technology to GMR MRAM is called tunneling magnetic resistance (TMR) or tunneling magnetic junction (TMJ) MRAM [20]. It is being pursued by the IBM-Infineon alliance, Motorola, and Hewlett-Packard, who believe that TMR is a more appropriate MRAM technology than GMR for large-scale stand-alone and embedded memories. The advantages of TMR over GMR stem from the higher achievable on-to-off resistance ratios (about 7% for SMR versus 20% or more in TMR). A TMR memory cell includes a stack of materials, with an ultra-thin (e.g. only a few atomic layers of aluminum oxide) insulating layer separating two very thin layers (also called plates) of ferromagnetic material (e.g. 10 to 20 atomic layers of iron-cobalt-nickel film) [14]. The magnetic spin orientation in one of the plates is pinned, while that of the other plate is free to be changed by the combined magnetic field produced by currents in digit and bit lines that intersect at cell positions. The current that is involved in sensing the state of a TMR cell flows vertically across the sandwich layers, unlike in a GMR cell, where the current flows horizontally. If the magnetic spins of electrons in the two magnetic plates are aligned, then electrons tunnel more readily from the upper layer across the insulator to the lower layer, and pass on through to a copper electrode under the TMR sandwich. If the spins of the two plates are anti-parallel, then tunneling electrons from the upper plate are repelled by the lower plate. Data is thus retained in the relative spin orientations of the two ferromagnetic plates, and data is sensed by measuring the tunneling current.

IBM has reported MRAM write times of as little as 2.3 ns and read access times of 3 ns, which is comparable to SRAM and is about 20 times faster than contemporary FeRAMs. What is more, MRAM read operations are reported to consume less than 1% of the energy of DRAM read operations. In 2001 IBM was characterizing a 1-Mb prototype MRAM made in a 180-nm process, and expected to introduce a 256-Mb commercial stand-alone MRAM by 2004. In 2001 Motorola described a 256-Kbit MRAM test vehicle built in a 0.6-μm, 3-V CMOS process and, like IBM, expects to move MRAM technology into production by 2004 [16]. Motorola has reported endurance of greater than 10^{10} write cycles with no significant degradation in cell resistance function. This is still inferior to the endurance of FeRAM.

Acceptance of MRAM faces several major challenges. Manufacturing the required cell structure on a mass scale will require very tight control over the chemical composition and thickness of the various layers of the stack structure, especially the ultra-thin insulating layer. The MRAM cell layers must also fit in economically into a standard CMOS flow. Questions have been raised as to the susceptibility of MRAM to magnetic interference, including write operations to nearby cells as well as stray magnetic fields emerging from off chip [9].

4. OVONIC UNIFIED MEMORY

Applications of the physical phenomenon underlying OVM have been studied for more than 30 years although relatively little has been published on the implementation of practical VLSI memories [10, 15]. OVN, as well as rewriteable CD and DVD optical disks, exploits thermally-induced reversible changes between amorphous (high resistance) and polycrystalline (low resistance) phases in a recording medium. In the OVMs being developed by Intel, Ovonyx Corp. (Santa Clara, CA), and Azalea Corp. (Santa Clara, CA), the medium is a thin film of chalcogenide glass, which has chemical structure $Ge_xSb_yTe_z$. If the programmed volume in a cell is small, the energy required to change the cell's state can be made very small, which is attractive for low-power applications. Ovonyx has additional strategic alliances to develop OVM along with STMicroelectronics and BAE Systems.

The 4-Mb OVM described in [10] was implemented in a modified 180-nm, 3-V CMOS process. Figure 3 shows the schematic of a storage cell. An 8-ns reset pulse applied with a ~5-ns falling edge produces local joule heating followed by a quenched change to an amorphous, higher resistance state (~85 kΩ) in the thin film region adjacent to the bottom cell electrode. A longer-duration (~85 ns) programming pulse produces sufficient local joule heating (up to under 600°C [10]) and more gradual cooling that induces a phase change to a lower resistance (~2 kΩ), crystalline state. The required reset voltage (<0.8 V) and the programming threshold voltage (~0.6V) allow the OVM to be accommodated within the 3-V power supply. The actively-driven read current is low enough to avoid disturbing the state of a cell, so reading is nondestructive, unlike FeRAM or DRAM. The cell's bipolar transistor should also facilitate scaling to smaller cells. Write cycle endurance of up to 10^{12} cycles at 5 MHz without significant degradation in the reset and programmed cell resistances is reported in [4]. Random access times are comparable to conventional DRAM (~100 ns). A data retention time of at least 10 years at 120°C is reported by Ovonyx [4]. Various cell designs have been built with sizes ranging from $5F^2$ to $8F^2$.

The weaknesses of OVM have not been presented in any detail in the open literature. There are clearly potential cell disturb mechanisms, such as unexpected thermally-induced programming when nearby cells are set. The reported write cycle endurance of 10^{12} cycles is far superior to that of flash memory (~10^5 cycles), but is inferior to the best reported endurance for FeRAM and would certainly not permit OVM to be considered as a replacement for conventional volatile memory. It is unclear how great the technical and economic challenges will be to integrating the required chalcogenide film and cell electrodes into standard CMOS logic processes.

5. CONCLUSIONS

This article briefly reviewed three promising emerging nonvolatile memory technologies. Table 1 compares three recent prototype NVMs. Note that the FeRAM chip is one year older than the other two (a 32-Mb part in 0.25-μm CMOS was revealed by the same team at ISSCC 2002). Each NVM candidate has the potential to provide an attractive combination of relatively high storage density, compatibility with standard CMOS processing, read and write cycle times, and endurance that is orders of magnitude superior to flash memory. It is too soon, however, to predict with any certainty which of these technologies has the best chance of mounting a successful challenge against steadily-improving flash memory and clever combinations of flash memory with conventional volatile memory. FeRAM appears to be the furthest along in development among the three NVM contenders, but has so far only achieved limited commercial success in a few niche applications. Significant progress has been made in tackling the key issues of fatigue/endurance and cell density.

Property	FeRAM [12]	MRAM [16]	OVM [10,4]
Storage capacity	4 Mbit	256 kbit	4 Mbit
CMOS technology	0.6-μm, 3.3 V	0.6-μm, 2.7 to 3.3 V	0.18-μm, 3 V
Cell size	9.38 μm^2	7.1 μm^2	< 0.26 μm^2
Sensing time	85 ns	< 50 ns	< 50 ns
Erase time	N/A	N/A	8 ns
Cycle/write/set time	150 ns	< 50 ns	< 50 ns
Retention	?	> 10 yrs	10 yrs at 120°C
Endurance	> 10^9 cycles	> 10^{10} cycles	10^{12} cycles

Table 1. Comparison of FeRAM, MRAM and OUM prototypes

However, it is unclear how easy it will be to include ferroelectric capacitors into conventional CMOS processes. MRAM is making impressive technical strides and is receiving high-profile support from major industrial players, but the manufacturing challenges posed by the ultra-thin insulating layer and the unconventional ferromagnetic films may prove to be formidable. Finally, OUM may turn out to be a dark horse that overtakes its competitors in what is both an economic contest as well as an engineering one

REFERENCES

[1]Anonymous, "Ramtron claims first nonvolatile memory with unlimited endurance," *Semiconductor Business News*, Mar. 27, 2001.

[2]W. D. Brown and J. E. Brewer (eds.), *Nonvolatile Semiconductor Memory Technology*, (IEEE Press, New York, 1998).

[3]A. Cataldo, "Japanese play catch-up on magnetic RAMs," *EE Times*, Apr. 30, 1999.

[4]A. Cataldo, "Intel finds allies for futuristic nonvolatile memory," *EE Times*, Nov. 13, 2001.

[5]P. Clarke and Y. Hara, "Infineon and Toshiba to collaborate on 32-Mbit FRAM," *EE Times*, Dec. 22, 2000.

[6]P. Clarke, "Infineon, IBM Hone Magnetic RAM Strategy," *EE Times*, Apr. 17, 2001.

[7]P. Clarke, "Motorola aims to sample 4-Mbit MRAM in 2003," *EE Times*, Apr. 17, 2001.

[8]M. Clendenin, "Taiwan company seeking MRAM foundry partner," *EE Times*, May 24, 2001.

[9]T. Costlow, "Magnetic RAM team feels the pull of joint work, " *EE Times*, Jan. 10, 2001.

[10]M. Gill, T. Lowrey, and J. Park, "Ovonic Unified Memory - A High-Performance Nonvolatile Memory Technology for Stand-Alone Memory and Embedded Applications," *ISSCC 2002*, pp. 202-203.

[11]S. S. Iyer and H. L. Kalter, "Embedded DRAM Technology: opportunities and challenges," *IEEE Spectrum*, vol. 36, no. 4, Apr. 1999, pp. 56-64.

[12]B.-G. Jeon., "A Nonvolatile Ferr oelectric RAM with Common Plate Folded Bit-line Cell and Enhanced Data Sensing Scheme," *ISSCC 2001,* pp. 38-39.

[13]R. E. Jones Jr., "Ferroelectric nonvolatile memories for embedded applications," *CICC 1998*, pp. 431-438.

[14]P. Kallender, "IBM says breakthrough will enable commercial MRAMs," *EE Times*, Dec. 8, 2000.

[15]S. Lai and T. Lowrey, "OUM - A 180 nm Nonvolatile Memory Cell Element Technology for Stand-alone and Embedded Applications," *IEDM 2001*, paper 36.5.

[16]P. K. Naji, M. Durlan, S. Tehrani, J. Ca lder, and M. F. DeHerrera, "A 256kb 3.0V 1T1MTJ Nonvolatile Magnetoresistive RAM," *ISSCC 2001*, pp. 122-123.

[17]J. F. Scott, *Ferroelectric Memories*, (Springer-Verlag, Heidelberg, 2000).

[18]Sheikholeslami and P. G. Gulak, "A Survey of Circuit Innovations in Ferroelectric Random-Access Memories," *Proc. IEEE*, vol. 88, no. 5, May 2000, pp. 667-689.

[19]S. Tanaka et al., "FRAM Cell Design with High Immunity to Fatigue and Imprint for 0.5 μm 3V 1T1C 1M bit FRAM," *IEDM 1998*, paper 13.4.

[20]S. Tehrani et al., "Recent Developments in Magnetic Tunnel Junction MRAM," *IEEE Trans. Magnetics*, vol. 36, no. 5 Sept. 2000, pp. 2752-7.

[21]J. Yamada et al., "A Self-Reference Read Scheme for a 1T/1C FeRAM," *1998 Symp. VLSI Circuits*, pp. 238-241

Section 6.2

Configurable Parallel Memory Implementation For System-on-Chip Designs

Jarno Vanne, Eero Aho, Kimmo Kuusilinna, and Timo Hämäläinen
Institute of Digital and Computer Systems, Tampere University of Technology
Korkeakoulunkatu 1, FIN-33720 Tampere, FINLAND

Abstract: System-on-chip (SoC) designs with parallel on-chip memories increase the memory bandwidth and feed the processor with only necessary data. This paper presents the implementation of a novel Configurable Parallel Memory Architecture (CPMA) that offers plenty of design and run-time configuration possibilities. CPMA is shown to decrease considerably the processor-memory bottleneck by widening the memory bandwidth, decreasing the number of memory accesses, and diminishing the significance of memory latency.

Key words: configurable parallel memory architecture, permutation, DRAM controller, pipelining

1. INTRODUCTION

Generally, System-on-Chip (SoC) design comprises a programmable processor, on-chip memory, and other functional units [2]. The development of memories has not followed that of processors [10] and the growing gap between memory access time and processor speed is becoming a severe limitation to the performance. In response to the performance gap, several attempts to improve the memory bandwidth and to reduce the memory latency have been made. Both technology oriented and architecture (memory bus) oriented approaches are used [4] [9]. For example, several new DRAM architectures have been created [3], of which embedded on-chip DRAM technology is very promising [7].

One of the potential architectural improvements is parallel memory architecture. Those have been shown to perform very effectively in specific applications. A module assignment function, an address function, and a scanning point are required to performa fully specified memory access. A collection of simultaneously accessed data items is called an access format and the scanning point determines the placement of the access format. The module assignment function determines the distribution of data to the modules, whereas the address function defines the distribution of data to addresses within the memory modules. The fundamental concepts are presented, for example, in [6].

The improvements of parallel memories compared to traditional memory architecture are twofold. Throughput between the processor and the memory is increased and the processor on the average gets more useful data. These properties are especially suitable for real-time and multimedia applications, which require large amounts of useful data being rapidly available. Furthermore, the aforementioned embedded DRAM is a technology that stands to profit from parallel memory implementations since it enables wide memory buses that can now be effectively used.

The Configurable Parallel Memory Architecture (CPMA) has been introduced in [1] and [8]. Figure 1 depicts a general high-level block diagram of CPMA. The core blocks of the architecture are the Configurable Address Computation Unit, application specific number of memory modules, and the Configurable Data Permutation Unit.

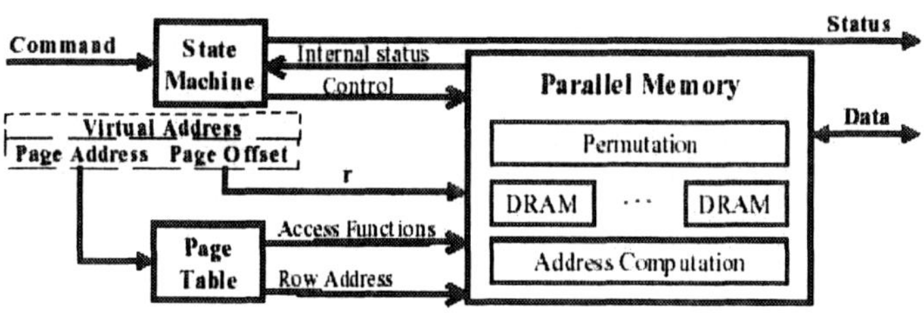

Figure 1. General Configurable Parallel Memory Architecture [1]

As shown in Figure 1, the inputs to CPMA are the virtual address and the operation command. The virtual address is divided into two sections, a page address and a page offset. The page address is a pointer to a page table. The page table transforms a page address suitable for parallel memory use. A page offset straightforwardly equals the scanning point (r) for CPMA.

Read and write operations are supported by CPMA. Data transmission between CPMA and other devices is performed via a bi-directional data bus. A Status signal determines the proper time for a new command to be fetched into CPMA. In addition, the State Machine controls the entire operation with the aid of internal status signals.

In this paper, the structure and the operation of CPMA are explained in detail. The main components of the architecture are described in Section 2 and the pipelined CPMA is presented in Section 3. Section 4 discusses the hardware implementation of the system. The performance comparison 3 between conventional memory and CPMA is examined in Section 5. Finally,Section 6 concludes the paper.

2. CONFIGURABLE PARALLEL MEMORYARCHITECTURE FOR SOC

The contemporary special implementations of parallel memories are extremely effective in particular applications. However, the configuration possibilities of these individual architectures tend to be inadequate for general purposes. CPMA meets the requirements of general-purpose architecture in the sense of being configurable at design and run time. The majority of the implementation parameters are VHDL generics, including the number and the size of memory modules and the width of the data buses as well as the input buses.

Configurable Address Computation Unit

The function of the Configurable Address Computation Unit is to convert the given access functions and the scanning point into a specific physical address. These addresses are used for accesses in each one of the memory modules. For a particular access, the unit produces the required memory module numbers and the corresponding addresses as an output. Moreover, this information is utilized by other components in the architecture. The implemented Configurable Address Computation Unit can utilize multiple access formats including horizontal and vertical straight lines, diagonals, rectangles, and special irregular access formats. A detailed description of the Configurable Address Computation Unit is presented in [1].

DRAM modules and DRAM Controllers

In principle, memory modules may be any random access memory elements. However, the typical maximum SRAM-module memory capacity is a quarter of the capacity of DRAM for the same chip size and process technology [11]. Due to the large capacity differences, SRAMs are about four times more expensive than DRAMs.

For the current implementation, conventional DRAM models are utilized with CPMA. As mentioned above, the number of the memory modules can vary between designs. The DRAM address bus is multiplexed into separate column and row addresses. At the design time, the width of the address bus can be determined to accommodate any DRAM size. The row address strobe (RAS) and the column address strobe (CAS) signals control the DRAM addressing.

A proper synchronous memory controller controls each of the asynchronous DRAMs. An internal state machine drives the utilized DRAM Controller, which is responsible for the delivery of signals to DRAM. The next operation to be performed is based on Valid, Refresh, and RW signals. If the Valid signal for a DRAM Controller is not active, controller remains in an idle state. The decision between read and write operations is based on the RW signal. Status information about the state of the DRAM Controllers is delivered to the main state machine of CPMA. According to the received information, the main state machine either enables or prevents new commands to be fetched for CPMA. A DRAM with a controller and the most significant signals are depicted in Figure 2.

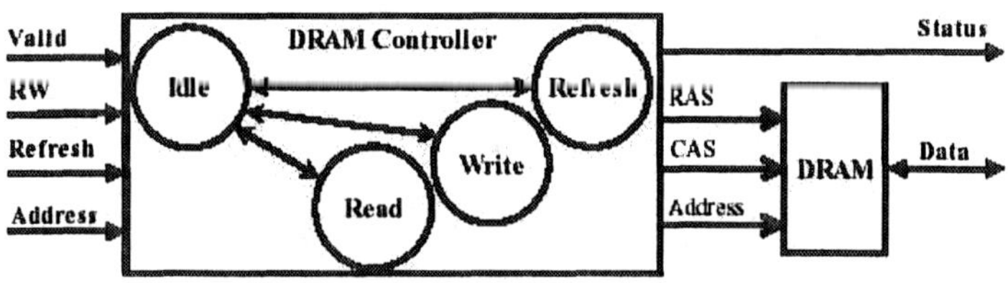

Figure 2. DRAM Controllerand DRAM

Currently, read operations can be performed in six clock cycles, while writing into memory takes five cycles. Memory controllers receive the address from the Address Computation Unit. During read and write operations, the RAS is set active after two and CAS is set active after four clock cycles from the operation activation. While reading, the accessed data is ready for further processing after five cycles, whereas the data is written into memory after five clock cycles after starting the write operation. At present, state machines are locked to a particular mode during operations and all non-accessed controllers remain in the idle state. For example, simultaneous read and write operations are not allowed. Enabling different concurrent operating modes between state machines and DRAM speed improvements utilizing the fast page mode DRAM model are our plans for the future.

The average memory access time can be significantly affected by the choice of the DRAM refresh mechanism. In some schemes, 50% of the total latency is due to waiting for DRAMs in a refresh mode [3]. In our design, the refresh mechanism is designed to be simultaneous for all DRAMs.

Configurable Data Permutation Unit

The final main component of the Parallel Memory is the Configurable Data Permutation Network. The physical memory module for a particular data item is determined by the module assignment function. Hence, the order of the data items differs from the required order determined by the used access format. The purpose of the permutation network is to shuffle the accessed data items into the correct order. Within a read operation, a conversion is made from physical memory module order to order of the access format, whereas during writing access ordered data items are organized according to physical location of the modules.

Different interconnection network topologies are presented in the literature, for example refer to [5] [12]. The chosen topology for used permutation unit is a single-stage interconnection network. The single-state interconnection network can be described as being dynamic, point-to-point, and non-hierarchical. More accurately, it is a crossbar network, which is able to provide a separate path between any input and output switch. The switches are implemented with multiplexers and demultiplexers.

The basic blocks of the Configurable Data Permutation Network are the Control Unit and the Crossbar Network. These components and the data flow around the permutation network are depicted in Figure 3.

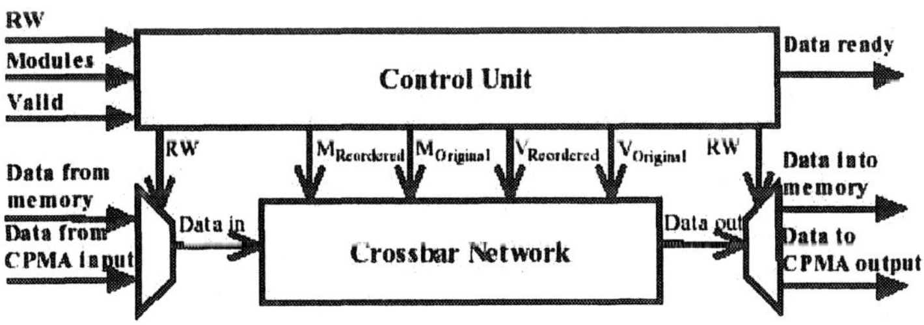

Figure 3. Components of the Configurable Data Permutation Network

As shown in Figure 3, the whole operation of the unit is controlled by a state machine driven Control Unit. The state machine of the Control Unit has two states: Reading and Writing. The selection of the state is made by the incoming RW signal. The required module vector (Modules) and valid vector (Valid) are generated in the Address Computation Unit. According to the current state, the correct module vectors and valid vectors are delivered to the Crossbar Network. During reading, inputs of the Crossbar Network receive reordered module ($M_{Reordered}$) and valid ($V_{Reordered}$) vectors, whereas outputs of the crossbar obtain the original module

($M_{Original}$) and valid ($V_{Original}$) vectors. In contrast, during writing the inputs of the crossbar are reordered and outputs remain unchanged.

The network is unidirectional regardless of the operation mode. During read operation, the data flows from DRAM memory via the permutation network to the permutation unit output. While writing, data is fed from permutation input and permuted data is delivered to DRAM Controllers. The input multiplexer and the output demultiplexer outside the proper permutation network are controlled by the RW signal.

At design time, the presented permutation network is fully configurable with respect to the number of memory modules and the bit width of a unique memory module. As an example, prime and Fibonacci numbers, advantageous in certain memory arrangements, are permitted. Naturally, the area size of the unit increases if the number of the memory modules or the data width of one memory module is increased.

3. PIPELINING

Pipelining is an implementation method in which commands are overlapped in execution [9]. The complex calculations of the Address Computation Unit are time-consuming tasks. To divide the Address Computation Unit into several stages is the main motivation for the pipelining. Beside address computation, many other operations of the system benefit from the pipelining.

CPMA pipeline overview

Currently, the CPMA implementation is divided into 16 pipeline stages. The pipeline is mainly controlled by the State Machine (Figure 1). The division of CPMA components into pipeline stages and the overall timing of the system are depicted in Figure 4.

The pipeline stages are defined as follows. The first stage is for the Page Table, during which the Page Address is converted into Row address and Access Functions. Stages from the second to the seventh are for Address Computation, which calculates proper module (Modules), address (Addresses), and valid (Valid) vectors. Within stages eight and nine, the calculated address vector is reordered (Reordered Addresses) to conform to DRAM inputs. In addition, module and valid vectors are delivered to the Data Permutation Unit.

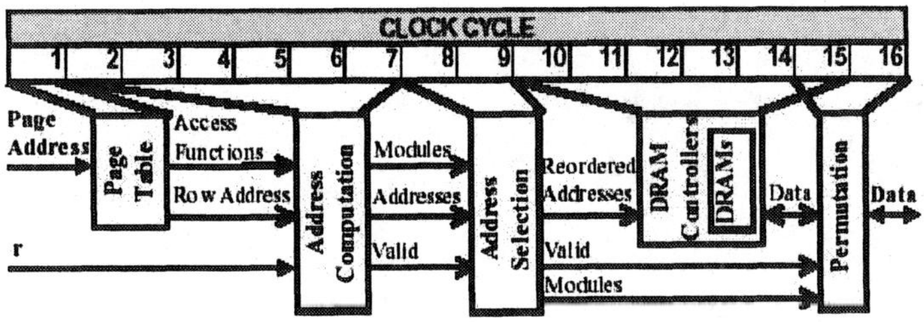

Figure 4. Timing of the CPMA pipeline within reading

During reading, the next five stages (stages 10-14) are taken by the DRAM Controllers, which access the addressed data from DRAM. The last two stages (stages 15 and 16) are for the Data Permutation Unit. The timing is somewhat different within write operation. An eight-stage buffer is required for the data in order to perform the first nine stages before permutation. The data permutation of the written data is made immediately after the Address Selection Unit has finished forming the module and valid vectors. Finally, the permuted data is written in memory.

State machine and pipeline control

The State Machine comprises two states: Operate and Stall. During the Operate state, a new instruction is fetched and previous instructions in the pipeline advance one stage in one clock cycle. The available commands are read, write, or no operation (NOP). After fetching the command, it is forwarded into the pipeline. If the pipeline becomes full, the State Machine changes to the Stall state. The Stall state halts the pipeline until the most advanced pipeline instruction is completed.

The first nine stages of the pipeline including the data buffer are completely controlled by the State Machine. Excluding the start time, the DRAM Controllers are State Machine independent. While reading, the final part of the pipeline is controlled by the DRAM Controllers. After the accessed data is delivered to data permutation, the data access will be completed in two cycles. However, data permutation is under the control of the State Machine during writing.

Pipeline characteristics and restrictions

The first part of the pipeline can accommodate up to nine access instructions until the pipeline is full. The bottleneck of the pipeline is the DRAM Controller phase. DRAM Controllers can only operate with one instruction simultaneously. The latency of the first instruction is 16 cycles, whereas the latency of the next instruction is 21 cycles, because the latter instruction has to wait 5 cycles before the previous instruction has completed the DRAM Controller stages. The worst latency for a command can be 56 cycles if nine instructions are fed into the pipeline during nine subsequent clock cycles. Within write operations the latency of the first instruction is 13 cycles, and in the worst-case latency is 45 cycles.

On the other hand, pipelining enhances the system throughput. In the case of a fully maintained pipeline, after completing the first instruction, every fifth clock cycle completes a CPMA instruction.

Only structural pipelining hazards are possible in our CPMA implementation. It is possible to start a read operation straight after a write operation. However, the write operation cannot be fetched until the last read instruction is completed due to the CPMA bi-directional data bus. There are no CPMA pipeline based data or control hazards.

4. IMPLEMENTATION

The described CPMA components are implemented with logic synthesis from VHDL. Being a configurable system, the area of CPMA is largely dependent on the selected implementation parameters. The number of DRAMs and the data width of an individual DRAM determine the total data bus width of a CPMA implementation. Variations in data bus width affect the size of the DRAM Controllers and the Data Permutation Unit.

The complexity of the Address Computation Unit is dependent on numerous design parameters including utilized module assignment functions, address functions, and access formats. The implementation areas of various components of the pipelined CPMA are tabulated in Table 1. The area estimations are based on the results obtained from logic synthesis. The area is declared as 2-input NAND -gates. The synthesized CPMA utilizes eight 8-bit wide memory modules. The memory capacities of the DRAMs are excluded from the estimation.

The area estimation of the Address Computation Unit is based on the implementation in [1]. At present, the pipeline stages are not uniformly distributed in the Address Computation Unit, but stages are located after the unit. The stage relocation (retiming) will be the next step in the CPMA development. It is evident from Table 1 that the Address Computation Unit dominates the implemented area. To radically minimize the area of the Address Computation Unit, operands limited in range to powers of two should be used.

Component	Area
Address Computation	47000
8 DRAM Controllers	5500
Data Permutation Unit	3500
Other pipelining parts	9000
Whole system	65000

Table 1. Gate count estimation for a CPMA implementation

5. TEST CASE

The following test case illustrates the benefits of a CPMA system. A performance evaluation is made by comparing a pipelined CPMA to a single non-pipelined conventional memory, which accesses data single pixel (8-bit) at a time. There are eight 8-bit wide memory blocks in the CPMA implementation. As a consequence, 64 bits can be accessed simultaneously. The evaluation is performed with QCIF (Quarter Common Intermediate Format) pictures. The standard size for a QCIF picture is 38016 pixels. The luminance of the picture is determined by the first 176 x 144 pixels, whereas roughly half as many pixels (88 x 72) determine the chrominance. The chrominance pixels are ignored in the evaluation.

Performed operations with CPMA

The evaluation process is divided into four phases. Figure 5 depicts the applied image during different phases of the evaluation. At the beginning of every phase, the whole picture (Original) is written in memory. The utilized access format is a horizontal straight line consisting of 8 pixels.

After writing the image in memory, it is read back using different access formats. Within the first phase, the picture is read with the original format resulting to the second picture from the left in Figure 5 (Whole). Next, the image is read with the same format, but the format is applied only to every second row of the original picture. As a result, an image is generated that is flattened in the j-direction (JFlattened). The zoomed image is created accessing the original image with a crumbled rectangle of 4 pixels (Zoom 4) and then with a crumbled rectangle of 8 pixels (Zoom 8).

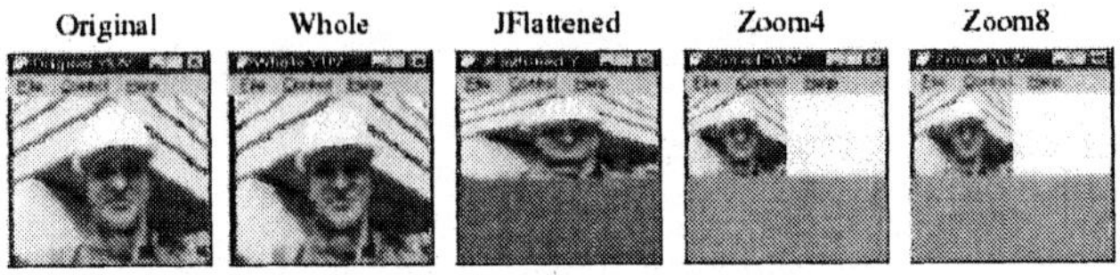

Figure 5. The applied image during the evaluation

Performance comparison

The aforementioned evaluation phases are performed and simulated with a register transfer level (RTL) CPMA. In addition, the execution cycles of the corresponding phases are calculated for ideal conventional memory and for ideal CPMA. The calculations are based on the following Equation:

$$\mathrm{Cycles}_{Total} = \mathrm{Cycles}_{Write} \cdot \left(\left\lceil \frac{\mathrm{Image}_{Length}}{\mathrm{Rate}_{L\,Write}} \right\rceil \cdot \left\lceil \frac{\mathrm{Image}_{Width}}{\mathrm{Rate}_{W\,Write}} \right\rceil \right) + \mathrm{Cycles}_{Read} \cdot \left(\left\lceil \frac{\mathrm{Image}_{Length}}{\mathrm{Rate}_{L\,Read}} \right\rceil \cdot \left\lceil \frac{\mathrm{Image}_{Width}}{\mathrm{Rate}_{W\,Read}} \right\rceil \right)$$

The values applied to the Equation are tabulated in Table 2. Used image values are the dimensions of the QCIF picture without chrominance pixels ($\mathrm{Image}_{Length} = 176$ and $\mathrm{Image}_{Width} = 144$). In each case, the execution times of read and write operations are 5 and 6 clock cycles, respectively. The Rate column identifies the stride between successive access scanning points. For example, $\mathrm{Rate}_{W\,Write} = 8$ signifies that every eight pixel is selected as a scanning point in the width direction during write operations.

As shown by the data in Table 2, the difference between conventional memory and CPMA is remarkable. In most cases, the conventional memory requires 8 times more clock cycles for similar operations than CPMA. The only exception is the case of crumbled rectangle of 4 pixels (Zoom 4), in which the speed-up factor is 6.5. It that case, a 4-pixel format instead of an 8-pixel format is applied during the read operation, which decreases the operation performance. The performance of CPMA would be slightly hindered, if the utilized format overlapped the picture boundaries. In that case, some useless pixels are accessed by CPMA.

Table 2. Performance evaluation results

Constants in all designs				
Memory	$Cycles_{Write}$	$Cycles_{Read}$	$Rate_{WWrite}$	$Rate_{LWrite}$
Conventiona	5	6	1	1
Ideal CPMA	5	6	8	1
RTL CPMA	5	6	8	1
Variables between designs and the results				
Memory	Picture	$Rate_{WRead}$	$Rate_{LRead}$	$Cycles_{Total}$
Conventiona	Whole	1	1	278784
Ideal CPMA	Whole	8	1	34848
RTL CPMA	Whole	8	1	34855
Conventiona	JFlattened	1	2	202752
Ideal CPMA	JFlattened	8	2	25344
RTL CPMA	JFlattened	8	2	25351
Conventiona	Zoom 4	2	2	164736
Ideal CPMA	Zoom 4	4	4	25344
RTL CPMA	Zoom 4	4	4	25351
Conventiona	Zoom 8	2	2	164736
Ideal CPMA	Zoom 8	8	4	20592
RTL CPMA	Zoom 8	8	4	20599

The slight difference between the total execution cycles of ideal and RTL CPMA can be explained with the external NOPs required by RTL CPMA. Memory refreshes cycles are not included in the evaluation. Refreshes increase the difference between memory performance values because conventional memory operates longer and therefore also needs more refresh cycles than CPMA.

6. CONCLUSIONS

Typical streaming multimedia and real-time applications are increasingly suffering from the memory access latency and the effects of sub-optimal cache refreshes. With the presented pipelined Configurable Parallel Memory Architecture high data throughput with a reasonable latency can be achieved. Especially SoC designs stand to profit from CPMA since it guarantees efficient usage of wide parallel memory structures and their wiring. Analytical results and simulations illustrate the considerable performance improvements of CPMA over the conventional memory architecture. The contemporary special implementations of parallel memories are extremely effective in particular applications. The configuration aspect enables a more general-purpose utilization of CPMA The complexity of CPMA is completely dependent on numerous design parameters including the number of memory modules and the various design options in the Address Computation Unit. However, this paper shows that these implementation need not to be excessively expensive in terms of silicon area.

Future research includes connecting the CPMA implementation with a processor core. The processor core will be developed to take full advantage of the CPMA operations. In addition, speed enhancements are necessary and they are achieved with further pipelining development and the use of more sophisticated DRAM modules.

REFERENCES

[1] E. Aho, J. Vanne, K. Kuusilinna, T. Hämäläinen, and J. Saarinen, "Configurable Address Computation in a Parallel Memory Architecture," *Advances in Signal Processing and Computer Technologies*, G. Antoniou, N. Mastorakis, and O. Panfilov, ed., WSES Press, Athens, Greece, 2001, pp. 390-395.

[2] H. Chang, L. Cooke, M. Hunt, G. Martin, A. McNelly, and L. Todd, *Surviving the SOC revolution: a guide to platform based design*, Kluwer, 1999, p. 235.

[3] V. Cuppu, B. Jacob, B. Davis, and T. Mudge, "A Performance Comparison of Contemporary DRAM Architectures," *Proc. Fifth International Symposium on Higherformance Computer architecture* (HPCA'99), Atlanta GA, May 1999, pp. 222-233.

[4] V. Cuppu and B. Jacob, "Concurrency, Latency, or System Overhead: Which Has the Largest Impact on Uniprocessor DRAM-System Performance?," *Proc. 28th International Symposium on Computer Architecture* (ISCA'01), Goteborg Sweden, June 2001, pp. 62-71.

[5] T. Feng, "A survey of interconnection networks," *IEEE Computer*, Vol. 14, No. 12, Dec. 1981, pp. 12-27.

[6] M. Gössel, B. Rebel, and R. Creuzburg, *Memory Ar chitecture & Parallel Access*, Elsevier Science B.V, Amsterdam, The Netherlands, 1994, p. 246.

[7] D. Keitel-Schultz and N. Wehn, "Embedded DRAM Development: Technology,Physical Design, and Application Issues," *IEEE Design and Test of Computers*, Vol. 18, No. 3, May-June 2001, pp. 7 –15.

[8] K. Kuusilinna, *Studies and Implementations of Bus-based Interconnections*, Doctoral Dissertation, Dept. Information Technology, Tampere Univ. of Technology, Tampere, Finland, 2001, p. 201.

[9] D. Patterson and J. Hennessy, *Computer Architecture A Quantitative Approach*, 2nd ed., Morgan Kaufmann, San Francisco, California, 1996, p. 760.

[10] D. Patterson, T. Anderson, N. Cardwell, R. Fromm, K. Keeton, C. Kozyrakis, R. Thomas, and K. Yelick, "A Case for Intelligent RAM," *IEEE Micro*, Vol. 17, No. 2, Mar. 1997, pp. 34-44.

[11] B. Prince, *High performance memories*, JohnWiley & Sons, New York, 1996.

[12] H. Siegel, *Interconnection Networks for Large-Scale Parallel Processing*, LexingtonBooks, Lexington, Massachusetts, 1986, p. 261.

Section 6.3

XOR-scheme Implementations In Configurable Parallel Memory

Eero Aho, Jarno Vanne, Kimmo Kuusilinna, and Timo Hämäläinen
Institute of Digital and Computer Systems, Tampere University of Technology
Korkeakoulunkatu 1, FIN-33720 Tampere, FINLAND

Abstract: Being able to continuously feed data to the processing nodes is one of primary problems in parallel data processing. Parallel memories increase memory bandwidth with several memory modules working in parallel and feed the processor with only necessary data. The Configurable Parallel Memory Architecture (CPMA) enables a multitude of access formats and module assignment functions to be used within a single hardware implementation, which has not been possible in prior embedded parallel memory systems. In this paper, a general XOR-scheme implementation for a previously implemented CPMA is presented. Any XOR-scheme can be used for the distribution of data to the memory modules with the introduced implementation. Results show significant reduction in the overall memory accesses, which can be viewed as one of the key approaches to solve the processor-memory performance bottleneck.

Key words: configurable parallel memory architecture, XOR-schemes, module assignment function, conflict free access

1. INTRODUCTION

The development of Dynamic Random Access Memories (DRAMs) has been exponential during the last decades. That is, the size of the memory cell has been shrinking and, thus, the capacity of the DRAMs has been increasing. However, the smaller feature s izes have not made the DRAMs significantly faster. The cell density has been the development driver and speed has partly been a secondary issue. Simultaneously, rapid changes have happened to the transistors in processors. This has allowed both speeds-ups and increased complexity in processors. For these reasons, the processor versus memory performance gap has been widening for a number of years. Several methods and architectures to enhance the memory system performance have been presented [11].

One solution to the problem is to place a cache between the main memory and the processor. Widening the bus between the memory and cache makes it possible to refresh the cache in larger blocks of data. Unfortunately, this can signify the fact that the cache contains a lot of currently unnecessary data.

Some real-time applications, e.g. motion estimation and discrete cosine transformation (DCT) in video processing, favor special data patterns that can be accessed in parallel [6]. This phenomenon is utilized in *parallel memories,* where the idea is to increase memory bandwidth with several memory blocks working in parallel and feed the processor with only necessary data. Traditionally the parallel memory development has been concentrated on *module assignment functions* (*skewing schemes*) that try to ensure *conflict free* parallel data accesses to as many *access formats (templates)* as possible. *Linear* module assignment functions were first studied with parallel memories [4]. *XOR-schemes* have been shown to allow several additional conflict free access formats [2], [5], [7], [9].

A generalized block diagram of a parallel memory architecture is shown in *Figure 1*. The Figure depicts an *address computation* unit, *N memory modules,* and a *permutation* unit [8]. Depending on the access format *F* and the address of the first element *r*, the address computation unit computes the right memory modules and addresses in the memory module. The permutation unit organizes the data into correct order.

A typical problem with previously presented parallel memory implementations is that they are application specific with fixed access formats implemented in hardware. This ensures the fastest speed, but requires redesign if other formats have to be included. In this work, the problem has been addressed by implementing the *Configurable Parallel Memory Architecture* (*CPMA*) [10], in which access formats and the needed address generation functions can be changed at run-time.

Time-to-market, increasing complexity, and system performance requirements have necessitated the increasing utilization of reusable intellectual property (IP) blocks and programmable platforms in system-onchip (SoC) designs. CPMA can be viewed as one step to this direction since it is able to cope with a multitude of algorithms needed in different applications without redesign of the memory system.

Promising technology development for CPMA includes embedded DRAM designs, where DRAM is embedded to the same die as the processor core. This enables the use of wide data buses, low latency, and vector processor like execution, which are all very attractive properties if data can easily be supplied from parallel memories. Especially real-time, streaming, applications are deemed to profit from this architecture because the cache memory performance bottleneck can be alleviated.

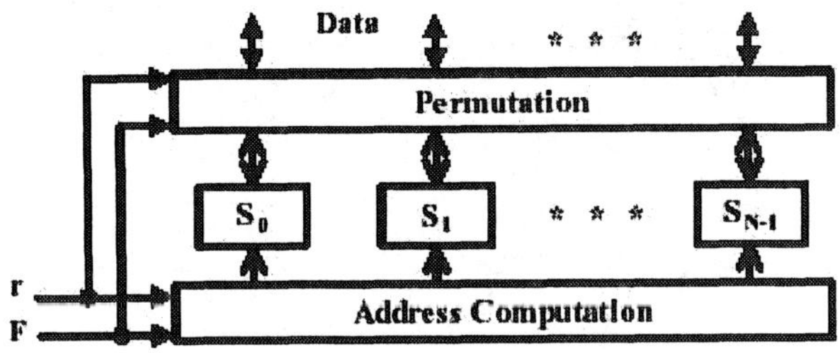

Figure 1. General parallel memory architecture.

Previous implementations of CPMA included linear module assignment functions and some special XOR-schemes [1]. In this paper, a generalized implementation of a XOR-scheme for the implemented CPMA is presented. An overview of the implemented CPMA is given in Section 2. XORschemes are introduced in Section 3, which is followed by the XOR-scheme implementation details in Section 4. Results are given in Section 5 and Section 6 concludes the paper.

2. CONFIGURABLE PARALLELMEMORY ARCHITECTURE

A block diagram of the implemented CPMA is depicted in *Figure 2*. The inputs to CPMA are the *virtual address* and the *command* (write, read). The virtual address includes a *page address*, which is a pointer to a *page table*. *Page offset* directly indicates the *scanning point r* for the parallel memory. The *row address* and the access functions are fed from the page table to the address computation unit. In addition, DRAM banks and a permutation unit are required for CPMA operation. [1]

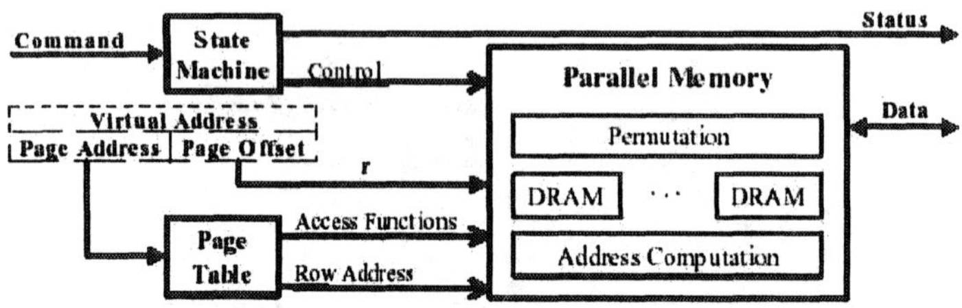

Figure 2. Configurable Parallel Memory Architecture.

CPMA address computation is implemented using multiple dedicated computation units in parallel. There is one unit for each class of access formats, module assignment functions, and address functions that the implementation supports. Moreover, a configurable data permutation unit is designed to conform to system specifications.

CPMA is designed to be configurable both at design time and at runtime. In the VHDL implementation, many parameters are generics, which means that the number of DRAMs, size of a DRAM module, width of the data bus, and width of input buses are adjustable. All the required blocks, logic, and buses are implemented with logic synthesis from VHDL. In the following, we describe some of the CPMA blocks in more detail.

Page Table

Our page table is a block of memory that is directly addressable from application software. The page address is a pointer to a specific row in the page table, which determines the functio nality in the CPMA during that memory access. Sample page table entries are tabulated in *Table 1* and the Table field contents are explained in the following.

Page Address	Row address	L_i	L_j	F	$S(i,j)$	$a(i,j)$
:						
:						

Table 1. Page table structure

Row address in the page table is like a page number in a physical memory. A single page should usually only carry one module assignment function $S(i,j)$.

Length of coordinates L_i and L_j indicate a memory area (dimensions i and j) to be accessed with the scheme defined in this page entry. The values are scaled according to the application at hand.

Parameters for access format function F. The data elements are read or stored in parallel with an access format F. Five example access formats are shown in *Figure 3*. Squares in an access format depict accessed data elements.

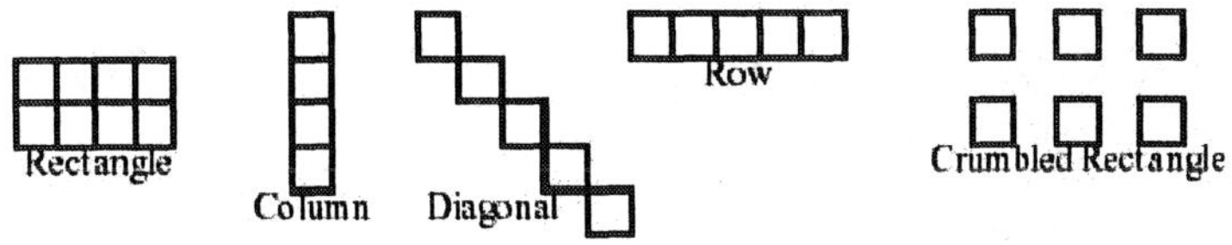

Figure 3. Access format examples.

Parameters for a module assignment function $S(i,j)$. The assignment of a data element to the memory module is determined by a module assignment function. The utilized module assignment function determines the access formats that can be used conflict free. There are two kinds of module assignment functions implemented in our system.

A *Linear* module assignment function is defined as

$$S(i, j) = (a \cdot i + b \cdot j) \bmod N, \tag{1}$$

where a, b and N are positive integers.

Most of the skewing schemes using dyadic functions (also called XOR schemes) are formulated as

$$S(i, j) = \pi(i \bmod N)_2 \oplus \tau(j \bmod N)_2, \tag{2}$$

where $N = 2^m, m \in \{0,1,2,...\}$. The operation denotes component-wise XOR. and | are permutations of the components of the binary numbers (i mod N) and (j mod N), respectively. That is $\pi(i \bmod N)_2 = \pi(i_{m-1}|..|i_0)$, where i_k is a binary number. [8]

Previously, only two hardwired permutations were implemented within the CPMA framework. These permutations are described by Equations (3) and (4) [8], [9].

$$\text{REVERSE}(i_{m-1} | ... | i_1 | i_0) = (i_0 | i_1 | ... | i_{m-1}) \tag{3}$$

and for $N = 2^{2m'}, m' \in \{0,1,2,...\}$

$$\text{SWAP}\ (i_{2m'-1} | ... | i_{m'} | i_{m'-1} | ... | i_0) = (i_{m'-1} | ... | i_0 | i_{2m'-1} | ... | i_{m'}) \tag{4}$$

Parameters for address function *a(i,j)*. Address function determines the address in a memory module for the accessed data element. The current system implements one address function $a(i, j) = b + (i + j \cdot L_{i)} / N$, where *b*, L_i and *N* are positive integers. L_i is specified in the L_i column of the page table.

When a user specifies an algorithm for CPMA, this process includes defining the contents of the associated page table row. First, the required access formats for the algorithm are evaluated. Then, a module assignment function for conflict free access has to be chosen. Moreover, a suitable address function for the module assignment has to be found. Finally, all this information is stored in the page table. Automation of this process requires further research.

Address Computation

A block diagram of the Address Computation unit is depicted in *Figure 4*. The inputs to the unit are the access functions and the scanning point *r* (*Figure 2*). The row address is directed straight to the DRAMs. The specific memory module addresses are the outputs from this unit.

Inside the unit, a *Format Control* unit receives the access format function, the scanning point *r*, and the length of coordinates L_i and L_j. This unit specifies the coordinates of the data elements that are accessed in the scanning field.

The Module Assignment unit computes the specific memory modules that contain the accessed data elements. The Address Decode unit, in turn, computes the physical addresses for each memory module. After that, the Address Select units direct the physical addresses for the proper memory modules.

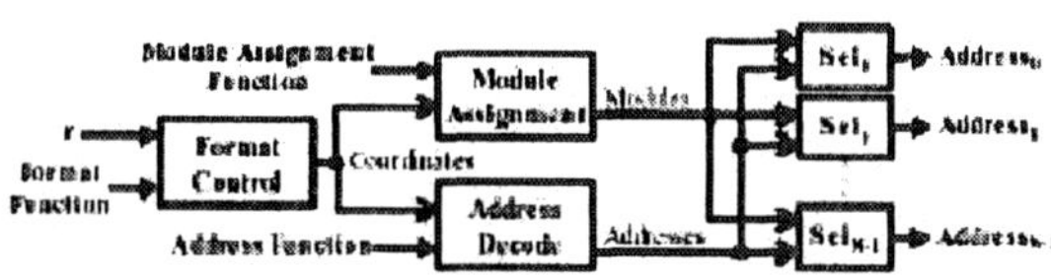

Figure 4. Address Computation unit.

3. XOR-SCHEMES

XOR-schemes always utilize a power of two ($N = 2^n$) memory modules.The dyadic function (2) can also be defined as:

$$S(i, j) = AI \oplus BJ\,, \tag{5}$$

where $I = (i \bmod 2^u)_2$ and $J = (j \bmod 2^v)_2$. **A** and **B** are $(n \times u)$ and $(n \times v)$ matrices containing binary numbers. The elements of **A** and **B** matrices are marked in an unordinary fashion. The top left element of matrix **A** is $a_{n-1,u-1}$ and the bottom right is $a_{0,0}$. The scanning field is decomposed into nonoverlapping $(2^u, 2^v)$ pages. In every page, the corresponding points are stored within the same memory module. The vector addition and multiplication in the function (5) are achieved by bit-wise logical operations XOR and AND, respectively. [7]

An example XOR-scheme is depicted in *Figure 5* where the number of memory modules N=8. A single page including 8*8 data elements is displayed. The data elements are stored in a memory module with a number shown in the corresponding table cell. The elements are indexed with (i,j) coordinates. E.g., a data element located at point (001,011) is stored in memory module 7. The values along the axels are shown in binary format since that number system is used with the XOR-scheme. The scheme is formed using function (6).

$$
\begin{aligned}
&S(i,j) - I \oplus RESERSE(J) \\
&= \begin{bmatrix} 1 & 0 & 0 \\ 0 & 1 & 0 \\ 0 & 0 & 1 \end{bmatrix} \cdot \begin{bmatrix} i_2 \\ i_1 \\ i_0 \end{bmatrix} \otimes \begin{bmatrix} 0 & 0 & 1 \\ 0 & 1 & 0 \\ 1 & 0 & 0 \end{bmatrix} \begin{bmatrix} j_2 \\ j_1 \\ j_0 \end{bmatrix} = \begin{bmatrix} i_2 \\ i_1 \\ i_0 \end{bmatrix} \otimes \begin{bmatrix} j_2 \\ j_1 \\ j_0 \end{bmatrix} \\
&= (i_2 \oplus j_0 \mid i_1 \oplus j_1 \mid i_0 \oplus j_2)
\end{aligned} \tag{6}
$$

j \ i	000	001	010	011	100	101	110	111
000	0	1	2	3	4	5	6	7
001	4	5	6	7	0	1	2	3
010	2	3	0	1	6	7	4	5
011	6	7	4	5	2	3	0	1
100	1	0	3	2	5	4	7	6
101	5	4	7	6	1	0	3	2
110	3	2	1	0	7	6	5	4
111	7	6	5	4	3	2	1	0

Figure 5. Example XOR-scheme.

The function (6) reorganizes each of the J binary vectors as formulated in (3). The scheme is called the EE (Exchange-Expansion) scheme in [9]. The depicted scheme allows row and column accesses to be conflict free in all the points, however, rectangle shaped accesses are restrictedly conflict free when the number of accessed elements M is equal to the number of memory modules N. E.g. a rectangle access format sized 4*2 is conflict free when the scanning point is located at $r(i,j)=(0,0)$ since all the accessed elements are located in different memory modules (*Figure 5*). However, the same access format is not conflict free when $r(i,j)=(1,1)$ because two of the data elements are accessed from the same memory modules 0 and 6. These data elements are marked with dots "•".

A more complicated XOR-scheme example is depicted in *Figure 6* where $N=8$. The function is formulated as follows

$$S(i,j)=\begin{bmatrix}1&1&1\\1&1&0\\0&1&1\end{bmatrix}\cdot\begin{bmatrix}i_2\\i_1\\i_0\end{bmatrix}\otimes\begin{bmatrix}0&0&1\\0&1&0\\1&0&0\end{bmatrix}\begin{bmatrix}j_2\\j_1\\j_0\end{bmatrix} \tag{7}$$

This formula is called the MG-EE scheme (MG stands for Modified-Gray). The scheme allows several conflict free access formats, e.g. rows, columns, diagonals, rectangles, crumbled rectangles, and chessboards. Some of the access formats are conflict free in arbitrary positions the rest can only be restrictedly placed. [9]

j \ i	000	001	010	011	100	101	110	111
000	0	5	7	2	6	3	1	4
001	4	1	3	6	2	7	5	0
010	2	7	5	0	4	1	3	6
011	6	3	1	4	0	5	7	2
100	1	4	6	3	7	2	0	5
101	5	0	2	7	3	6	4	1
110	3	6	4	1	5	0	2	7
111	7	2	0	5	1	4	6	3

Figure 6. 8*8 mapping matrix for the MG-EE scheme.

4. GENERALIZED XOR-SCHEME IMPLEMENTATION

A and **B** in the XOR-scheme Equation (5) are (n x u) and (n x v) matrices when the number of memory modules is $N=2^n$. The vector addition and multiplication are achieved by logical operations bitwise XOR and bitwise AND respectively. Therefore, the calculations are done as follows:

$$AI = \begin{bmatrix} a_{n-1,u-1} & a_{n-1,u-2} & \cdots & a_{n-1,0} \\ a_{n-2,u-1} & a_{n-2,u-2} & \cdots & a_{n-2,0} \\ \vdots & \vdots & \cdots & \vdots \\ a_{0,u-1} & a_{0,u-2} & \cdots & a_{0,0} \end{bmatrix} \begin{bmatrix} i_{u-1} \\ i_{u-2} \\ \vdots \\ i_0 \end{bmatrix}$$

$$= (a_{n-1,u-1} \cdot i_{u-1} \oplus a_{n-1,u-2} \cdot i_{u-2} \oplus \ldots \oplus a_{n-1,0} \cdot i_{u-1} \mid a_{n-2,u-1} \cdot i_{u-1} \oplus \quad (8)$$

$$a_{n-2,u-2} \cdot i_{u-2} \oplus \ldots \oplus a_{n-2,0} \cdot i_0 \mid \ldots \mid a_{0,u-1} \cdot i_{u-1} \oplus a_{0,u-2} \cdot i_{u-2} \oplus a_{0,0} \cdot i_0)$$

The operations . and depicts logical AND and XOR, respectively. The calculation of **B***J* is done using a respective notation where a symbol u is replaced with a symbol v.

Let us denote

$$AI = (a_{n-1} \mid a_{n-2} \mid \ldots \mid a_0),$$
$$BJ = (b_{n-1} \mid b_{n-2} \mid \ldots \mid b_0)$$

Then, the final calculation can be expressed as

$$AI \oplus BJ = (a_{n-1} \oplus b_{n-1} \mid a_{n-2} \oplus b_{n-2} \mid \ldots \mid a_0 \oplus b_0). \quad (9)$$

A block diagram of the implementation of the XOR-scheme (5) is depicted in *Figure 7*. Matrices **A** and **B** are supplied from a page table. The block **A***I* executes the logical operations in function (8) and the respective operations are done in the block **B***J*. Finally, logical bit-wise XOR operations are done as indicated in (9) to get the module number $S(i,j)$.

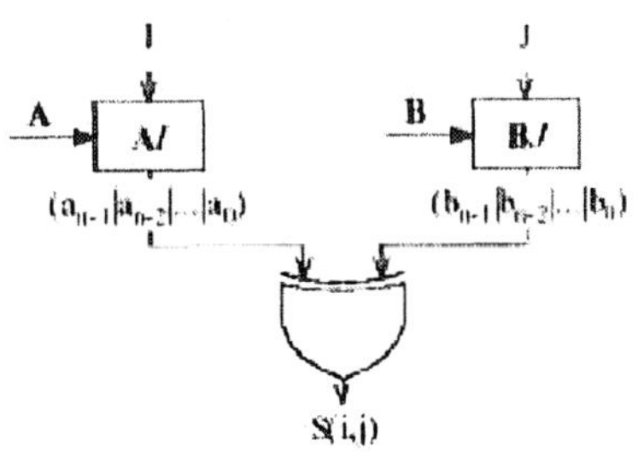

Figure 7. Generic implementation of the XOR-scheme.

The sizes $(n \times u)$ and $(n \times v)$ of matrices **A** and **B** have to be determined before synthesizing the presented XOR-scheme implementation. $(n*u)$ units of 2 -input AND ports and n units of u -input XOR ports are required in the **AJ** block. Respectively, $(n*v)$ units of 2 -input AND ports and n units of v - input XOR ports are required in the **BJ** block. The final bitwise XOR operations need n units of 2 - input XOR ports. Altogether, the required ports in the implementation of the generalized XOR-scheme (5) are tabulated in *Table 2*.

The values of the binary matrices **A** and **B** are required to be stored in the page table. Synthesis only defines the maximum size $(n \times u)$ and $(n \times v)$ for the matrices. In the case of smaller sized matrices, the elements of the matrices **A** and **B** have to be placed to the right down corner and the left up corner values are set to zero

.

Port type	Number of ports
2 input AND ports	$(n*u) + (n*v)$
u -input XOR ports	n
v -input XOR ports	n
2 -input XOR ports	n

Table 2. Required resources in the implementation

An overview of configuring a XOR-scheme is depicted in *Figure 8*. Before a memory access, the values for the matrices **A** and **B** in Equation (5) are stored in the page table. When a data pattern is read from CPMA, the Central Processing Unit (CPU) references the page table with a page address to these predetermined matrix values. The memory module number $S(i,j)$ for each of the accessed data elements is determined with the help of the matrix values and the (i_0,j_0), (i_1,j_1)... (i_{N-1},j_{N-1}) coordinates from the Format Control unit (*Figure 4*).

All the XOR-schemes using the Equation (5) can be formed using the presented XOR-scheme implementation. These kinds of schemes have been introduced e.g. in [2], [5], [7], and [9]. Those schemes enable the use of a multitude of conflict free access formats. For instance, the XOR-scheme introduced in [7] enables accesses for rows, columns, rectangles, crumbled rectangles and chessboards. The XOR-scheme was also used in the STARAN computer architecture [3]. STARAN was able to access many kinds of templates including rows, columns, *folded lines*, *partial-row-pairs*, and *partial-column-pairs*. The last three access formats are useful in pattern analysis [9].

5. RESULTS

The introduced XOR-scheme is implemented with synthesizable register transfer level VHDL. The sizes (n x u) and (n x v) of matrices **A** and **B** have to be determined before synthesizing the design. A test case design was synthesized to confirm the number of required logical ports. The parameters were set as $n=v=u=3$. That makes the number of memory modules $N=2^n=8$. This implementation can form, e.g. the schemes depicted in *Figure 5* and *Figure 6*. The synthesis tool estimates an area that equals to 390 2 –input NAND ports.

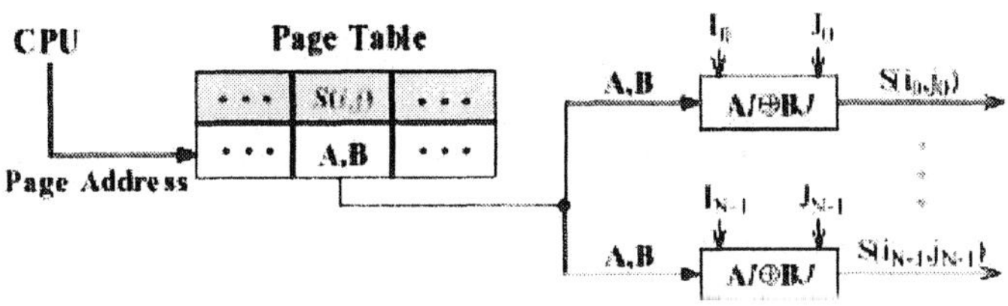

Figure 8. Configuring a XOR-scheme.

The required theoretical resources are tabulated in *Table 3* for the test case design when $N=8$. The area estimation for the above ports from the synthesis tool is equivalent to 442 NAND ports. The difference between the synthesized implementation and the theoretical 442 ports indicates that the synthesis tool has managed to optimize the test case design.

Table 3. Required theoretical resources when N=8.

Port type	Number of ports
2 -input AND ports	8*18=144
3 -input XOR ports	8*6=48
2 -input XOR ports	8*3=24

The advantage from using the implemented Configurable ParallelMemory Architecture is apparent if access times are compared when processing a picture with different access formats. Crumbled rectangle access formats can be utilized in video processing (e.g. DCT/IDCT), motion estimation). An example of such an access format is shown in *Figure 9*. The access format can be used conflict free within e.g. the XOR-scheme depicted in *Figure 6*.

If a QCIF (Quarter Common Intermediate Format) picture containing 176*144 pixels, is written to memory a single pixel at a time it takes 176*144 = 25344 write operations. (Assuming every pixel is written individually.) When writing the same picture with the access format shown in *Figure 9*, it takes 3168 write operations. That is 8 times less than with a single pixel access format.

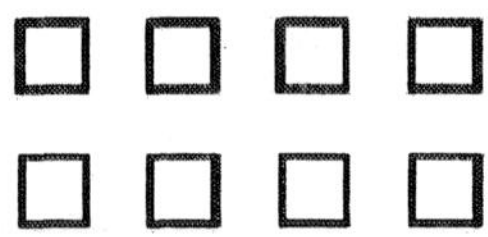

Figure 9. Access format used in the example.

6. CONCLUSION

The discussed Configurable Parallel Memory Architecture enables a system-on-chip processor to use a multitude of memory access templates. These access formats can guarantee the usefulness of data and data refresh operations are required less frequently. In addition, the configurability of the CPMA facilitates running several different computation threads, each with a unique skewing scheme, in a single system. This has not been possible in the prior embedded parallel memory systems.

In this paper, a general implementation of a XOR-scheme for the implemented CPMA was presented. The implemented schemes allow several additional conflict free access formats to be used. Parameters for constructing a XOR-scheme can flexibly be changed at run-time without a need to redesign the whole hardware.

In the future, our research will focus on clarifying the access formats that are required in special applications. Especially templates that can be utilized with algorithms used in video and graphics are studied. Moreover, other module assignment functions are researched for their implementation feasibility.

REFERENCES

[1] E. Aho, J. Vanne, K. Kuusilinna, T. Hämäläinen, and J. Saarinen, "Configurable Address Computation in a Parallel Memory Architecture," *Advances in Signal Processing and Computer Technologies*, G. Antoniou, N. Mastorakis, and O. Panfilov, ed., WSES Press, Athens, Greece, 2001, pp. 390-395.

[2] M. A. Al-Mouhamed and S. S. Seiden, "Minimization of Memory and Network Contention for Accessing Arbitrary Data Patterns in SIMD Systems," *IEEE Trans. Computers*, Vol. 45, No. 6, June 1996, pp. 757-762.

[3] K. E. Batcher, "The Multidimensional Access Memory in STARAN," *IEEE Trans. Computers*, Feb. 1977, pp. 174-177.

[4] P. Budnik and D. J. Kuck, "The Organization and Use of Parallel Memories," *IEEE Trans. Computers*, Vol. C-20, No. 12, Dec. 1971, pp. 1566-1569.

[5] S. Chen, A. Postula, and L. Jozwiak, "Synthesis of XOR Storage Schemes with Different Cost for Minimization of Memory Contention," *Proc. Euromicro Conf.*, Sep. 1999, pp. 170-177.

[6] S. Dutta, W. Wolf, and A. Wolfe, "A Methodology to Evaluate Memory Architecture Design Tradeoffs for Video Signal Processors," *IEEE Trans. Circuits and Systems for Video Technology*, Vol. 8, No. 1, Feb. 1998, pp. 36-53.

[7] J. M. Frailong, W. Jalby, and J. Lenfant, "XOR-Schemes: A Flexible Data Organization in Parallel Memories," *Proc. Int'l Conf. Parallel Processing*, Washington, DC, USA, Aug. 1985, pp. 276-283.

[8] M. Gössel, B. Rebel, and R. Creuzburg, *Memory Architecture & Parallel Access*, Elsevier Science B.V, Amsterdam, The Netherlands, 1994, p. 246.

[9] Z. Liu and X. Li, "XOR Storage Schemes for Frequently Used Data Patterns," *Journal of Parallel and Distributed Computing*, Vol. 25, No. 2, Mar. 1995, pp. 162-173.

[10] K. Kuusilinna, *Studies and Implementations of Bus-based Interconnections*, Doctoral Dissertation, Dept. Information Technology, Tampere Univ. of Technology, Tampere, Finland, 2001, p. 201.

[11] D. A. Patterson and J. L. Hennessy, *Computer organization and design: the hardware/software interface*, Morgan Kaufmann, San Francisco, California, 1994, p. 648.

Section 6.4

An Novel Low Power Embedded Memory Architecture for MPEG-4 Applications with Mobile Devices

Mohammed Sayed and Wael Badawy
Department of Electrical and Computer Engineering
University of Calgary, Calgary, Alberta, Canada T2N 1N4

Abstract: This paper presents a novel embedded memory architecture for MPEG-4 synthetic video object decoding that is used in a new chipset for third generation mobile phones. The embedded memory architecture stores video frame and motion vectors, then it computes in memory the predicted frame, which generates the MPEG-4 video sequence. The embedded memory is optimized with a parallel reading and writing operations. The architecture has been prototyped in 0.35 ì m CMOS technology using TSMC standard cells and the Digital Video Processing and Digital VLSI Prototyping (DVP^2) group's custom cells.

Key words: System on chip, embedded memory, digital video processing, MPEG-4 applications.

1. INTRODUCTION

In the last two decades, the industries of mobile devices increased rapidly. These mobile devices contain sophisticated digital circuits to perform the complicated computations required for speech compression algorithms and the advanced functions carried out by these devices. Power consumption of these mobile devices increases as their capabilities increase. Moreover, with increasing video applications over mobile networks, the power consumption increases dramatically. Since, these mobile devices are battery powered and as a result of the recent batteries' limited power capabilities, low power designs are required.

Reducing power consumption can be achieved at different levels of the design. At CMOS circuit level, reducing the supply voltage, reducing the clock frequency or minimizing the number of transitions occurs [1] are commonly used to reduce power consumption. At the architecture level, clock gating techniques [2], decreasing the number of memory access (i.e. decreasing the bus activity), parallel and/or pipelined design, and using embedded memories all are possible techniques for decreasing the power consumption and increasing the performance. Architecture level approaches are more promising for reducing power consumption [3].

The object-based video representation, as specified by MPEG-4 standard [4], is one of the most important data compression techniques. Moreover, the affine transformation is a good candidate for motion compensation in MPEG-4 compression technique, but the high computational cost of the affine transformation is not suitable for the MPEG-4 applications on mobile devices.

This paper we presents novel embedded memory architecture for video object decoding from the MPEG-4 bit stream. The architecture uses a multiplication free algorithm to compute the affine transformation [5]. Moreover, using an embedded memory enables us to optimize the memory structure to decrease the number of memory accesses and performing the reading and writing operations in parallel, which decreases the bus activity and enhance the computation performance.

The rest of the paper is organized as follows: Section 2 outlines an overview of reducing power consumption by using paralle l implementation, Section 3 presents the proposed embedded memory architecture, Section 4 presents the architecture prototyping, Section 5 concludes this paper.

2. POWER CONSUMPTION IN CMOS TECHNOLOGY

In digital CMOS systems there are three major sources of power dissipation. These sources are shown in the following equation [3]:

$$P_{total} = p_t(C_L V_{dd}^2 f_{clk}) + I_{sc} V_{dd} + I_{leakage} V_{dd} \tag{1}$$

The first term represents the switching power component, where P_t is the transition probability, C_L is the loading capacitance, V_{dd} is the supply voltage, and f_{clk} is the clock frequency. The second term ($I_{sc}V_{dd}$)is the short circuit power component, where I_{sc} is the short circuit current, which arises when both the NMOS and PMOS transistors are simultaneously active. The third term ($I_{leakage}V_{dd}$) is due to the leakage current $I_{leakage}$. Switching power is the dominant source of power dissipation in digital CMOS systems.

Reducing the supply voltage is the most effective way to reduce power consumption as shown in equation 1. We can have quadratic improvement in power consumption by reducing the supply voltage but this comes at the cost of increasing the delay [3]. Parallel implementation is one of the techniques used to increase the throughput. By using parallel implementation we can reduce the supply voltage while maintaining the same throughput, which mean reducing the power consumption at the cost of increasing the area instead of increasing the delay.

In video applications, memory transfer operations occur very frequently, which mean the I/O power, due to transfer data between the processor and the off chip memory, will be tremendous. Using an embedded memory will reduce the power consumption so far. With embedded memory we can avoid using the memory off chip driver with its large capacitances [6] and the bus capacitances on PCB boards, which mean reducing the power consumption as shown in equation 1. Moreover, the embedded memory can work with the same low supply voltage as the remaining logic parts, which mean more power reduction.

Using embedded memory not only decreasing the power consumption but also decreases the latency and so improves the performance. The embedded memory is a suitable choice for the parallel architectures due to the flexibility in its design as will be shown in the next section.

3. THE PROPOSED EMBEDDED MEMORY

In MPEG-4 simple profile, the video object can be defined using a rectangular 2D mesh with triangular patches, where the deformation of the 2D mesh define the dynamics of video object. Motion compensation is done using the affine transformation to decode the video object.

As shown in Figure 1 the image is divided into right angle triangles with edge length equal to C, in this paper we chose C=16 pixel and an image of 32x32 pixels for simplicity. The affine's computations and video object decoding will be done using 32 processing elements (one processing element per column) working in parallel. These processing elements are divided into 4 main processing elements, working on the edges of the triangle patches, and 28 sub-processing elements, working on the internal pixels of the triangle patches, as shown in Figure 2.

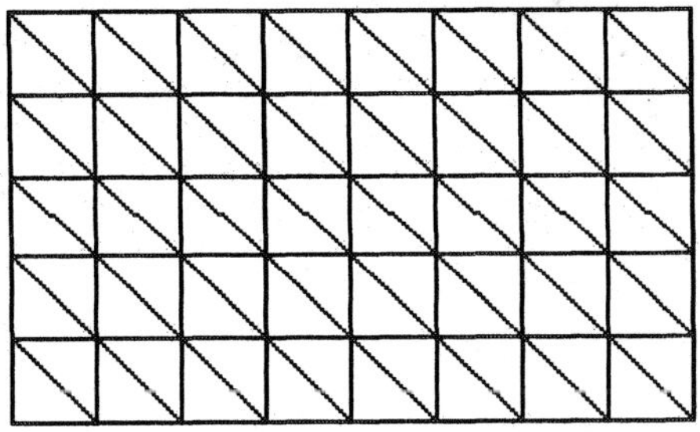

Figure 1: The 2D mesh

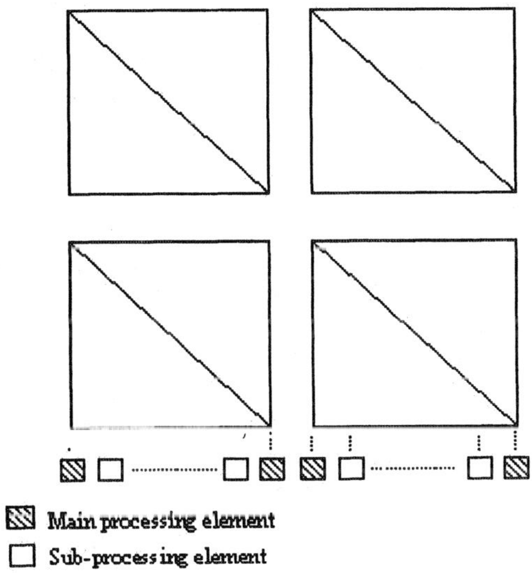

Figure 2: The proposed parallel architecture for texture mapping using affine transformation

Each processing element generates the addresses needed for texture mapping for the pixels in its column using the affine transformation [7] as shown in equations (2):

$$\begin{aligned} x &= a_1 X + a_2 Y + a_3 \\ y &= a_4 X + a_5 Y + a_6 \end{aligned} \qquad (2)$$

With the scan line algorithm [5], the affine transformation can be evaluated using the values of neighboring pixel as shown in equations (3). The number of coefficients needed to compute the affine transformation is reduced to four coefficients (a_1, a_2, a_4, a_5) instead of the six coefficients (a_1, a_2, a_3, a_4, a_5, a_6) needed in the original affine transform.

$$
\begin{aligned}
x(X+1,Y) &= x(X,Y) + a_1 \\
y(X+1,Y) &= y(X,Y) + a_4 \\
x(X,Y+1) &= x(X,Y) + a_2 \\
y(X,Y+1) &= y(X,Y) + a_5
\end{aligned}
\qquad (3)
$$

Moreover, using the multiplication free algorithm proposed in [5], the affine transformation's complexity can be reduced so far.

The main processing elements need only to compute a_2 and a_5 for the processed triangular patch. The sub-processing elements need only to compute a_1 and a_4 for the processed patch and they use a_2 and a_5, which are computed by the main processing elements. The block diagrams of the main processing elements and the sub-processing elements are shown in Figures 3 and 4 respectively.

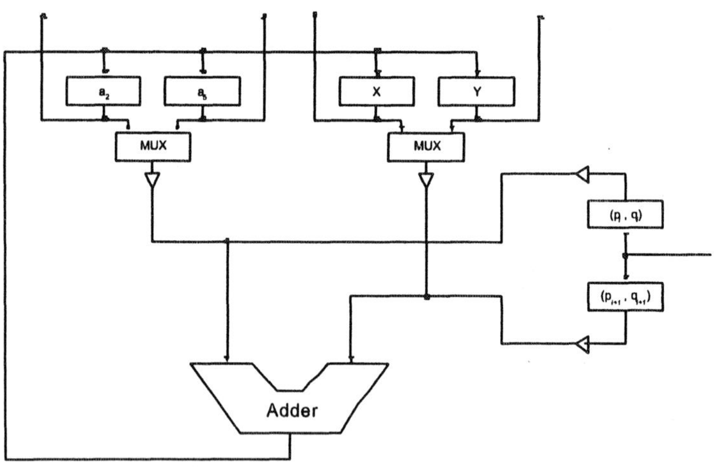

Figure 3: The block diagram of the main processing element

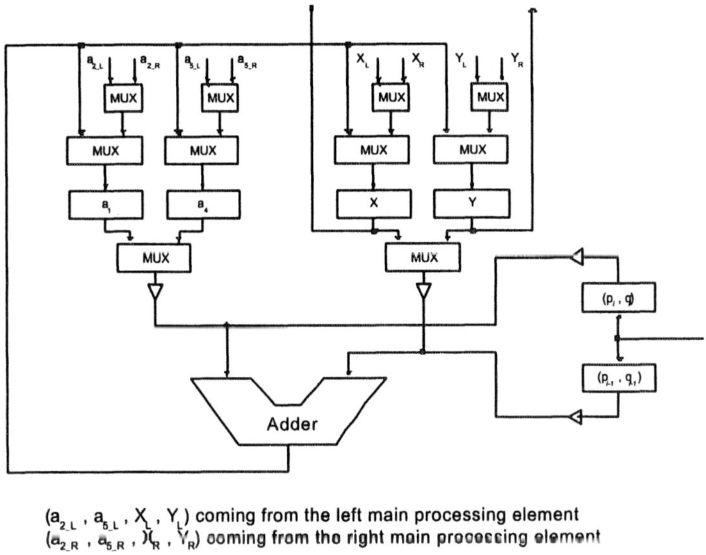

Figure 4: The block diagram of the sub-processing element

The proposed architecture consists of two memories each of size 32x32 bytes (one for the given texture and the other for the new texture), motion vectors' memory, and the 32 processing elements. The processing elements generate the required memory address (X,Y) using the affine transformation and the main control unit transfers the texture from (X,Y) location in the given frame to the (x,y) location in the new frame, this operation is done row by row. The block diagram for the whole architecture is shown in Figure 5.

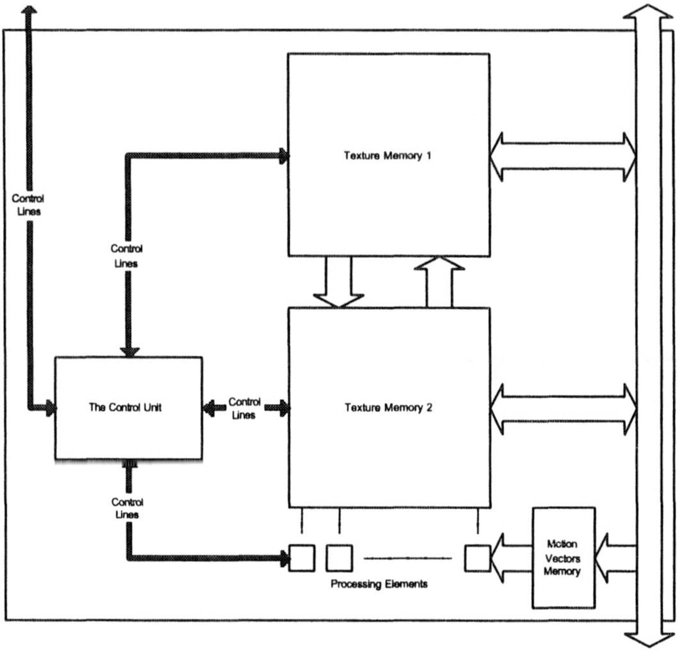

Figure 5: The block diagram of the proposed texture mapping architecture

The bottleneck in any memory-based architecture is the memory read/write time. With the normal memory architecture's implementations the read and write operations are done one byte (i.e. one pixel) per memory read/write cycle. In the techniques used for MPEG-4 motion estimation the value of any motion vector lay between –7 and 7 [8], so any point i (x_i,y_i) in the new frame will be mapped to a point j (X_j,Y_j) in the given frame, where (xj-7 < Xj < xj+7) and (yi–7 < Yj < yi+7). Dividing our memory into N blocks, where the width of each block > 14, and with repeating the overlap memory area as shown in Figure 6, we can read/write N bytes per memory read/write cycle. In the prototype of the proposed architecture we choose N=2 (2 blocks) so 2 bytes will be read/write per each memory read/write cycle and C=16 (the smallest 2^n number > 14).

Moreover, the number of blocks can easily increased without increasing the number of memory read/write cycles.

The block diagram of one memory block is shown in Figure 7. Each memory block consists of 32 rows and 23 columns, 7 columns of these columns are repeated texture from the adjacent memory block. Each memory block has one reading bus connected to the 23 columns and has two writing buses, one is connected to the 16 columns, which will contain local data, and the other is connected to the 7 columns, which will contain repeated data from the adjacent memory block. Each memo ry block uses two address decoders for horizontal and vertical addresses. The flow chart for the control unit is shown in Figure 8.

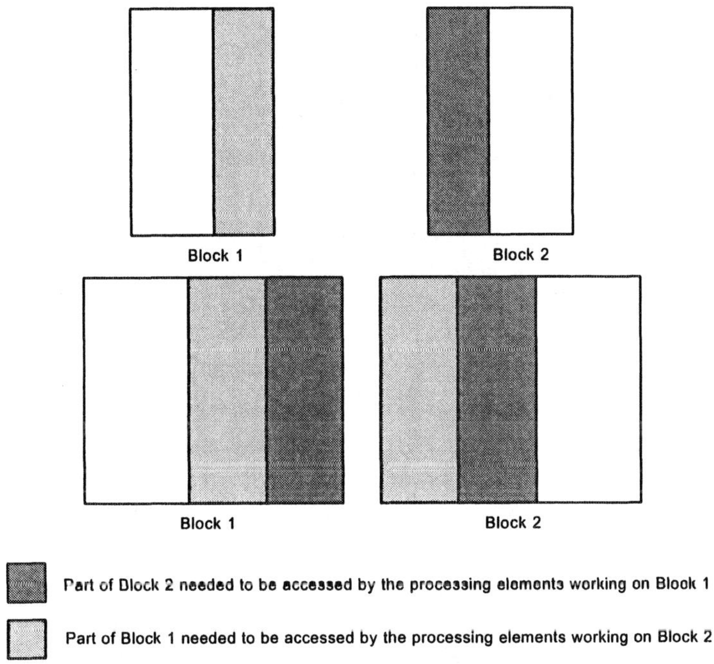

Figure 6: The proposed embedded memory

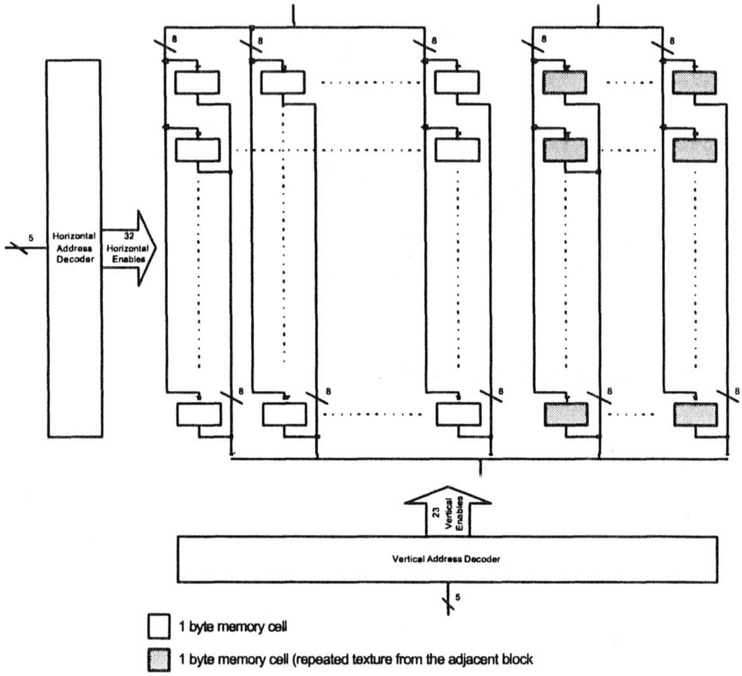

Figure 7: The block diagram of one memory block

4. THE ARCHITECTURE PROTOTYPING

The architecture has been prototyped using CMOS 0.35 ìm technology with TSMC standard cells. Synopsys tools are used for the "front-end" steps and Cadance tools are used for the "back-end" steps. The design has been done using Verilog HDL. The Verilog code is simulated, analyzed, and synthesized using Synopsys tools.

The synthesized prototype takes an area of 21 mm2 and consumes 288 mW.

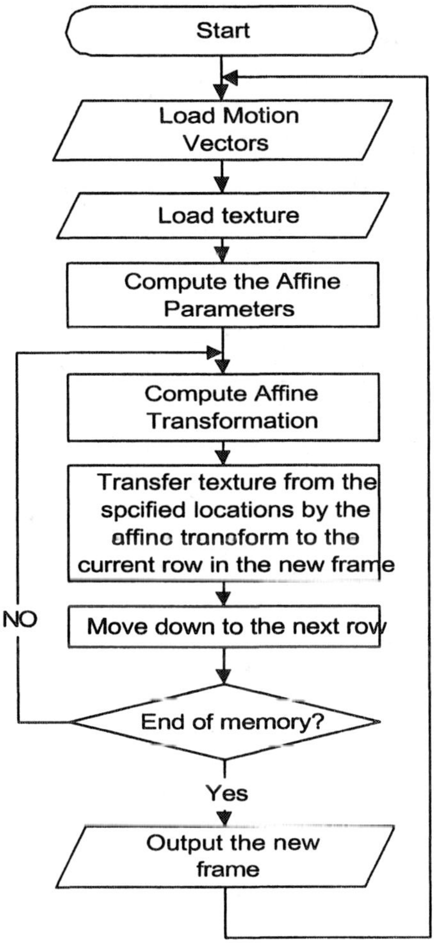

Figure 8: The control unit flow chart

5. CONCLUSION

This paper presents a novel low power embedded memory architecture for MPEG-4 video object decoding. The proposed architecture uses parallel implementation and embedded memory techniques to reduce the power consumption and enhance the performance. Moreover, a new memory structure is proposed that decreases the memory read/write by reading and writing N bytes per memory access time. The prototype of the proposed architecture is designed using Verilog HDL then synthesized using TSMC standard cell libraries. The synthesized prototype takes an area of 21 mm2 and consumes 288 mW.

ACKNOWLEDGEMENT

The authors would like to thank Natural Sciences and Engineering Research Council of Canada (NSERC), Canadian Microelectronics Corporation (CMC), Micronet, Human Res ource Canada (HRC), University Research Grant Committee (UGRC) and Department of Electrical Engineering at University of Calgary for supporting this research.

REFERENCES

[1]Kaushik Roy and Sharat C. Prasad "Low power CMOS VLSI circuit design," 2000.

[2]Ga ry K. Yeap "Practical low power digital VLSI design," 1997.

[3]Anantha P. Chandrakasan, Samuel Sheng, and Robert W. Brodersen "Low power CMOS digital design," IEEE journal of solid state circuits, Vol. 27, No. 4, April 1992.

[4]ISO/IEC JTC/SC29/WGll N1730, "MPEG-4 Overview," July 1997, http://drogo.cselt.it/mpeg

[5]Wael Badawy and Magdy Bayoumi, "A Multiplication-Free Parallel Architecture for Affine Transformation," the IEEE International Conference on Application-specific Systems, Architectures and Proces sors, Boston, MA, July 10-12, 2000 pp.25-34.

[6]Keitel-Schulz, D.; Wehn, N. "Embedded DRAM development: technology, physical design, and application issues," IEEE Design & Test of Computers, Vol 18 Issue 3 , May-June 2001, pp 7 -15.

[7]G. Wolberg, Digital Image Warping, Los Alamitos, CA: Computer Society Press, 1990.

[8]Wael Badawy and Magdy Bayoumi, "On Minimizing Hierarchical Mesh Coding Overhead: (HASM) Hierarchical Adaptive Structured Mesh Approach," IEEE International Conference on Acoustics, Speech, a nd Signal Processing, Istanbul, Turkey, June 5-9, 2000, pp. 1923-1926.

Section 6.5

Assessment of MPEG-4 VTC and JPEG2000 Dynamic Memory Requirements

Gauthier Lafruit, Jan Bormans
IMEC, DESICS, Kapeldreef 75, B-3001 Leuven, Belgium

Abstract: The dynamic memory requirements for ASAP schedules of MPEG-4 and JPEG2000 wavelet codecs are studied, using a queueing model. The straightforward assumption that the tight data dependencies of the MPEG-4 parent-children tree coding induce higher memory requirements than the attractive, independent block coding approach of JPEG2000 is demonstrated to be wrong. In particular, an MPEG-4 wavelet decoder is always less memory hungry than a JPEG2000 decoder. The memory requirements for an MPEG-4 wavelet encoder are often lower than, and in exceptional cases similar to those of its JPEG2000 counterpart.

Key words: Wavelet Still Image Coding, Memory requirements, MPEG-4, JPEG2000.

1. INTRODUCTION

MPEG-4 Visual Texture Coding (VTC) [1,2] and JPEG2000 [3,4,5] are two standardized still image compression tools that use a similar wavelet transform approach, but a very different entropy coding method. On the one hand, VTC is based on a zero-tree coding technique, using wavelet coefficients that are grouped as parent-children trees over the subbands. On the other hand, JPEG2000 codes equally-sized blocks from each subband independently of each other. Consequently, since JPEG2000 does not have to extract a full parent-children tree for starting the coding process, it is often taken for granted that JPEG2000 has lower memory requirements than VTC. However, it is often overlooked that (i) any optimal ASAP schedule for calculating the Wavelet Transform inevitably creates parent-children trees [6,7], and (ii) JPEG2000 requires blocks of BxB wavelet coefficients in the highest level of the Wavelet Transform for the coding process [4,5], while VTC uses only one single parent coefficient [2]. Because of (i), VTC's zero-tree coding approach does not necessarily introduce memory size penalties. Furthermore, as a consequence of (i) and (ii), JPEG2000 has inevitably to create several parent-children trees for collecting enough parent wavelet coefficients (BxB) for starting the coding process in the highest level. As a consequence, we question that JPEG2000 has lower memory requirements than VTC. This paper therefore presents an analysis, answering the prickly question of which of both wavelet coding tools has the lowest memory requirements in an optimal ASAP schedule. A full system perspective, i.e. including the encoder, buffering memory and decoder, is followed. Compression issues are not tackled in this study. The only assumption taken is that the JPEG2000 block size B is typically set to 32 or 64 for good compression performances [4].

The paper is organized as follows. In Section 2, the data dependencies influencing the memory requirements of VTC and JPEG2000 are analyzed. Section 3 describes a queueing model for tracking the required memory sizes. Section 4 presents simulation results that are discussed in Section 5. Conclusions are drawn in Section 6.

2. DATA PROCESSING DEPENDENCIES

In the Wavelet Transform, each wavelet coefficient is calculated from a bounded region of lower level wavelet coefficients, through a filtering process. By recursively applying this data dependency down to the input image level, one can draw a Data Dependency Pyramid (DDP) for each wavelet coefficient in the highest level of the Wavelet Transform. Such a DDP collects through all wavelet levels, those wavelet coefficients that are actively involved in the calculation of one particular wavelet coefficient in the highest level [6].

In VTC, the entropy coding module processes the wavelet data in packets of parent-children trees, which are typically included in one DDP. Therefore, in VTC, the data dependencies are fully covered by a DDP.

In JPEG2000, a block of BxB wavelet coefficients in the highest wavelet level covers a much larger region of BxB DDPs, suggesting that JPEG2000 has more severe (larger) data dependencies than VTC. However, within this data dependency region, some BxB blocks can be formed in intermediate levels of the Wavelet Transform, prior to the calculation of any parent coefficient. These blocks can be instantaneously processed by JPEG2000 entropy coding unit and thus prematurely freed up from memory. Consequently, JPEG2000 memory requirements could be much lower than those suggested by the BxB DDPs. The simulation model of Section 3 tackles this issue.

3. MEMORY QUEUEING MODEL

This section describes a simulation model for tracking VTC/JPEG2000 dynamic memory requirements accurately, while hiding low-level data processing details. The queueing model of Fig 1 corresponds to a full VTC/JPEG2000 codec system, with a wavelet-based encoder (WT-EC), a bitstream buffer and a wavelet-based decoder (ED-IWT). The bitstream buffer can possibly be partitioned over the encoder and decoder, depending on the application requirements. Once a queue contains the required data to start its succeeding data processing module, the latter is activated, reads the queue, processes the data instantaneously and fills the next queue in the processing chain. In either a software or hardware implementation, each queue can be mapped to one single memory with a particular buffering functionality. For instance, in VTC, the Transform Memory (TM) collects wavelet coefficients, created by the Wavele t Transform (WT) data processing module, as units of parent-children trees. Once a parent-children tree is fully created, it is transferred to the Entropy Coder (EC). In JPEG2000, the Transform Memory groups the wavelet coefficients in BxB blocks for each level of the Wavelet Transform. Partial BxB blocks are not released from TM, while fully completed BxB blocks are immediately sent to the Entropy Coder.

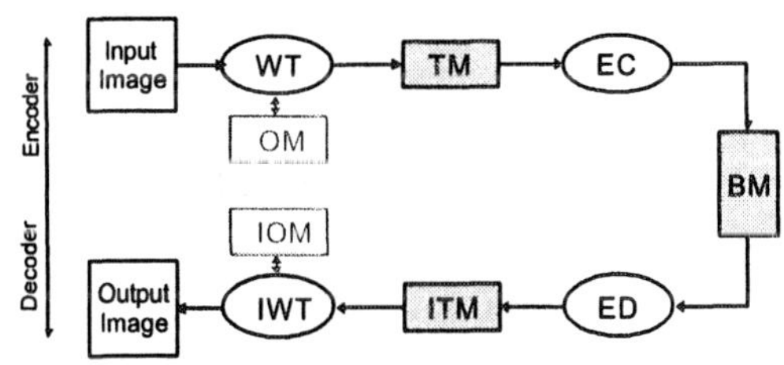

Fig. 1 Queueing simulation model of the wavelet codec (coder + bitstream buffer + decoder). The shaded boxes are Queues, the ellipses are Data Processing modules.

The Entropy Coder stores the coded information into the Buffer Memory (BM), as long as an insufficient amount of information is available for starting the decoding process. Since the Buffer Memory contains (possibly highly) compressed data, it is from the global memory requirements' point of view more appropriate to keep data into this buffer, as long as the activation of *all, successive* decoding processes does not create additional output pixe ls. More specifically, there is no reason to trigger the Entropy Decoding (ED) module if its results cannot be (possibly partially) used immediately by the Inverse Wavelet Transform (IWT), for creating new image pixels. From all the entropy decoded wavelet coefficients, only a portion of them is actually participating to the filtering operations of the Inverse Wavelet Transform. All other decoded wavelet coefficients are held into a buffer memory between the Entropy Decoder and the Inverse Wavelet Transform module: the Inverse Transform Memory (ITM).

Note that deviating from the above described scenario - i.e. triggering the Entropy Decoding module as soon as possible, instead of taking into account the processing capability of the Inverse Wavelet Transform - would reduce the Buffer Memory requirements, but inevitably increases the size of the Inverse Transform Memory.

It is obvious that the implementation style (i.e. calculation sequence) of the (Inverse) Wavelet Transform has a large impact on the simulated memory requirements. Since it is our objective to draw conclusions for the most optimal conditions, we choose a well-known ASAP (As Soon As Possible) calculation schedule: the RPA algorithm [8]. Since wavelet coefficients are calculated recursively over all levels, the RPA exploits this inter-level data dependencies optimally for creating the wavelet coefficients over all levels ASAP. This RPA algorithm has been adapted by several authors [6,7,9,10] to optimize functionality (e.g. Region of Interest coding) and/or performance (e.g. memory bandwidth), while the memory size requirements remain essentially identical (theoretical bounds can be found in [6]). The simulations and theoretical derivations of the present paper are therefore derived from the RPA.

An additional remark is in place here. In VTC, the Transfo rm Memory is exactly equal to the so-called Tree Memory, described in [6,7]. The memory required to seamlessly glue the filtering operations of the (Inverse) Wavelet Transform for adjacent rows (memory required in the RPA algorithm in [5,8]) or blocks (the so-called Overlap Memory in [6,7,9,10]) is not counted in the (Inverse) Transform Memory. The reason for this is that the size of this Overlap Memory (OM-IOM) is independent of the entropy coding algorithm and is thus reflected in the same amount for the VTC and JPEG2000 simulations. Since the main goal of this paper is to establish a comparison between VTC and JPEG2000, a common offset in the memory size graphs does not influence the final conclusions. As a consequence, the Overlap Memory has been discarded from the analysis.

4. SIMULATION RESULTS

In Subsections 0 and 0, the RPA schedule has been injected in the queueing model of Fig 1 for a one dimensional Wavelet Transform, and the memory consumption has been tracked over time. The results are extrapolated for a 2D Wavelet Transform in Subsection 0.

1D Dynamic Memory Requirements

During the simulation of the 1D-RPA, the memory consumption patterns are typically obtained. Observe that the data movements between the different queues (memories) are footprinted by the peaks in the graphs. In these example snapshots, a one-dimensional, 4-level, 7/5-tap Wavelet Transform is followed by a 1D JPEG2000 32-samples block coding or a 1D parent-children tree VTC coding. The main difference between these coding modes is that VTC waits for a *complete* parent-children tree to be created before transferring it to the Entropy Coder, while JPEG2000 transfers independent BxB blocks to the Entropy Coder, whenever they become available.

Fig 2 shows that the ITM requirements are of the same order of magnitude for VTC and JPEG2000. However, the average TM and BM requirements (thus, discarding the peaks) are higher for JPEG2000 than for VTC. This can be explained by JPEG2000 relatively large block size (typically 32 or 64 samples), which requires collecting more data than the number of samples in a parent-children tree of VTC. For the same reason, the peaks in the graphs of JPEG2000 are higher than VTC's.

Fig 3 reflects the same phenomenon in the delay between the insertion of a sample in the encoder (horizontal axis) and the reconstruction time stamp of the same sample at the decoder (vertical axis). For the same working conditions as in Fig 2, VTC creates the first output sample after having processed the first 93 input samples in the encoder, while JPEG2000 must have coded-decoded 541 input samples, before a first chunk of 451 samples can be reconstructed. From that moment on, VTC reconstructs the output in packets of 16 samples, for each 16 new input samples at the encoder. On the other hand, JPEG2000 reconstructs the output in packets of 512 samples, which corresponds exactly to $32x2^4$ samples, i.e. (block size)$x2^{(\text{number of levels})}$.

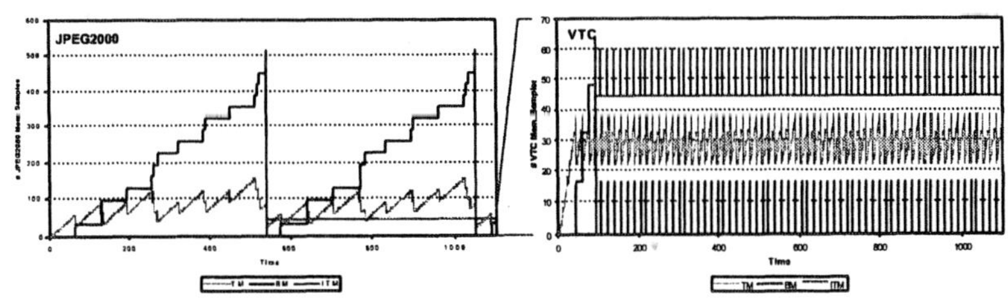

Fig 2: Memory occupation pattern in the 4-level, 7/5-tap wavelet filter, 1D-RPA for JPEG2000 with 32-samples blocks (Left), and VTC (Right). The green connection lines refer to equal memory sizes in JPEG2000 and VTC.

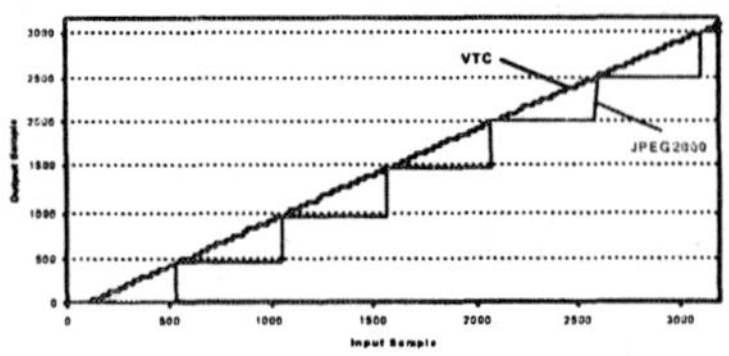

Fig. 3: Coding-Decoding delay of 4-level, 7/5-tap wavelet filter, 1D-RPA for VTC and JPEG2000 with 32-sample blocks.

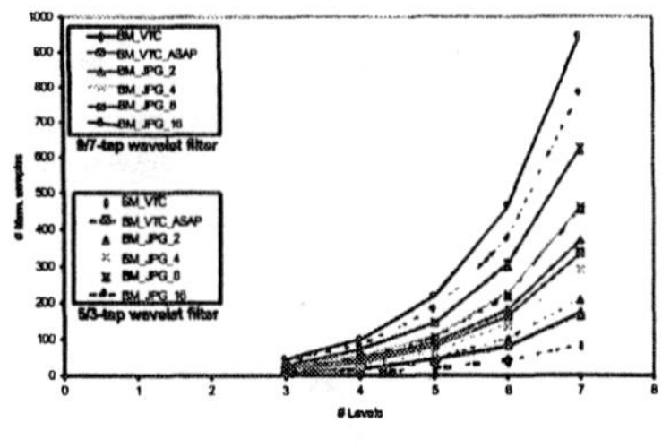

Fig. 4: Variation of the Buffer Memory requirements for different 2M+1 filter sizes, i.e. 5/3-tap and 9/7-tap wavelet filters.

Observe that JPEG2000 is never able to reconstruct samples faster than VTC, which already suggests that the parent-children tree extraction of VTC does not penalize its coding speed and memory requirements, compared to the independent block coding of JPEG2000. Moreover, JPEG2000 independent block coding does not represent a real advantage from the decoding point of view either. Indeed, Fig 2 (top) shows that when JPEG2000 creates a new output packet of samples, the BM is completely emptied and the ITM is directly filled with some entropy decoded wavelet coefficients. This means that all so-far coded blocks have to be decoded for providing the Inverse Wavelet Transform enough information in order to process it down to the lowest level and create a new output packet of reconstructed samples. Some of the entropy decoded wavelet coefficients cannot be consumed by the Inverse Wavelet Transform and are therefore stored in the ITM. The surprising fact that BM is dropping to zero at each new packet of reconstructed samples clearly indicates that at least one wavelet coefficient in *each* of the so-far coded/decoded blocks is actually actively involved in the decoding chain.

1D Maximum Memory Requirements

The maximum memory sizes are now tracked over time *t*, in different configurations of the 1D-RPA.

It is observed that the VTC results highly depend on the filter sizes used in the (Inverse) Wavelet Transform. In contrast, the variations of JPEG2000 results with these filter sizes (e.g. BM: see Fig 4) do not alter the main conclusions about the relative performances of VTC and JPEG2000. Therefore, in VTC, a parameter M is added for identifying the (2M+1 Lowpass / 2M-1 Highpass)- biorthogonal filter pairs of the Wavelet Transform, while JPEG2000 results are averaged over all M-values (M = 2, 3, 4).

The results are summarized in Fig. 5, with graphs labeled as X_Y_Z_[M], with following conventions:

X corresponds to the memory that is tracked, i.e. TM, BM or ITM.

Y identifies the compression mode, i.e. VTC stands for MPEG-4 VTC and JPG stands for JPEG2000.

If Y=JPG, then Z indicates the one-dimensional size of the independent blocks to be coded, i.e. Z wavelet coefficients (ZxZ in 2D) are transferred as an indivisible unit to the Entropy Coder. With Y=VTC, Z=nil, since a zero-tree coding is used, with one parent per parent-children tree to code.

Even if in the Y=VTC mode, a full parent-children tree should be coded as one indivisible unit, it should be noted that many coding schemes use a two-step approach [11]. In the first step, each wavelet coefficient is preprocessed as soon as it is created. The second step then processes the intermediate symbols to the final zero-tree symbols. In such a scenario, wavelet coefficients are individually trans ferred to the Entropy Coder in an ASAP schedule and the coding scheme is then virtually identical to a JPEG2000 coding with one sample blocks. Such a VTC coding scheme is then referred to as Z=ASAP.

2D Maximum Memory Requirements

For the 1D-RPA no ambiguity exists with respect to its implementation. However, as already mentioned in Section 3, a multitude of different improved implementations of the RPA exist in 2D [5,6,7,9,10], therefore requiring another approach for obtaining the 2D memory requirements of Fig 6.

In order to ensure independence from the implementation style, optimal 2D-RPA results are obtained by assuming that each 1D memory of the 1D-RPA stores wavelet coefficients over one column of a subband image of the wavelet pyramid. This is consistent with the analysis in [6]. Consequently, the maximum size of a 2D-memory (TM, BM or ITM) over time t for processing an input image of width W with the 2D-RPA (or variant) is obtained by summation of the individual 1D-RPA memory contributions $memory_{i,j}$ of subimage i, having width $W/2^j$ at level j (j=1..*Max_Level*) of the wavelet pyramid, as follows:

$$\max_t[memory(t)] = \max_t\left\{\sum_{i,j} memory_{i,j}(t).\frac{W}{2^j}\right\} \qquad (1)$$

For a memory containing non-compressed information (thus, TM and ITM), its size only depends on the level j (i.e. it is independent of the LH, HL or HH type of the wavelet subimage). This assumption may not be true for BM that contains compressed wavelet coefficients, for which the compression performance may differ between the LH, HL and HH wavelet subimages of the same level. Nevertheless, for convenience, we assume that there is never a dependency with the wavelet image type. In this case, we may drop the index i dependency and just perform the summation over the levels:

$$\max_t[memory(t)] \leq 3.W.\sum_j\left\{\max_t\left[memory_j(t)\right].\frac{1}{2^j}\right\} \qquad (2)$$

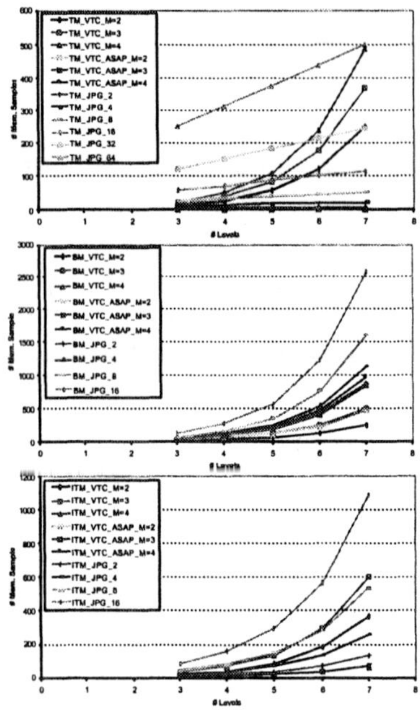

Fig. 5: 1D Memory requirements of MPEG-4 VTC and J PEG2000 codec. The memory is subdivided into the Transform Memory – TM (Top), the Buffer Memory – BM (Middle) and the Inverse Transform Memory – ITM (Bottom).

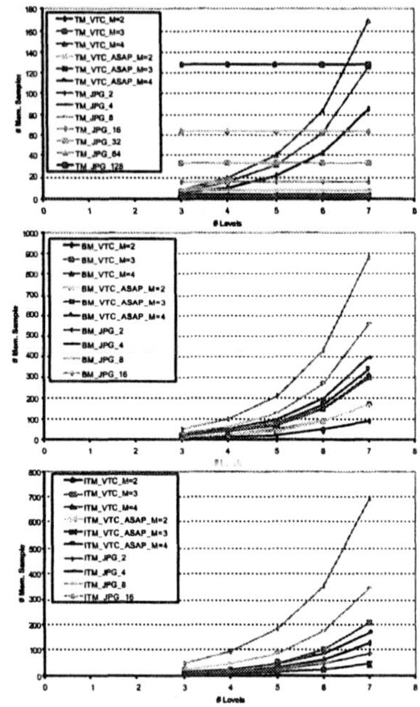

Fig. 6: 2D Memory requirements of MPEG-4 VTC and JPEG2000 codec. All values should be multiplied by three times the Input_Image_Width to get the final memory size. The memory is subdivided into the Transform Memory – TM (Top), the Buffer Memory – BM (Middle) and the Inverse Transform Memory – ITM (Bottom).

An upper bound of the memory requirements is thus obtained by tracking the maxima of the 1D-RPA memory requirements and calculating the weighted sum of eq. (2), yielding the results of Fig 6 (the constant factor 3 W has been discarded from the graphs).

Using this approach, one question remains to be solved, viz. "How far does the upper bound deviate from the real maximum value ?". In the 1D-RPA, we have observed for the VTC working conditions that there is less than 1% deviation between summing the maxima over each level j individually and tracking the maximum over the full system, i.e.:

$$\max_t[memory(t)] \approx \sum_j \max_t[memory_j(t)] \qquad (3)$$

This suggests that for the VTC coding performed on two-dimensional images, the procedure of eq. (2) yields a result, very near to reality.

When checking the relationship between the two terms of eq. (3) for the 1D-RPA in the JPEG2000 working conditions, a variation of a factor of respectively 1.5 and 1.9 is consistently observed over the large range of parameter settings for the TM and ITM memories. By extrapolation, this suggests that eq. (3) yields an upper bound which may deviate with at most the same factors from the real value in the two-dimensional JPEG2000 implementation. The reader may verify that these *uncertainty* factors for JPEG2000 do not influence the main conclusions that will be drawn throughout this paper.

5. MPEG-4 VTC VERSUS JPEG2000 MEMORY REQUIREMENTS

Asymmetry between encoder and decoder

The VTC memory requirements increase exponentially with the number of levels in the Wavelet Transform both at the encoder and decoder, while in JPEG2000 this phenomenon only occurs at the decoder (see Fig. 5 and Fig 6). Obviously, there exists a possible high asymmetry between the memory requirements of the encoder (TM) and decoder (ITM), which is in contrast with the straightforward assumption that wavelet encoders and decoders have symmetrical memory characteristics. Consequently, the TM and ITM memory requirements are analyzed separately in the following two subsections.

Transform Memory

Interestingly, the linear dependency of the TM size with the number of levels in the JPEG2000 1D-RPA (Fig.5 (top)) is totally compensated by the weighting factors of Eq. (1) in the JPEG2000 2D-RPA of Fig 6 (top). Consequently, while for VTC the TM requirements change exponentially with the number of Wavelet Transform levels, the TM size is independent of the number of levels in JPEG2000. However, the JPEG2000 TM size increases with the size of the blocks to be coded. For the default size of 64x64 coefficient blocks (TM_JPG_64 in Fig 6 (top)), the VTC coding (TM_VTC_M=2→4) requires less memory, as long as the number of wavelet levels does not exceed 5. At 6 wavelet levels, the 9/7-tap wavelet filter VTC coding requires slightly more memory than JPEG2000, but the other VTC modes are less expensive. However, when using 128x128 blocks instead of 64x64 blocks, VTC is less expensive in TM size up to a 7-level 5/3- and 7/5-tap Wavelet Transform.

Taking the uncertainty factor 1.5 of Subsection 0 into account, the conclusions change slightly. At 6 wavelet levels, only the 5/3-tap wavelet filter VTC coding requires less memory than JPEG2000. Compared to a 128x128 blocks JPEG2000 coding, VTC is then less expensive in TM size up to a 6-level Wavelet Transform, and even up to a 7-level Wavelet Transform if the wavelet filter is limited to 5/3-taps.

In summary: up to a 5-level Wavelet Transform, a VTC encoder requires less memory than its JPEG2000 counterpart. Above 5 levels, working conditions should be checked to draw accurate conclusions.

Inverse Transform Memory

Fig 6 (bottom) shows that once JPEG2000 uses larger blocks than 8x8 samples, its ITM is always larger (for any number of levels) than in VTC coding. With the uncertainty factor of 1.9, mentioned of Subsection 0, this observation remains valid, except for exceptional cases, for which the JPEG2000 block size must be increased to 16x16, in order to make VTC coding less memory hungry than JPEG2000. Moreover, Fig 6 (bottom) shows that ITM increases drastically with the block size in JPEG2000. For JPEG2000 default block size of 64x64, the ITM size can grow up to 2700 units (of 3. *W* samples) for a 7-level Wavelet Transform, while VTC is limited to 210 units. This indicates that a VTC decoder can have a memory size, which is up to one order of magnitude smaller than a JPEG2000 decoder.

Buffer Memory

Observe that the Buffer Memory (BM) between the encoder and decoder contains a larger number of (compressed) samples than TM and ITM. This is a consequence of the phenomenon described in section 3, of not starting *any* decoding process as long as no new output pixels can be created.

The Buffer Memory always increases with the number of wavelet levels, both in VTC and JPEG2000. Similar evolutions occur in the 1D and 2D graphs of Fig.5 (middle) and Fig 6 (middle).

For any wavelet filter, the VTC coding requires less Buffer Memory than JPEG2000 in any configuration, even for as little as 2x2 block sizes in JPEG2000. This suggests that VTC's parent-children tree processing schedule is probably the most appropriate from the point of view of memory size.

6. CONCLUSION

We have shown with a queueing simulation model that from the memory requirements point of view, an optimal implementation of an MPEG-4 wavelet decoder (VTC) is always less memory expensive than an optimal implementation of a JPEG2000 wavelet decoder with 16x16 (or larger) blocks. At the encoder side, a VTC encoder with up to 5 levels of the Wavelet Transform, requires less memory than a JPEG2000 encoder using a default block size of 64x64. Above 5 levels, the same conclusion may not necessarily hold, but in realistic conditions (up to 7 levels of the Wavelet Transform and up to 9/7-tap wavelet filter), the VTC encoder has memory requirements which remain in the same order of magnitude as the JPEG2000 encoder. The straightforward assumption that JPEG2000 is less memory hungry than VTC, because of its independent block coding, is thus wrong. Obviously, JPEG2000 pays a high penalty because of its constraint of processing equally sized blocks of BxB ($B \neq 1$, typically B=32 or 64) wavelet coefficients over all wavelet levels. In fact, the parent-children tree processing of VTC appears to be a very natural and probably the most appropriate schedule for wavelet data processing.

Finally, we also showed that there exists no strict symmetry in the memory requirements of encoder and decoder.

REFERENCES

[1] —, "Information technology - Coding of audio-visual objects - Part 2: Visual," ISO/IEC JTC 1/SC 29/WG 11 N 3056 Maui, December 1999.

[2] I. Sodagar, H.J. Lee, P. Hatrack, Y.Q. Zhang, "Scalable Wavelet Coding for Synthetic/Natural Hybrid Images," *IEEE Transactions on Circuits and Systems for Video Technology*, Vol. 9, No. 2, pp. 244-254, March 1999.

[3] —, "JPEG 2000 Image Coding System, " ISO/IEC JTC 1/SC29/WG1, FCD 15444-1.

[4] A. Skodras, C. Christopoulos, T. Ebrahimi, "The JPEG2000 still image compression standard," *IEEE Signal Processing Magazine*, Vol. 18, No. 5, pp. 36-58, September 2001.

[5] D. Taubman, "High Performance Scalable Image Compression with EBCOT," *IEEE Transactions on Image Processing*, vol. 9, no. 7, pp. 1158-1170, July 2000.

[6] G. Lafruit, L. Nachtergaele, J. Bormans, M. Engels, I. Bolsens, "Optimal Memory Organization for scalable texture codecs in MPEG-4," *IEEE Transactions on Circuits and Systems for Video Technology*, Vol. 9, No. 2, pp. 218-243, March 1999.

[7] G. Lafruit, L. Nachtergaele, B. Vanhoof, F. Catthoor, "The Local Wavelet Transform: a memory-efficient, high-speed architecture optimized to a Region-Oriented ZeroTree coder," *Journal of Integrated Computer-Aided Engineering*, Vol. 7, No. 2, pp. 89-103, March 2000.

[8] M. Vishwanath, "The Recursive Pyramid Algorithm for the Discrete Wavelet Transform," *IEEE Transactions on Signal Processing*, Vol. 42, No. 3, pp. 673-676, March 1994.

[9] Y. Andreopoulos, N. Zervas, G. Lafruit, P. Schelkens, T. Stouraitis, C. Goutis, J. Cornelis, "A Local Wavelet Transform implementation versus an optimal Row-Column algorithm for the 2D multilevel decomposition," *Proceedings ICIP*, Vol. 1, pp. 330-333, Oct. 2001.

[10] Yiliang Bao, C.-C. Jay Kuo, "Design of Wavelet-Based Image Codec in Memory-Constrained Environment," *IEEE Transactions on Circuits and Systems for Video Technology*, Vol. 11, No. 5, pp. 642-650, May 2001.

[11] B. Vanhoof, M. Peon, G. Lafruit, J. Bormans, M. Engels, I. Bolsens, "A Scalable Architecture for MPEG-4 Embedded Zero Tree Coding," *Custom Integrated Circuits Conference*, pp. 65-68, San Diego, May 1999

7

CIRCUIT TECHNIQUES

Although much of current SOC research is focused on the higher levels of abstraction for system design, the properties of the final integrated system are very dependent upon the circuitry that is used to build the hard IP used in the final design fabrication. This IP may be at the level of a reasonably complex core, or can be at a lower level where soft or firm IP generators and synthesizers are used to assemble the circuits into system level components. In this chapter we will see four different approaches and levels of circuit design for both digital and analog circuit blocks.

In the first section, "Modified Distributed Arithmetic Architecture for Adiabatic DSP Systems, by Dusan Suvakovic and C. Andre T. Salama of the University of Toronto, an approach to building adiabatic distributed arithmetic architectures (DAA) is presented. A modified DAA, is optimized for adiabatic operation, by replacing the ROM and the adder used in the conventional DAA, by parallel counters implemented using complex adiabatic logic gates. The design of the adiabatic modified DAA is fully automated by a set of CAD tool capable of generating the final circuit layout based on a high level specification. The energy efficiency of the modified 9-input adiabatic DAA exceeds that of the conventional DAA by more than five times, by eliminating the dissipation associated with ROM bit lines. In addition, the physical implementation of the modified architecture is significantly more compact than that of the conventional DAA.

In section 7.2, "Design of a CMOS Wide Range Logarithmic Amplifier with a Modified Parallel Architecture", by Sanghoon Joo, Minkyu Song, Hyunyeon Cho, and Sangki, of Dongguk University, Seoul, Korea, a circuit for a Logarithmic Amplifier (LA) is described for applications in a wide range and high sensitivity radar system. A new mixed type of series and parallel architecture is proposed for the logarithmic stage, and to decrease the input range error in the front of the LA, a novel input stage is designed. The amplifier is designed into a target O.5μm standard CMOS technology; the effective chip area is 131 μm x 154μum, and has a power consumption of 90mW at a 3 .3V supply voltage. Simulation and measurements, demonstrates that the amplifier is capable of 60dB dynamic range and 5Ons fall time.

Chaotic generators are finding increased use, for example, in CDMA and secure communications systems, that are prime examples of potential System-on-Chip applications. In the next section, "Digital Hardware Implementation of Continuous And Discrete Chaotic Generators", by Mohamed I. Sobhy, Mohammed A. Aseeri and Alaa E. R. Shehata, of the University of Kent at Canterbury, UK, a method to design chaotic generator models in real time is developed which is capable of implementing chaotic systems that are given by state equations in real time. A scaling factor to control the frequency band is introduced, and the method is useful for the implementation of chaotic generators at low frequencies. Both continuous and discrete chaotic generators can be implemented even if the system cannot be represented by a physical electronic circuit. Modification of any system is a simple change in the block diagram or the parameter values within the block.

Arithmetic represents one class of basic processing operation that is found in almost all digital systems. Addition is the most basic arithmetic operation and efficient construction of adder circuits is a constant quest by circuit designers. In the last section, 7.4 "New 1-Bit Full Adder Cells For Low-Power System-On-Chip Applications", by Mohammed Sayed and Wael Badawy, of the University of Calgary, three new 1-bit full adder cells are presented. A performance analysis of power consumption and delay for these cells is presented, and the cells are prototyped, simulated and tested in three different CMOS technologies: 0.18μm CMOS, 0.25μm CMOS, and 0.35 μm CMOS. The performance analysis of the proposed cells and six commonly used 1-bit full adder cells shows that the proposed cells outperform the other cells in both power consumption and power-delay product in all three technologies.

Section 7.1

Modified Distributed Arithmetic Architecture for Adiabatic DSP Systems

Dusan Suvakovic, C. Andre, T. Salama
Edward S. Rogers Sr. Department of Electrical and Computer Engineering, University of Toronto, 10 King's College Rd., Toronto, Ontario M5S 3G4, phone: *(416) 978-8658*

Abstract: This paper presents a modified distributed arithmetic architecture (DAA), optimized for adiabatic operation. In this novel architecture, the ROM and the adder used in the conventional DAA, are replaced by parallel counters implemented using complex adiabatic logic gates. Design of the adiabatic modified DAA is fully automated by a set of CAD tool capable of generating the final circuit layout based on a high level specification. The energy efficiency of the modified 9-input DAA, in adiabatic operation, exceeds that of the conventional DAA by more than five times. In addition, the physical implementation of the modified architecture is significantly more compact than that of the conventional DAA.

Key words : adiabatic, distributed arithmetic architecture, DSP

1. INTRODUCTION

Adiabatic techniques provide a potential for significant improvement in energy efficiency of digital circuits and systems and represent a potential enabling technology for new generations of battery powered and self-powered applications. These applications, such as digital hearing aids and various biomedical devices, are expected to involve extensive digital signal processing (DSP) while operating with microwatt dissipation. Therefore, adiabatic implementation of typical DSP building blocks represents a major research challenge.

This paper describes an adiabatic implementation of the distributed arithmetic architecture (DAA) [1], frequently used to compute a sum of products, in which either a multiplicand or a multiplier is constant. Such computation is performed as part of numerous digital signal processing (DSP) algorithms such as: finite impulse response (FIR) filtering with constant coefficients, discrete Fourier transform (DFT) and discrete cosine transform (DCT). DAA includes a lookup table storing partial results for a particular computation, a carry-propagate adder (CPA) and an accumulator register. Since it does not include multipliers, the DAA is considered to be compact, fast and energy efficient compared to alternative DSP building blocks capable of performing the same operation.

This paper describes an adiabatic implementation of the DAA, based on a recently developed adiabatic logic style OBDDL (Ordered Binary Decision Diagram Logic) [2]. In the implementation, a modification of the DAA is made, in which the conventionally used lookup table and the adder are replaced by an array of parallel counters [3] implemented using complex OBDDL gates. This modification is performed in order to minimize the amount of energy unavailable for recovery during computation, thus boosting the energy efficiency.

The complexity and variety of complex gates necessitates an automated design of the modified DAA. To that end, a set of CAD tools, which generates the physical layout of the modified DAA based on a high level specification, was developed.

The rest of the paper is organized as follows. In Section 2, the structure and operation of the modified DAA are explained. In Section 3, the circuit design of the OBDDL counter gates is described. Custom CAD tools developed to generate the complete adiabatic DAA based on a high level specification, are described in Section 4. The achieved energy efficiency and design compaction, compared to the conventional DAA, are discussed in Section 5. Conclusions are given in Section 6.

2. DAA IMPLEMENTATION USING OBDDL GATES

Conventional DAA

A block diagram of the conventional DAA implementation is shown in Fig. 1. Data inputs $x_0,\ldots x_{N-1}$ are applied to it in the bit-serial fashion, starting with the least significant bit. For a chosen L-bit output precision, this architecture includes a $2^N \times L$ - bit ROM, accessed by the inputs .x0,....xN-1. The DAA adds the pre-calculated sums of products of constant coefficients and one-bit input data read from the ROM, to the value in the accumulator wire-shifted right by one position.

Modified DAA for OBDDL implementation

In the modified DAA (MDAA), described here, the ROM and the adder are replaced by OBDDL counters, resulting in an architecture that is significantly more compact in silicon implementations and allows efficient energy recovery. The architecture involves one OBDDL counter per bit position. Each counter has direct data inputs $x_0,\ldots x_{N-1}$ and feedback inputs driven by one of its own outputs or by a counter output from an adjacent bit position. The feedback connections play the roll equivalent to that of the adder-accumulator-shifter block in the conventional DAA.

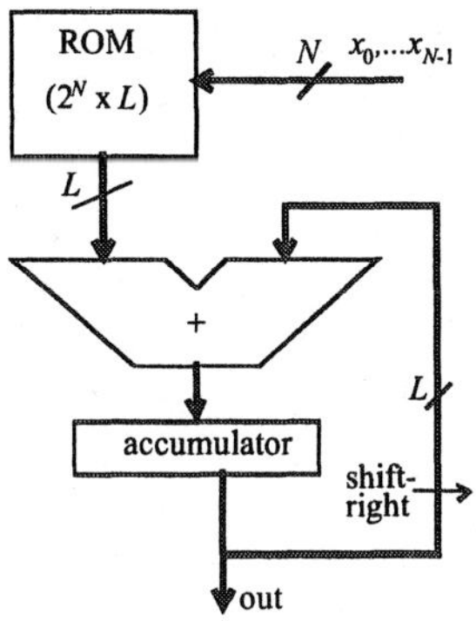

Figure 1. Conventional DAA

An example of MDAA for $N = 9$, is illustrated in Fig. 2. Each bit position of the MDAA includes a 13-input combinational counter. Unlike the conventional DAA, which implicitly uses the floating point precision of the multiplication constants, the MDAA assumes their fixed point representation. Each counter performs binary encoding of the number of logic "ones" for each bit position of the multiplication constants. For a sum of 9 products

$$S = \sum_{i=0}^{8} A_i \cdot x_i \qquad (1)$$

the constants A_i can be expressed as

$$A_i = \sum_{l=0}^{L-1} A_{i,1} \cdot 2^l \qquad (2)$$

and the input data as

$$x_i = \sum_{m=0}^{M-1} x_{i,m} \cdot 2^m \qquad (3)$$

where xi,m, Ai,1 belongs to {0,1}and L>=M

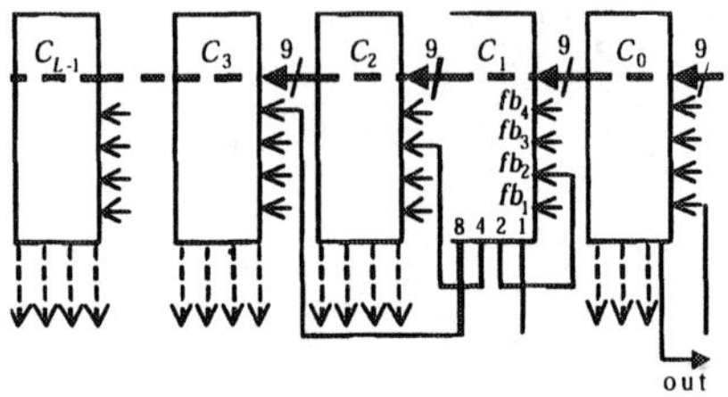

Figure 2. Modified DAA

The processing unit consists of L counters, one for each bit of A_i. Each counter is implemented as a collection of up to four OBDDL counter gates. The network size is reduced by customizing the counter at bit position l such as to count the input bit $x_{i,m}$ only if $A_{i,l} = 1$, whereas the remaining input bits are not used by the particular counter. The output C_l of the counter at the bit position l in the clock cycle m, can be expressed as

$$C_{l,m} = fb_1 + fb_2 + fb_3 + fb_4 + \sum_{\substack{0 \le i \le 8 \\ A_{i,l} = 1}} A_{i,l} \cdot x_{i,m} \tag{4}$$

Where f_{b1}, f_{b2}, f_{b3} and f_{b4} represent the feedback bits as illustrated in Fig. 2. In the physical imple mentation, the OBDDL gates are designed such as to provide a feedthrough path for the inputs multiplied by bit '0'. Further reduction in the counter size is achievable by scaling all constants A_i by a constant factor that maximizes the total number of zeros in their binary representation for all values of i. The correct handling of negative numbers for 2's complement binary representation is achieved by a weighted counter network at the highest bit position, which subtracts the direct inputs and adds the we ighted feedback inputs.

3. CIRCUIT DESIGN

Each counter output bit is driven by an adiabatic OBDDL gate. As shown in Fig. 3, an OBDDL gate consists of a differential NMOS transistor network implementing the particular logic function and a sense amplifier. The circuit is operated by two non-overlapping adiabatic clock signals V_{ck1} and V_{ck2}. The operation involves precharging of nodes F and FB and subsequent detection of a small differential voltage between these outputs [2].

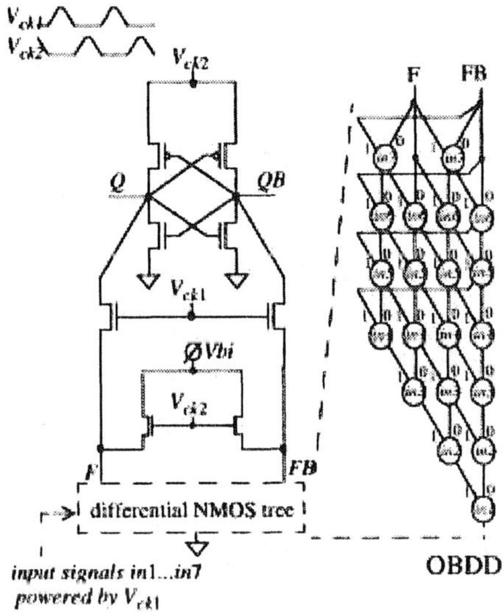

Figure 3. Structure of OBDDL gates

The topology of the NMOS tree corresponds to an ordered binary decision diagram (OBDD) [4] and performs the logic function of one of the counter outputs. The particular OBDD illustrated in Fig. 3, evaluates the middle output bit of a 7-input counter. For efficient energy recovery, adiabatic drivers [5] should be added at the gate outputs *Q* and *QB.*

4. DESIGN AUTOMATION

The complexity and variety of complex counter gates necessitates automatic generation and silicon compilation of the modified DAA. The set of CAD tools developed for this purpose includes two major parts: the OBDD network synthesizer and the layout generator. The former group of tools are implemented in MATLAB and the latter ones in SKILL. The flow chart of the overall MDAA generator algorithm is shown in Fig. 4.

The input for the MDAA generator is specified in the ASCII format and includes the constant coefficients A_i , the number of bits in the input data x_i and the number of output data bits. Where applicable, analytical specification of the coefficients, supported by MATLAB, is used. In the first algorithm step, the OBDD synthesis tool finds the number of '1's for each bit position of the constant coefficients and locates the feedthroughs for each counter. Subsequently, an OBDD netlist is generated for each output bit of each identified counter size, using Bryant's algorithm for OBDD synthesis [4]. In the final step of the synthesis part, separate OBDD netlists, in which feedthrough locations are indicated, are generated for each logic gate.

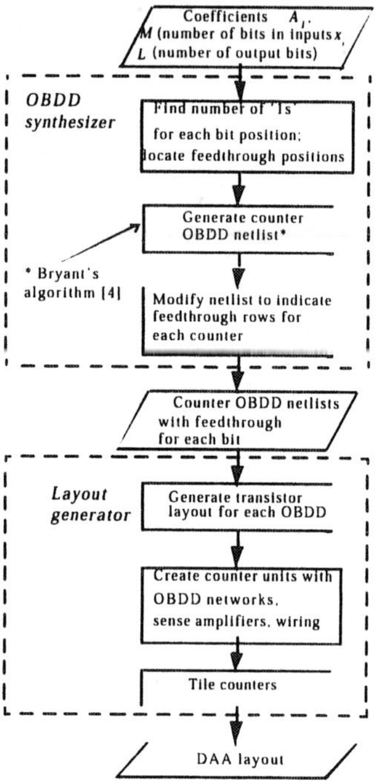

Figure 4. MDAA generation algorithm

Using the output from the synthesis tool, the layout generator creates OBDD-style networks with feedthroughs for each bit of each counter. The layout of one such network is shown in Fig. 5. Subsequent ly, counter layouts, including OBDD-style networks, sense amplifiers and interconnecting wires are generated. Finally, the individual counter layouts are tiled in an array to form the final MDAA layout.

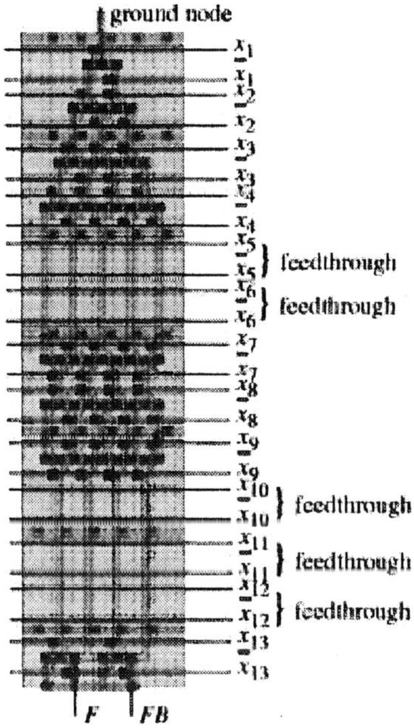

Figure 5. NMOS OBDD-style counter network layout

5. COMPARISON WITH CONVENTIONAL DAA

A number of adiabatic MDAA units with 8 and 9 inputs and 16 output bits have been implemented in a 0.25μm CMOS process and their operation was experimentally verified. The average energy consumption per computation for an adiabatic supply voltage of 1V and an operating frequency of 25MHz, is 29pJ for the 9-input DAA and 21pJ for the 8-input DAA. Whereas the operating frequency was limited by the experimental setup, SPICE simulations indicated that the maximum operating frequency for the implemented MDAA should be above 100MHz.

The described modified DAA, operated adiabatically, is significantly more energy efficient than an equivalent conventional, ROM-based DAA, also operated adiabatically. This results from the inherently high dissipation at ROM bit lines, which cannot be reduced by energy recovery, as compared to a smaller amount of irreducible dissipation associated with output detection in OBDDL counter gates in the OBDDL-based DAA.

For example, if the number of DAA inputs is 9 and the number of output bits is L, the described MDAA includes between $3L$ and $4L$ OBDDL gates, whereas the equivalent ROM-based DAA includes L bit lines exhibiting a total dissipation equal to the dissipation of 20 L average OBDDL gates, in the same 0.25μm CMOS process implementation. Such high dissipation in ROM is caused by a large total transistor drain junction capacitance attached to each bit line. Therefore, even without considering other sources of dissipation in the conventional DAA, it is clear that the proposed DAA modification reduces the energy dissipation in adiabatic operation, at least by a factor of five.

In addition, the MDAA implementation is significantly more compact than that of the conventional DAA. For a 9-input DAA, the area occupied by the ROM-based version is estimated to be at least four times greater than that occupied by the MDAA.

6. CONCLUSION

In this paper, a modified distributed arithmetic architecture, suitable for adiabatic implementation, was presented. A set of CAD tools, developed for silicon compilation of the proposed architecture, was described. The modified architecture allows a more efficient energy recovery than the conventional DAA, by eliminating the dissipation associated with ROM bit lines. Also, the new architecture occupies a significantly smaller area than the conventional one, suggesting that it may represent an attractive alternative to the conventional DAA, not only in adiabatic designs but in the conventional designs as well.

ACKNOWLEDGMENTS

The work was supported by NSERC, Micronet, Gennum, Mitel, Nortel and PMC Sierra.

REFERENCES

[1] A.S. White, "Applications of distributed arithmetic to digital signal processing: A tutorial review," *IEEE ASSP Magazine*, pp. 4-19, July 1989

[2] D. Suvakovic, C. A. T. Salama, "Energy-Efficient Arithmetic Architectures for Adiabatic DSP Systems", *accepted for publication in Journal on VLSI Signal Processing*

[3] E. E. Swartzlander, Jr., "Parallel Counters", *IEEE Transactions on Computers*, vol. c-22, pp. 1021-1024, 1973

[4] R. E. Bryant, "Graph-Based Algorithms for Boolean Function Manipulation", *IEEE Transactions on Computers*, vol. C-35, pp. 677-691, 1986

[5] W. Athas, L. J. Svensson, J. G. Koller, N. Tzartzanis, E. Chou, "Low-Power Digital Systems Based on Adiabatic Switching Principles", *IEEE Transaction on VLSI Systems*, vol. 2, pp. 398-407, 1994

Section 7.2

Design of a CMOS Wide Range Logarithmic Amplifier with a Modified Parallel Architecture

Sanghoon Joo, Minkyu Song, Hyunyeon Cho, and Sangki Kim
Dept. of Semiconductor Science, Dongguk University, Seoul, KOREA, RE Lab., Gamma-Nu Inc., Kyunggi-Do, KOREAE*

Abstract: In this paper, we discuss a VLSI design of a Logarithmic Amplifier (LA) for wide range and high sensitivity radar system. In general, LA consists of an input stage, a logarithmic stage, and an output stage. In order to make a much wider dynamic region and a higher speed than the conventional LA, a new mixed type of series and parallel architecture, namely a modified parallel architecture is proposed in the logarithmic stage. Further, to decrease an input range error in the front of LA, a novel input stage is designed. It is fabricated on the basis *of O.5um* standard CMOS technology. Effective chip area is 131 Oum x 1 *540um,* and shows the power consumption of 90mW at 3 .3V supply voltage. Through the simulation and measurements, it is verified that it shows the characteristics of 60dB dynamic range and 5Ons falling time

Key words: logarithmic amplifier modified parallel architecture, input stage, logarithmic stage, and dynamic range

1. INTRODUCTION

Logarithmic Amplifier (LA) converts the wide range signals into linear range signals. For example, as shown in Fig. 1, a 1 OmV peak-to-peak voltage signal is amplified into lVpp, a lOOmVpp into 2Vpp, and a lVpp into 3Vpp. When we handle a wide range signal like a radar system, the system covers wide range dynamic signal over 30dB. Thus, it is necessary to design a logarithmic amplifier that converts the wide range signals into linear range signals, when we make a radar system or those kinds of systems. Among the traditional logarithmic amplifiers *[1]-[5],* however, most of them is designed on the basis of Bipolar Junction Transistors, and have a narrow input signal range under 40dB. Nowadays, we need a wide input range logarithmic amplifier over 60dB, according to the development of new technology. Further, we also need a CMOS logarithmic amplifier, since most of the systems are integrated into CMOS one-chip solution.

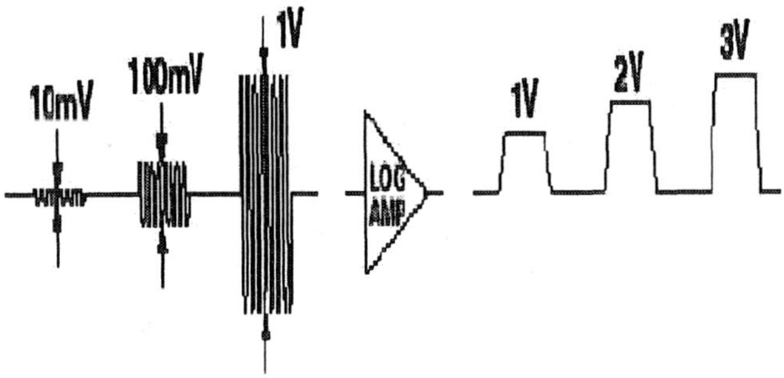

Figure1. Principle of a Logarithmic Amplifier

Therefore, a wide input range over 60dB is proposed in this paper with a CMOS technology. The contents of the paper are as follows. In Section II, we discuss the conventional logarithmic amplifiers and the proposed architecture. Then, the concrete circuit design is described in Section III. The experimental results and conclusions are shown in Section IV, and V.

2. ARCHITECURE

Fig.2 shows a conventional architecture of linear limiting logarithmic amplifier (LLLA) [1]-[3]. When a small power signal is driven at the input stage, the signal is amplified by certain times at the first amplifier. Then the amplified signal is also amplified at the second stage. However, the maximum output of each amplifier must be limited at certain voltage. After it continues until the last stage, each amplified signal is added at the adder block. Thus we obtain a logarithmic output of the small input signal at the output stage

In comparison with the linear limiting logarithmic amplifier, a linear summing logarithmic amplifier (LSLA) is shown in Fig.3 *[4][5]*. The input signal is driven at all the amplifiers. Then the output is added at the final stage. Compared to the linear limiting logarithmic amplifier, it has a small offset voltage. However, it is difficult to design many kinds of amplifiers.

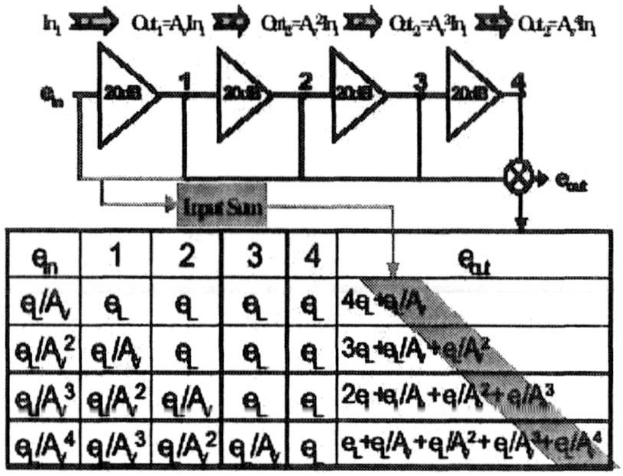

Figure 2 Liner Limiting Logarithmic Amplifier

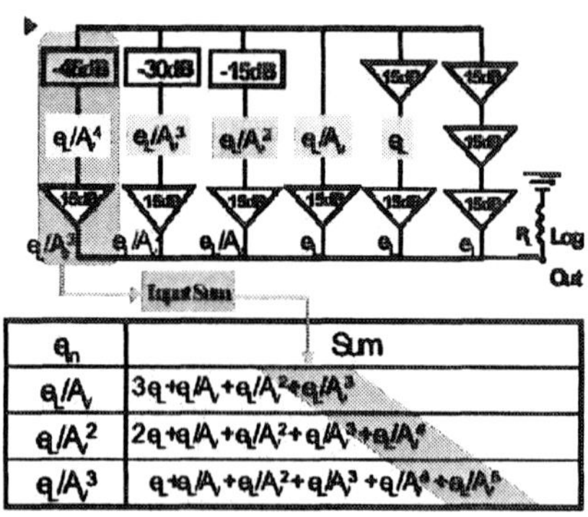

Figure 3 Linear Summing Logarithmic Amplifier

Fig.4 shows the conceptual block diagram of the proposed logarithmic amplifier. When we design a conventional parallel summation logarithmic amplifier, it is necessary to design many kinds of amplifier. If we need a wider dynamic range, the number of amplifiers is also increased. However, we can reduce the number of amplifiers, if we use the concept of linear limiting logarithmic amplifier. From the figure, we can reduce the number of amplifiers and advantage of lower offset voltage

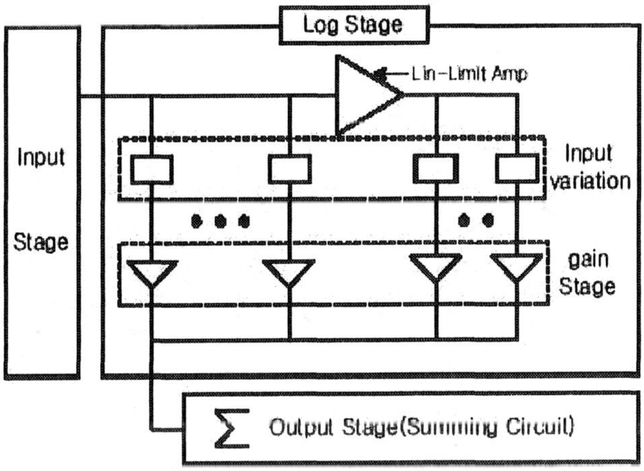

Figure 4 Modified Parallel Architecture

3. CIRCUIT DESIGN

Input stage

Fig.5 shows the transfer curve for detector diode located at the front of logarithmic amplifier. Since the transfer curve of the diode is not linear, it is necessary to design an input buffer that compensates the non-linearity. In general, an ordinary operation amplifier is sufficient to compensate the no linearity.

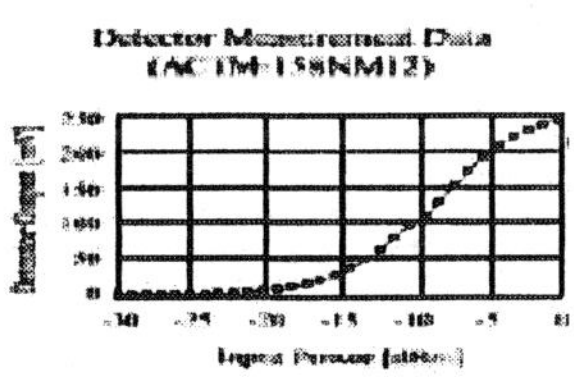

Figure 5 Transfer Curve for Detector Diode at Input Stage

Log Stage

In the middle of the proposed logarithmic amplifier, the most important part is the design of log stage and the distribution of input voltage. Fig.6 shows the block diagram of the log stage composed of several amplifiers. According to the design of amplifiers, the input voltage is amplified by 30dB, and transfers to the second stage. Fig.7 shows the circuit diagram of distribution technique.

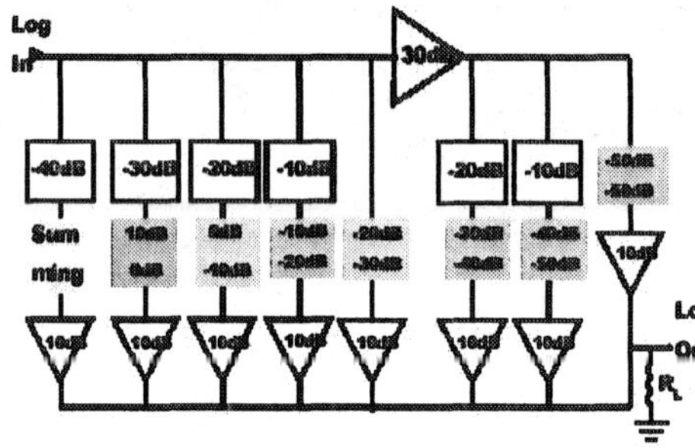

Figure 6 Block Diagram for Log Stage

4. EXPERIMENTAL RESULTS

Fig.8 shows the chip microphotograph of the proposed logarithmic amplifier. The proposed one has been fabricated with a 3-metal *0.Sum* CMOS technology and the chip area is about 1.31 xl *.54* mm^2 with 90mW power dissipation at 3.3V power supply. The measured waveforms and frequency res ponse are shown in Fig.9 (a) and (b), respectively. According to the input voltage, the output is measured as shown in Fig.9 (a). The 3dB frequency of the logarithmic amplifier is about 700kHz, as shown in (b).

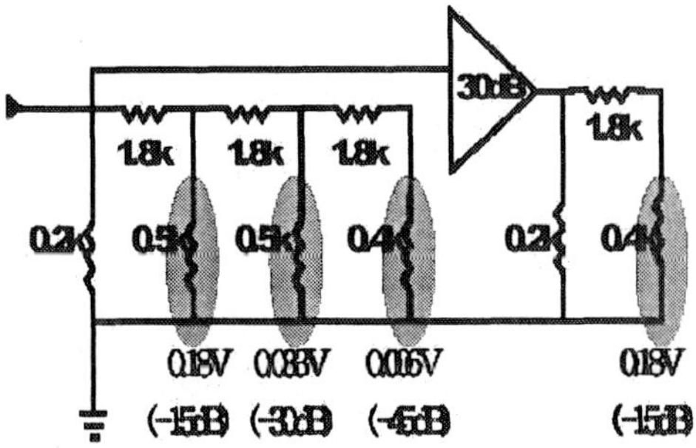

Figure 7 Distribution of Input Voltage by resistor

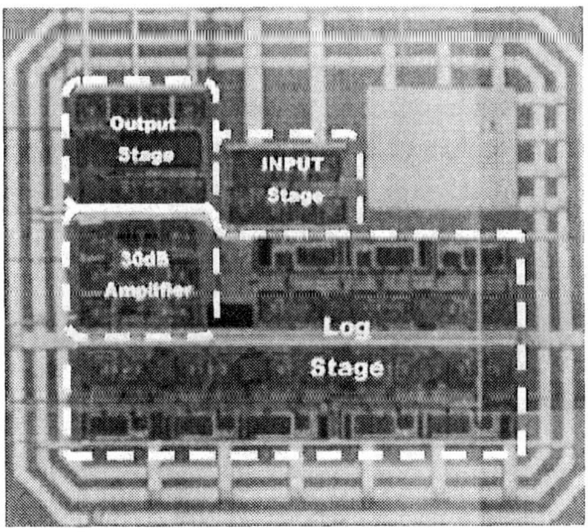

Figure 8 Chip Microphotograph of Logarithmic Amplifier

Fig. 10 shows the DC transfer curve of the proposed logarithmic amplifier. According to the input voltage, optimized output voltage is obtained. Fig. 11 shows pulse response of the proposed logarithmic amplifier. The falling time of the LA is about SOns. This is much faster than that of other ones, because the modified parallel architecture has a fast discharge time.

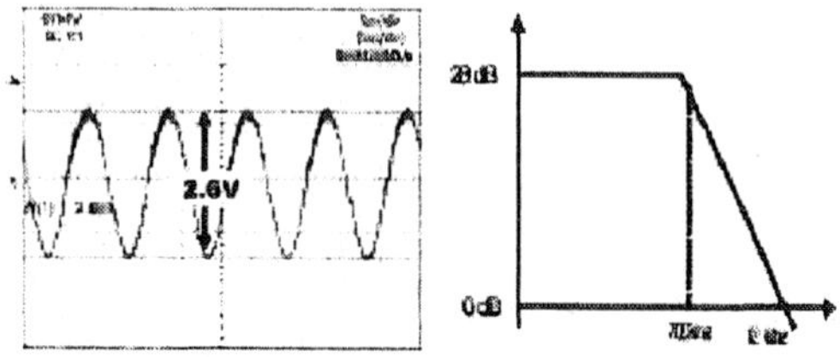

(a) (b)

Figure 9 Experimental Results
Measured Waveforms (b) Frequency Response

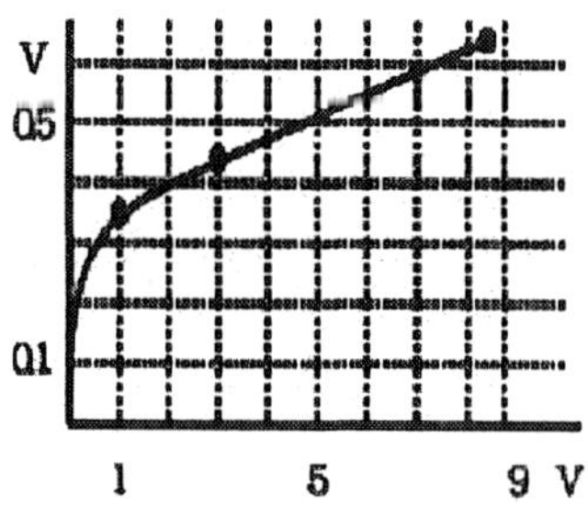

Figure 10 DC Transfer Curve of Logarithmic Amplifier

5. CONCLUSION

In this paper, we discussed a novel modified parallel architecture of a Logarithmic Amplifier for wide range and high sensitivity radar system. In order to make a much wider dynamic region and a higher speed than the conventional LA, a new mixed type of series and parallel architecture, namely a modified parallel architecture was proposed in the logarithmic stage. It has been fabricated on the basis of 0.Sum standard CMOS technology. Effective chip area was 131 Oum x 1 540um, and had the power consumption of 90mW at 3.3V supply voltage. Through the simulation and measurements, it was verified that it had the characteristics of 60dB dynamic range and SOns falling time. The experimental results are summarized in Tab 1

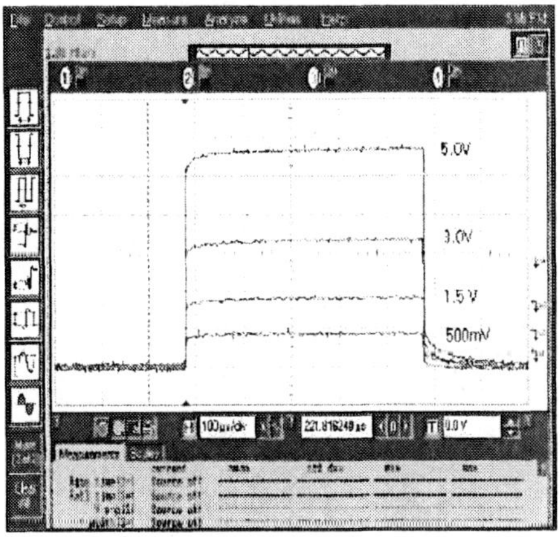

Figure 11 Measured Results for Pulse Response

Input Frequency	0.5 – 150 MHz
Log Linearity	± 0.2 dB
Power Supply	3.3V
Technology	0.5um n-well CMOS
Dynamic Range	60dB
Recovery Time	50 ns
Power Consumption	90mW
Effective Chip Area	1310um x 1540 um

Table 1 Summary of the Measured Results

REFERENCES

[1]Richard Smith Hughes "Logarithmic Amplification with Application to Radar and EW", Artech House,1986

[2] N. Scheinberg and R. Michels, "A Monolithic GaAs Low Power L Band Successive Detection Logarithmic Amplifier", IEEE Journal of Solid-State Circuit., vol 29, pp 151-154, Feb.1994

[3]MA-COM mc, Logarithmic IF Amplifiers, MA-COM mc, Application527, pp 2-134 ---2-137, 1994

[4]Stephen E. Lipsky, "Log Amps Improve Wideband Direction Finding:, Microwaves, pp. *58-65* . May 1973

[5] L.W. Chua, "A GaAs MMIC for A 2 7 GHz Successive Detection Logarithmic Amplifier", IEEE Microwave and Millimeter Wave Monolithic Circuit Symposium Digest, pp.145-148, 1992

Section 7.3

Digital Hardware Implementation OF Continuous And Discrete Chaotic Generators

Mohamed I. Sobhy, Mohammed A. Aseeri and Alaa E. R. Shehata.
Department of Electronics, The University of Kent at Canterbury, Canterbury, Kent, CT2 7NT, UK.

Abstract: The paper describes how chaotic generators can be implemented using digital hardware. A continuous chaotic generator based on the Rössler model and a discrete generator based on the Henon map models are described. A complete implementation procedure is given and the results from the simulated and hardware generators are compared. The developed hardware can be used directly in chaotic communication systems.

1. INTRODUCTION

Analogue chaotic generators have been used for communication systems [1]-[6]. Recovery of the information signal depends on how well the receiver is synchronised with the transmitter. This requires that the parameters of both receiver and transmitter be matched to a high degree of accuracy. This requirement is difficult to achieve in analogue systems especially that the values of analogue circuit component are functions of age and temperature. The most obvious solution is to implement the generators using digital hardware. The generators are first represented by a set of non-linear equations and a system-based model is developed to represent these equations directly. The algorithm is then converted to a real time version, which is linked to an I/O card. This way we can adjust the generator parameters to a very high accuracy in order to achieve good synchronization. The method can be applied to continuous or discrete chaotic generators described by state equations even if there is no electronic circuit representation for the system. All the required procedures to achieve this, are toolboxes within MATLAB [7]. The limitation on the frequency of the output is mainly due to the speed of the I/O card.

2. CONTINUOUS CHAOTIC GENERATORS

The steps of the real time design are as follows:

- First a system model for the state equations is developed using MATLAB Simulink Software.
- The model is then simulated to check the output and adjust the required frequency.
- The real time program is created and linked to the I/O card using the Real Time Windows Target.

The state equations of the Rössler system was used as example of a continuous chaotic generator. The equations are given by [8]

$$\begin{aligned} \dot{x} &= -y - z \\ \dot{y} &= x + Ay \\ \dot{z} &= B + z(x - C) \end{aligned} \qquad (1)$$

where A, B and C are constants and A=0.398, B=2 and C=4. The real time Rössler model is shown in Figure 1.

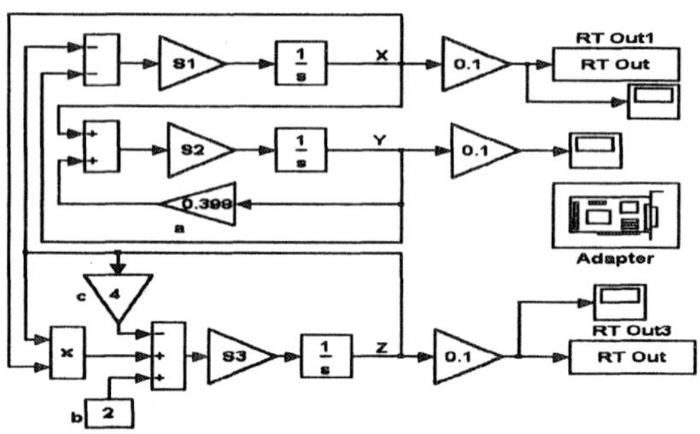

Figure 1 Real time model of the Rössler generator.

The scaling factors S_1, S_2 and S_3 are used before every integrator in the model to control the frequency. Figures 2 and 3 show the results of the simulation and the x-z attractor of the model is shown in Figure 4.

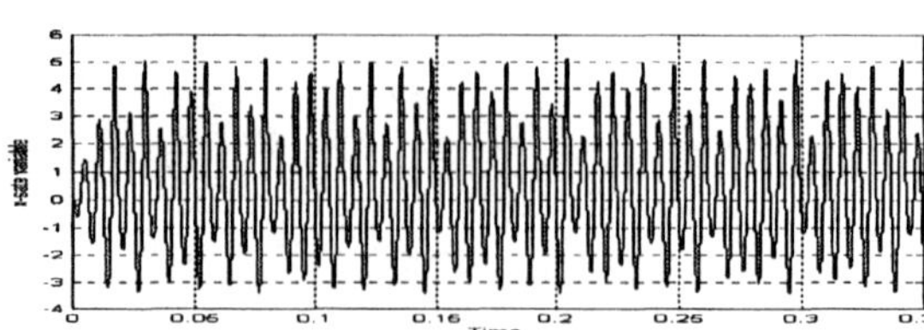

Figure 2 The x-state variable output from simulation model.

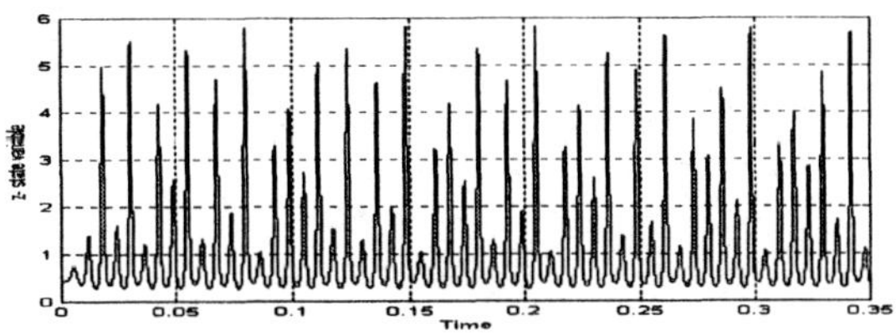

Figure 3 The *z*-state variable output from simulation model.

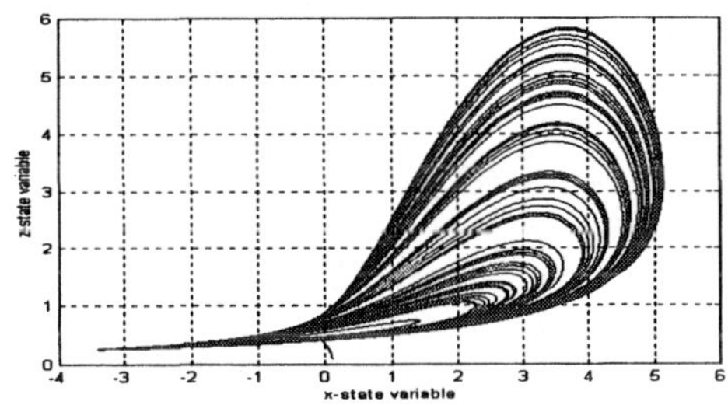

Figure 4 The x-y Attractor of the simulation.

In the simulation model, we choose a clock time step of $5x10^{10}$, and the scaling factors (S_1, S_2 and S_3) were $10x10^7$ to achieve an equivalent maximum frequency of 100 MHz. The spectrum of the output signal is shown in Figure 5.

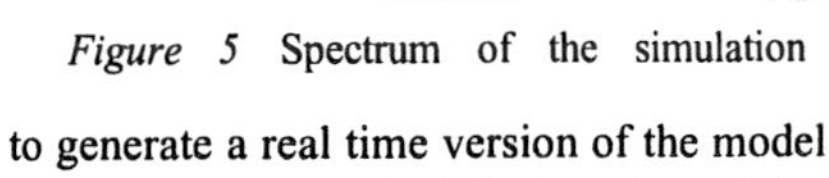

Figure 5 Spectrum of the simulation

The next step is to generate a real time version of the model shown in Figure 1, which is also interfaced with an I/O card. This is achieved through the Real Time Windows Target (rtwintgt)', which is available in MATLAB. The model allows the control of the frequency and amplitude of the chaotic signal. The model parameters are now determined to an accuracy of 10^{15}. The Measured results of the real time model are shown in Figures 6 and 7 and Figure 8 shows the real time attractor for the Rössler Model.

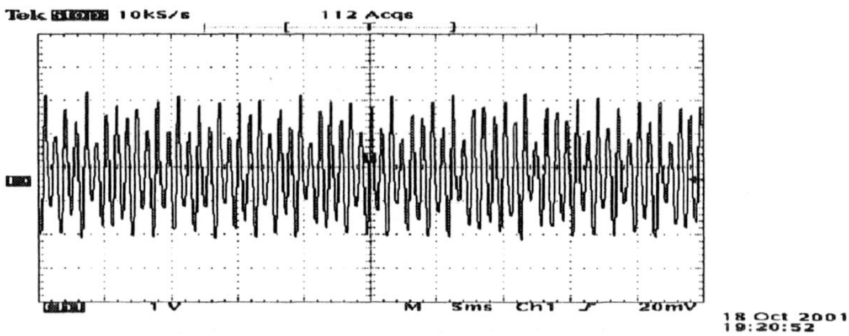

Figure 6 The Measured x-state variable in real time.

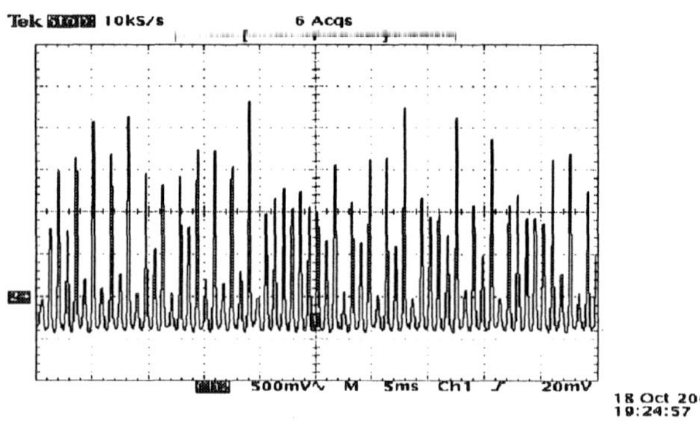

Figure 7 The measured z-state variable in real time.

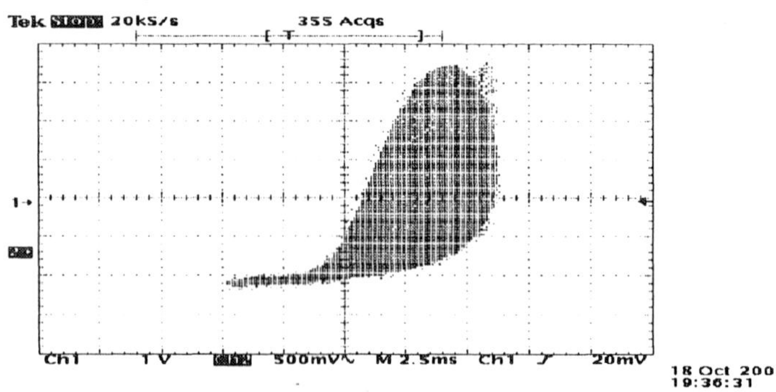

Figure 8 The real time attractor of Rössler model.

To achieve the maximum frequency when using I/O Lab-PC-1200 card we decrease the step size to $4x10^{-6}$ and increase the scaling factor to $8x10^{3}$. The maximum frequency that can be achieved is 12.5 KHz as shown in Figure 9. The results show very good agreement between the performance of the simulated model and the real time implementation, except for the absolute value of the output frequency, which is scaled to the capabilities of the I/O card.

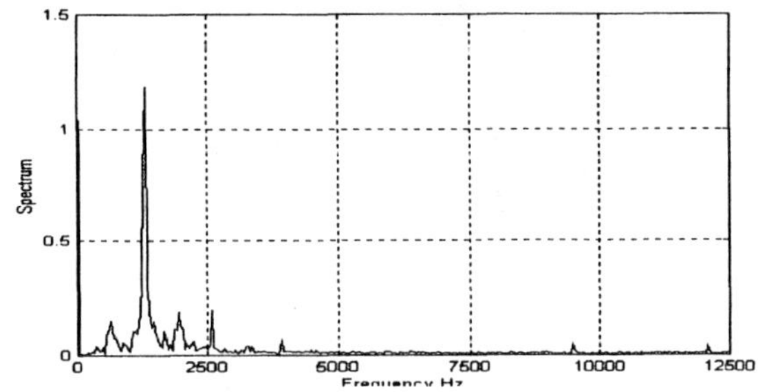

Figure 9 Spectrum of the real time system.

3. DISCRETE CHAOTIC GENERATOR

The state equations of the Henon map was used as a discrete chaotic generator and are given by [9]

$$\begin{aligned} x_{n+1} &= 1 + y_n - a x_n^2 \\ y_{n+1} &= b x_n \end{aligned} \tag{2}$$

where a and b are constants and a=-1.4 and b=0.3. The system model converts the state equations to a Simulink model and the output frequency is controlled by the step size. Figure 10 shows the real time model of the Henon map chaotic generator.

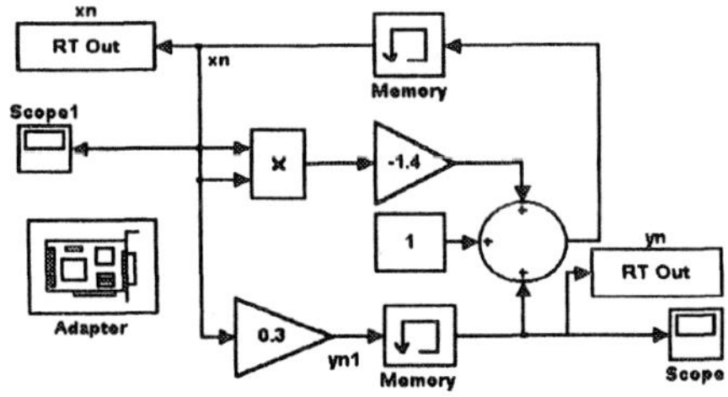

Figure 10 Real time model of Henon map generator

In this example we choose the clock time step $dt = 0.5x10^{10}$, to achieve an equivalent maximum frequency of 1 GHz. The output results of the simulation model are shown in Figures 11 and 12. Figure 13 shows the attractor of the simulation model. The spectrum of the output signal x_n state variable is shown in Figure 14.

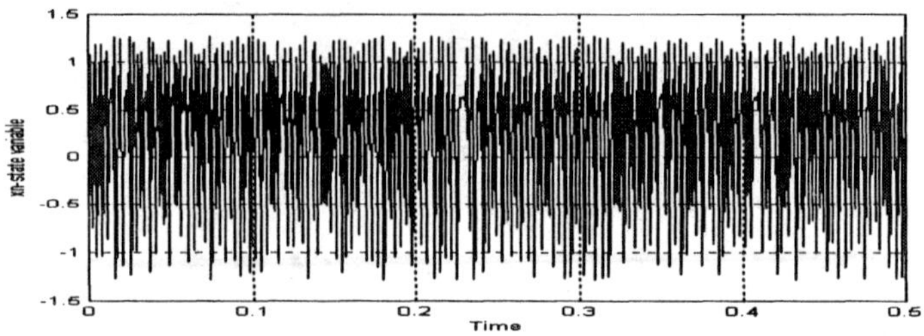

Figure 11 The xn state variable from the simulation

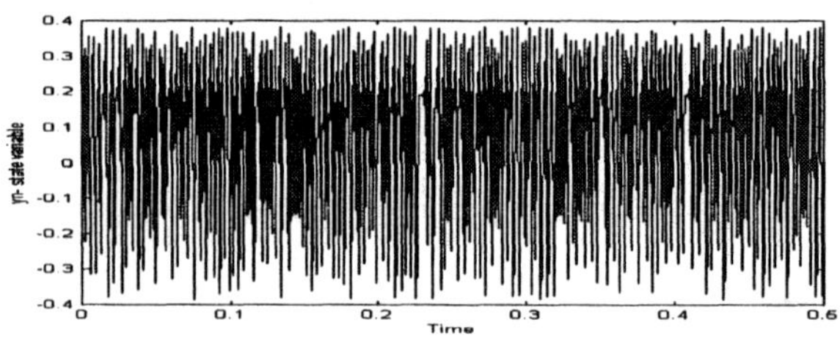

Figure 12 The y_n state variable from the simulation

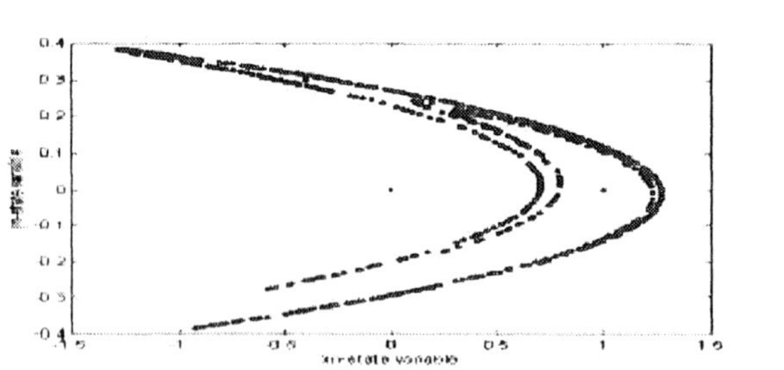

Figure 13 The attractor of the Henon map simulation.

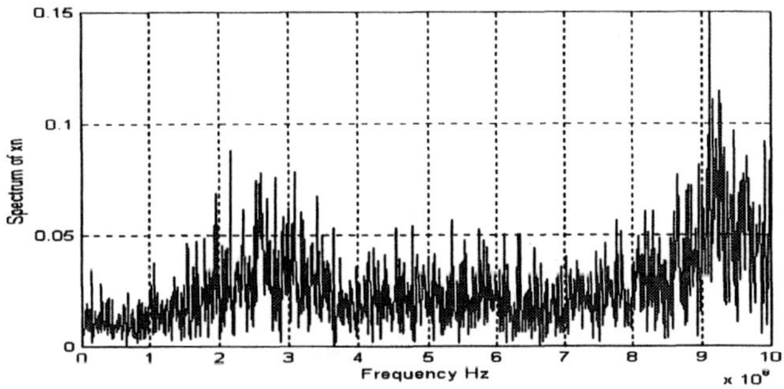

Figure 14 The spectrum of the Henon map simulation.

Similar to the Rössler chaotic generator, the next step is to generate a real time version of the model shown in Figure 10, which is also interfaced with an I/O card. The Measured results of the real time mode are shown in Figures 15 and 16.

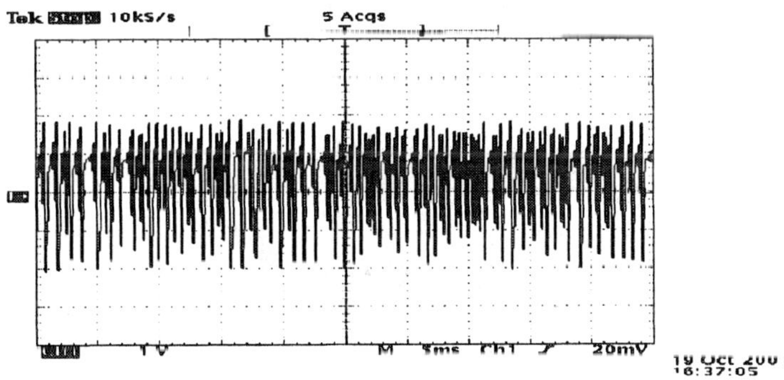

Figure 15 The x_n state variable in real time.

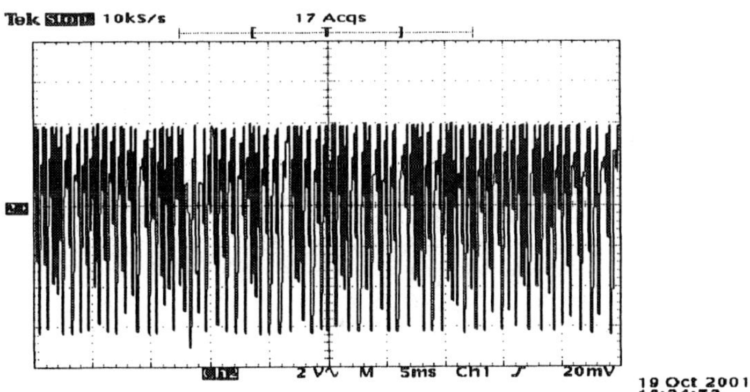

Figure 16 The y_n state variable in real time.

Figure 17 shows the x_n-y_n attractor of the real time generator using a Tektronix TDS 360 oscilloscope. The maximum frequency is 20 KHz when using a Lab-PC-1200 I/O card when the step size is 10×10^{-5}. The spectrum of the output is shown in Figure 18.

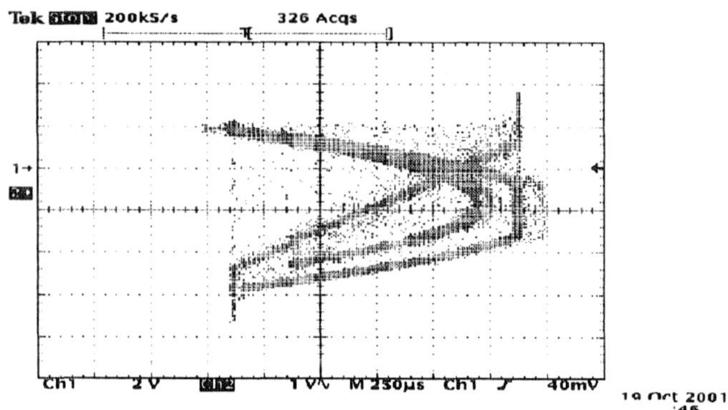

Figure 17 The real time attractor of the Henon map.

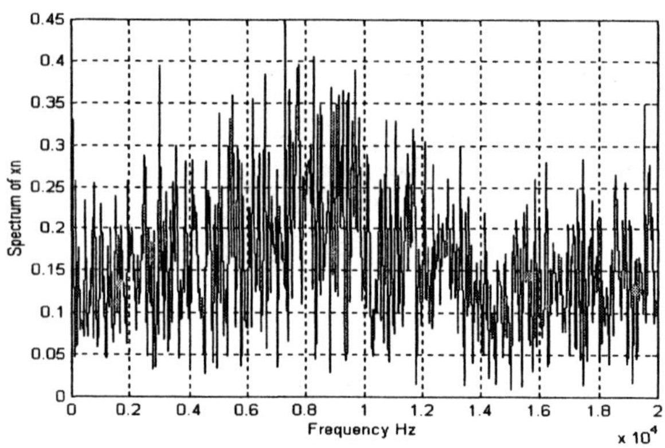

Figure 18 The real time spectrum of the Henon map.

4. CONCLUSION

A method to design chaotic generator models in real time is developed which is capable of implementing the chaotic systems that are given by state equations in real time. The method is implemented by MATLAB, SIMULINK, real time workshop, real time target window and the data acquisition card Lab PC 1200. A scaling factor to control the frequency band is introduced. The method is useful for the implementation of chaotic generators at low frequencies because the limitation of the sampling rate for the data acquisition card Lap PC 1200. Both continuous and discrete chaotic generators can be implemented even if the system cannot be represented by a physical electronic circuit. Modification of any system is a simple change in the block diagram or the parameter values within the block.

REFERENCES

[1] C. A. Murphy and M. P. Kennedy, "Chaos controller for autonomous circuits," *Proc. 3 rd Int. Specialist Workshop on Nonlinear Dynamics of Electronic Systems NDES'95*, pp.225-228, 1995.

[2] M. Storace, M. Parodi and D. Robatto, "A hysteresis-based chaotic circuit: Dynamics and applications, " *Int. J. of Circuit. Theory appl.*, pp. 527-542, 1999.

[3] A. S. Elwakil and M. P. Kennedy, "Chaotic oscillator configuration using a frequency dependent negative resistor," *Int. J. of Circuit. Theory appl.*, pp. 69-76, 2000.

[4] N. J. Carron and D. W. Hahs, "A new approach to communications using chaotic signals," *IEEE Trans. Circuits Systems. I*, vol. CAS-44, No. 5, pp. 373-382, Oct. 1993.

[5] H. Kamata, T. Endo and Y. Ishida, "Secure communication system using chaos via DSP implementation," *IEEE Int. Symposium. On Circuit and System (ISCAS 95)*, pp. 112-115, 1995.

[6] H. Kamata, T. Endo and Y. Ishida, "Communication with chaos via DSP implementation," *IEEE Int. Symposium. on Circuit and System ISCAS 97*, pp. 1069-1072, 1997.

[7] P. Marchand, *Graphics and GUIs with MATLAB*: CRC Press, 1999.

[8] T. Yoahinaga, H. Kitajima and H. Kawakami, "Bifurcations in a coupled Rössler system," *IEICE Trans. Fundamentals*, vol. E78-A, No. 10, Oct. 1995.

[9] J. M. H. Elmirghani, "Data transmission through chaotic perturbation and associated security issues," *SPIE special issue on Chaotic Circuits for Communication*, vol. 2612, pp.76-85, 1994.

Section 7.4

Novel 1-bit Full Adder Cells for Low-Power System-On-Chip Applications

Mohammed Sayed and Wael Badawy
Department of Electrical and Computer Engineering
University of Calgary, Calgary, Alberta, Canada T2N 1N4

Abstract: This paper presents three novel 1-bit full adder cells. The performance of the proposed cells is analyzed in terms of power consumption and delay for 0.18 and 0.35 μm CMOS technologies. The performance analysis shows that the proposed adder cells outperform the commonly used adder cells in terms of power consumption and delay components.

Key words: System-on-chip, low power, adder cells.

1. INTRODUCTION

The computationally intensive nature of the real time applications emerges the need for more lower power high performance systems. Using low power block is one of the promising directions to reduce the power consumption. The increase of the circuit density in System-On-Chip applications demands the use of low power building blocks so that the System-On-Chip platform consumes lower power.

Addition operation is the nucleus of all the computation units. It highlights the importance of the 1-bit full adder cell as a building block. Adder cells are used to determine the throughput of a processor's executive unit, floating-point unit, and memory address generation [1]. The 1-bit full adder cell can be implemented in many different logic structures such as Complementary Pass-Transistor Logic (CPL) [2], Dual Pass-Transistor Logic (DPL) [3], Complementary Pass Transistor-Transmission Gate (CPL-TG) [4], or Transmission Gate logic (TG) plus variant [5].

This paper presents three novel adder cells. The cells are prototyped and simulated using 0.35 and 0.18 ìm CMOS technologies. Their performance is compared in terms of power and delay with six different 1-bit full adder cells, which are commonly used CMOS designs for a 1-

bit full adder cell [6]. The performance analysis was done using 0.35 and 0.18 ìm CMOS technology.

The rest of the paper is organized as follows: Section 2 reviews the designs of the 1-bit full adder cells; Section 3 introduces the proposed 1-bit full adder cells; Section 4 discusses the performance analysis of the proposed cells in terms of power consumption and delay parameters; and finally; Section 5 concludes this paper.

2. THE 1-BIT FULL ADDER STRUCTURES

A 1-bit full adder is a logic circuit that has three inputs (A, B, and Cin) and two outputs (Sum and Cout). The adder functionality can be described as in equation (1):

$$Sum = (A \oplus B) \oplus Cin$$
$$Cout = A \bullet B + Cin \bullet (A \oplus B)$$
Equation (1)

Figure 1 shows a direct implementation of 1 bit full adder in CMOS. It will be called "Conventional CMOS Design" (C. CMOS) in the rest of this paper. The C. CMOS is a direct implementation of equation (1).

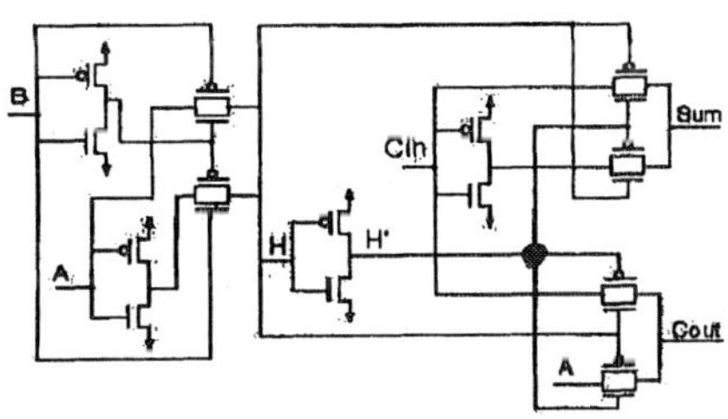

Figure 1. Conventional CMOS (C. CMOS) 1-bit full adder cell

Let $H = A \oplus B$ (H is called the half sum of A and B [2]), Equation (1) can be rewritten as follows:

$$Sum = H \oplus Cin = \overline{H} \bullet Cin + H \bullet \overline{Cin}$$
$$Cout = A \bullet \overline{H} + Cin \bullet H$$
Equation (2)

Equation (2) yields a new equation to construct a 1-bit full adder. The XOR function becomes a building block in various designs of these 1-bit full adders [7], some of these designs are shown in Figure 2.

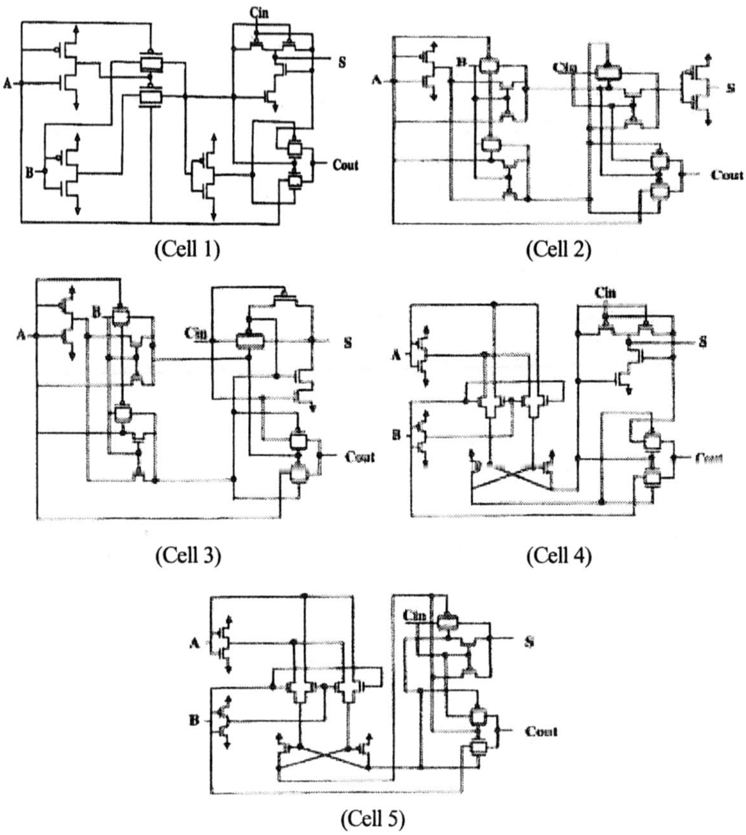

Figure 2. 1-bit full adder cells

3. THE NOVEL 1-BIT FULL ADDERS

Equation (2) shows that the half-sum and the inverted half-sum blocks are the main blocks to calculate Sum and Cout. A new circuit that combine XOR and XNOR functions using pass logic implementation has been introduced in [8].

The main idea in the new XOR and XNOR circuit is that referring to the XOR and XNOR k-maps in Figures 3a and b respectively, it may be noted that the XOR can be represented by three pass implicants ($\bar{A}$(B), $\bar{B}$(A), and AB(0)). Similarly, the XNOR can be represented by three pass implicants (A(B), B(A), and $\bar{A}\bar{B}$(1)).

The first and the second pass implicants in both of the XOR and XNOR function can be implemented by the circuits shown in Figures 3c and d respectively. Moreover, when A=B=LOW, the cross-coupled pMOS transistors circuit produces a logic LOW at its output as shown in Figure 3c and when A=B=HIGH, the cross-coupled nMOS transistors circuit produces a logic HIGH at its output as shown in Figure 3d. The circuit in Figure 3e is constructed to generate the LOW output, corresponding to A=B=HIGH, for the XOR function and to generate the HIGH output, corresponding to A=B=LOW, for the XNOR function. The output of the cross-coupled section in Figure 3c is used as gating signal to pass logic HIGH to the output of the cross-coupled section in Figures 3d and the output of the cross-coupled section in Figure 3d is used as gating signal to pass logic LOW to the output of the cross-coupled section in Figures 3d as shown in Figure 3e. This configuration reduces the number of transistors needed to implement XOR and XNOR circuit [8].

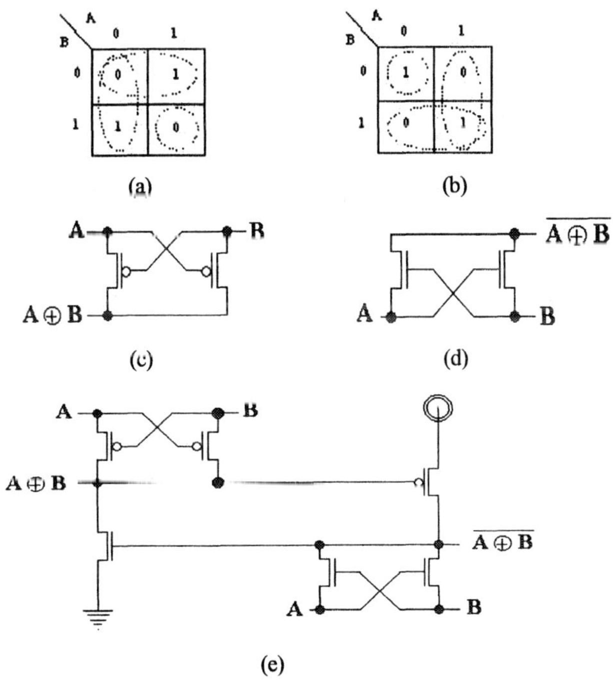

Figure 3. The new XOR and XNOR circuit

The novel adder cells are designed based on this new XOR/XNOR design along with three different circuitries needed to produce the required outputs as shown in Figure 4.

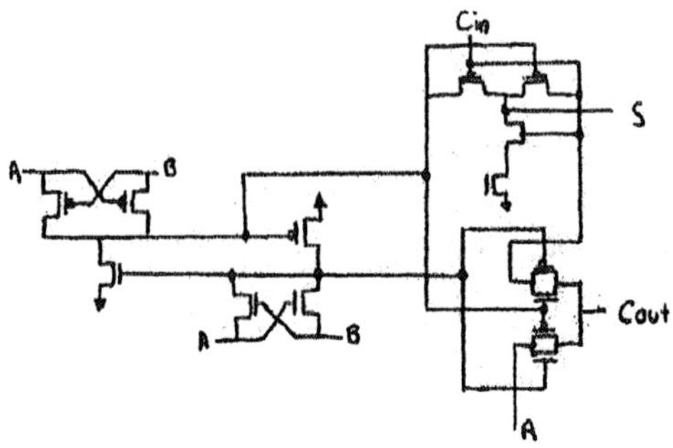

(N-Cell 1)

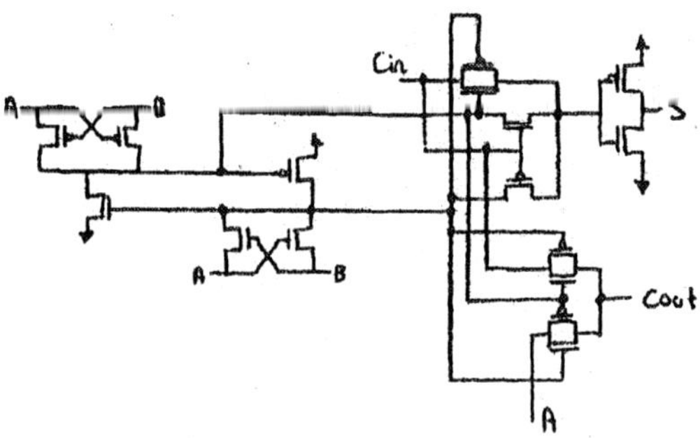

(N-Cell 2)

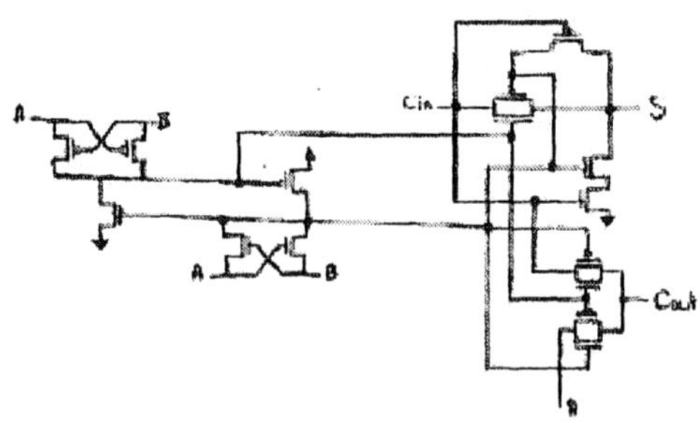

(N-Cell 3)

Figure 4. The three proposed 1-bit full adder cells

4. THE PERFORMANCE ANALYSIS

The performance of the proposed adder cells with six different adder cells is analyzed in terms of power consumption and delay. All adder cells are shown in Figures 1, 2, and 4.

The 1-bit full adder has three inputs (A, B, and Cin). Each input has four transitions ('LOW' to 'LOW', 'LOW' to 'HIGH', 'HIGH' to 'HIGH', and 'HIGH' to 'LOW'). Therefore, there are a maximum of sixty-four possible transitions. A pattern of sixty-four different transitions is being used to test the circuits as shown in Figure 5. A power-delay component is used to project the impact of the power consumption and delay of a 1-bit full adder cells.

The adder cells are tested at the minimum size feature that can generate the correct output signal. An inverter has been used to load the output of each cell (i.e. Sum and Cin). This inverter will help deciding on the smallest size of the transistors that will be able to generate the proper output for a given input signal. To simulate the output of cascaded circuit blocks a buffer is used to generate the input signals. The performance of the adder cells is measured without the additional inverters and buffers.

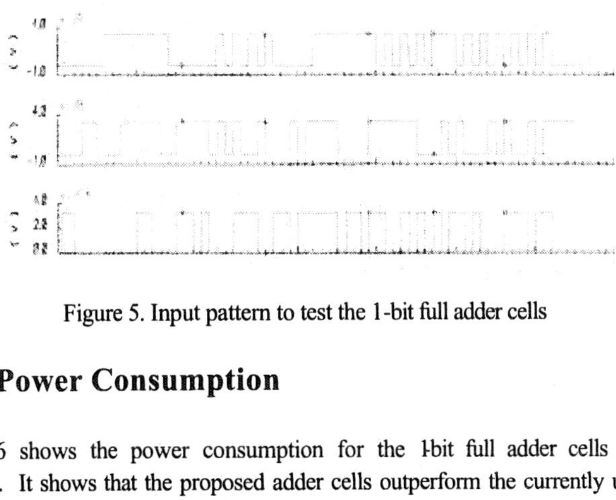

Figure 5. Input pattern to test the 1-bit full adder cells

4.1 Power Consumption

Figure 6 shows the power consumption for the 1-bit full adder cells in different technologies. It shows that the proposed adder cells outperform the currently used cells in 0.18 and 0.35 ìm technologies. Table 1 includes the power consumption numerical values for the 1-bit full adder cells in ascending order (for each technology). Note that voltage was reduced to its normal level in the 0.18 ìm technology, which is 1.8 V.

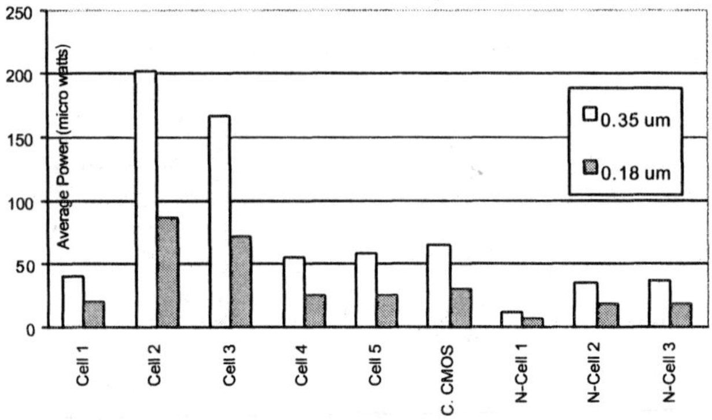

Figure 6: Power consumption of the 1-bit full adder cells using different technologies

Power(uW)			
0.18 um Tech.		**0.35 um Tech.**	
N-Cell 1	5.942	N-Cell 1	12.1
N-Cell 2	18.02	N-Cell 2	35.07
N-Cell 3	19.06	N-Cell 3	37.39
Cell 1	19.2	Cell 1	40.56
Cell 4	24.34	Cell 4	54.49
Cell 5	25.75	Cell 5	57.93
C. CMOS	29.65	C. CMOS	64.97
Cell 3	70.93	Cell 3	166.3
Cell 2	86.93	Cell 2	201.7

Table 1. The power consumption of the 1-bit full adder cells in ascending order

4.2 Delay

Figure 7 indicates the maximum delay of the 1-bit full adder cells. To compute the delay, the critical path is used to measure the maximum delay for the nodes to switch and generate the output. The longest delay is measured when the signal Sum became stable since Sum represents the longest critical path for the adder cells.

It can be noted from Figure 7 that N-Cell 1 outperforms the other cells in 0.35 ìm technology. But N-Cell 2 and N-Cell 3 have less delay than N-Cell 1 in the 0.18 ìm technology. Table 2 includes the maximum delay numerical values for the adder cells in ascending order (for each technology).

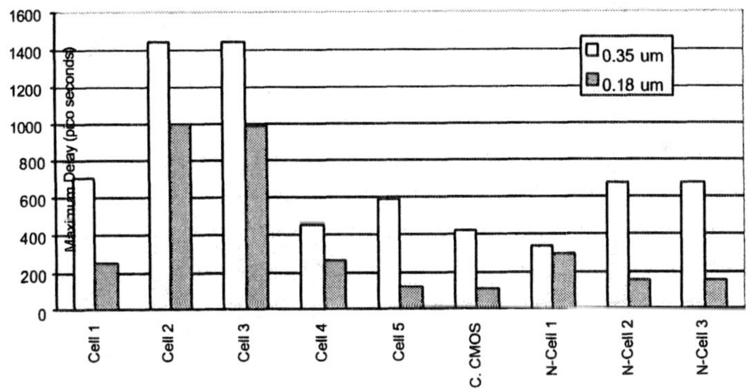

Figure 7. Maximum delay bars of the 1-bit full adder cells using different technologies

Delay(pS)			
0.18 um Tech.		**0.35 um Tech.**	
C. CMOS	110.887	N-Cell 1	327.044
Cell 5	117.903	C. CMOS	412.04
N-Cell 2	154.1	Cell 4	446.384
N-Cell 3	154.431	Cell 5	587.463
Cell 1	251.127	N-Cell 3	665.082
Cell 4	258.2	N-Cell 2	670.103
N-Cell 1	284.97	Cell 1	702.278
Cell 3	986	Cell 2	1445.53
Cell 2	997.59	Cell 3	1446.95

Table 2. The maximum delay for the 1-bit full adder cells in ascending order

4.3 Power-Delay Product

Power and delays are usually trade-offs, meaning that the fastest circuit may not be the most efficient power user. Therefore, power-delay parameter simply combines the two by multiplying power consumption value by delay value of the same design. Figure 8 compares the proposed cells with the other cells.

Table 3 includes the power-delay product for the adder cells in ascending order (for each technology). It shows that N-Cell 1 and N-Cell 2 outperform the other adder cells in all technologies.

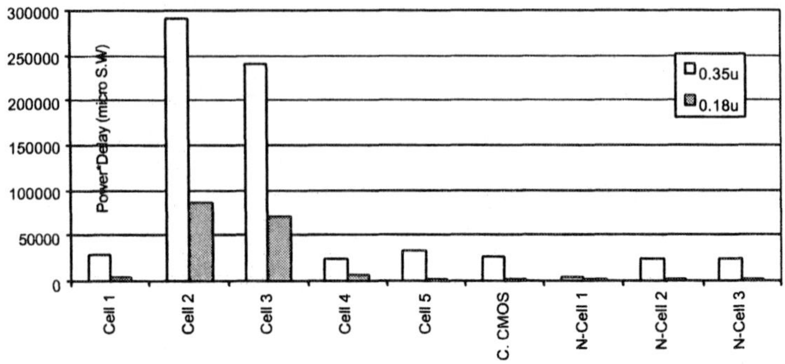

Figure 8. Power-delay product of the 1-bit full adder cells using different technologies

Power-Delay product			
0.18 um Tech.		**0.35 um Tech.**	
N-Cell 1	1693.2917	N-Cell 1	3957.2324
N-Cell 2	2776.882	N-Cell 2	23500.5122
N-Cell 3	2943.4549	Cell 4	24323.464
Cell 5	3036.0023	N-Cell 3	24867.416
C. CMOS	3287.7996	C. CMOS	26770.238
Cell 1	4821.6384	Cell 1	28484.3957
Cell 4	6284.588	Cell 5	34031.731
Cell 3	69936.98	Cell 3	240627.785
Cell 2	86720.499	Cell 2	291563.401

Table 3. The power-delay product for the 1-bit full adder cells in ascending order

5. CONCLUSION

This paper presents three novel 1-bit full adder cells and the performance analysis for these cells in terms of power consumption and delay. The cells are prototyped, simulated and tested in two different CMOS technologies 0.18 and 0.35 μm. The performance analysis of the proposed cells and six commonly used 1-bit full adder cells shows that the proposed cells outperform the other cells in power consumption and in power-delay product in all technologies.

6. ACKNOWLEDGEMENT

The authors would like to thank Natural Sciences and Engineering Research Council of Canada (NSERC), Canadian Microelectronics Corporation (CMC), Micronet, Human

Resource Canada (HRC), University Research Grant Committee (UGRC) and Department of Electrical Engineering at University of Calgary for supporting this research.

Moreover, The authors would like to thank Prof. Graham Jullien for his constructed remarks on the performance analysis, Mr. Arash Shoarinejad and Ms. Sue Ann Ung for their help in simulating the adder cells.

7. REFERENCES

[1] Uming Ko, Poras T. Balsara, and Wai Lee, "Low-power design techniques for high-performance CMOS adders," IEEE Transactions on Very Large Scale Integration Systems, vol. 3, no. 2 June 1995, pp. 327-333.

[2] K. Yano, T. Yamanaka, T. Nishida, M. Saito, K. Shimohigashi, and A. Shimizu, "A 3.8-ns CMOS 16x16-bit multiplier using complementary pass-transistor logic," IEEE Journal of Solid-State Circuits, vol. 25, 1990, pp.388-395.

[3] M. Suzuki, K. Shinbo, T. Yamanaka, A. Shimizu, K. Sasaki, and Y. Nakagome, "A 1.5-ns 32b CMOS ALU in dual pass-transistor logic," IEEE Journal of Solid-State Circuits, vol. 28, 1993, pp. 1145-1151.

[4] Issam S. Abu-Khater, Abdellatif Bellaouar, and M. I. Elmasry, "Circuits techniques for CMOS low-power high-performance multipliers," IEEE Journal of Solid-State Circuits, vol. 31, no. 10, October 1996, pp. 1535-1546.

[5] Nan Zhuang and Haomin Wu, "A new design of the CMOS full Adder," IEEE Journal of Solid-State Circuits, vol. 27, no. 5, May 1992, pp. 840-844.

[6] M. Shams, "High speed and low-power architectures and circuit techniques for a hybrid mesh-based/block-based video coder," Ph.D. dissertation, University of Louisiana, Louisiana Lafayette, LA, USA, spring 2000.

[7] J. Wang, S. Fang, and W. Feng, "New efficient designs for XOR and XNOR functions on the transistor level," IEEE Journal of Solid-State Circuits, Vol. 29, No. 7, July 1994.

[8] D. Radhakrishnan, "Low-voltage low-power CMOS full adder," IEE Proceedings Circuits, Devices, and Systems, Vol. 148, No. 1, February 2001.

8

LOW POWER

SoC technologies produce very dense circuitry that requires careful attention to power dissipation. Along with low-power are attendant problems with interconnect networks and other dense technology issues. In this chapter we present three sections that deal with current research work in the area of low-power and interconnect.

In the first section, "A New Logic Design Method for Considering Low Power and High Testability" by Yoon-Sik Son and Jong-Wha Chong of Hanyang University, Seoul, Korea. In this work, a new logic transformation method to achieve both low power consumption and high testability is proposed. The method is based on the redundancy insertion approach, and uses a structure of redundant connections that operate as test points in the test mode. Under the condition that a FFR of the circuit is unobservable at all primary outputs, the proposed method reduces the switching activity of the circuit while maintaining low area overhead. The effects that the logic transformation method has on the testability of the logic circuit are described, and the structure of redundant connections, to improve the testability of the circuit, is described. From the experiments on the MCNC benchmark circuits, the results show that the proposed method simultaneously improves the power consumption and testability of circuits.

In the second section, "System Synthesis for Optically-Connected, Multiprocessors On-Chip", by Neal K. Bambha and Shuvra S. Bhattacharyya of the University of Maryland, the authors discuss the use of optical techniques as a possible solution to the problems of scaling in VLSI interconnects. Along with enabling higher speed interconnects, optical connections allow the construction of highly connected and irregular networks that are streamlined for particular applications. Using these networks, it is possible to implement application mappings that allow flexible, single-hop commu nication patterns between processors. This has advantages for reduced system latency and power. Such optically connected multiprocessors are particularly promising for embedded digital signal processing (DSP) applications, which are highly parallel, and typically have tight constraints on latency and power consump tion. This work addresses novel trade-offs involving communication routing flexibility, power consumption, and performance that arise in the context of system synthesis of optically -interconnected multiprocessors. The authors report on experimental results that expose these trade-offs, and propose systematic techniques to address them efficiently. They demonstrate the performance of these techniques on several benchmark examples, and demonstrate that incorporating a flexibility metric into existing scheduling algorithms improves the schedule makespan, and that single -hop schedules can significantly reduce the power consumption of the system.

In the final section, "Low Power System-on-Chip Platform Architecture for High Performance Applications", by W.-C. Lo, A.T. Erdogan and T. Arslan of the Institute for System Level Integration, Livingston, Scotland, and the University of Edinburgh, Scotland, work is described on the development of a scheme for implementation of low power high performance Digital Signal Processing intensive AMBA based System-On-Chip platforms. The scheme is based on a novel interfacing technique which utilises the bus hierarchy within AMBA in order to allow single and multiple high performance DSP Intellectual Property cores to be integrated into the SoC platform. The paper describes the overall AMBA SoC architecture, the integration scheme, and an SoC generic wrapper which allows multiple low power high performance DSPs to be connected to the platform. Initial results for timing and power consumption of the whole SoC platform are provided for a low power FIR filter core demonstrating the impact of power dissipation on various sections of the platform.

Section 8.1

A New Logic Design Method for Considering Low Power and High Testability

Yoon-Sik Son and Jong-Wha Chong
Departmet of Electronics Engineering, Hanyang University, 17 Haengdang-Dong, Seongdong-Ku, Seoul, Korea, 133-791.

Key words: Logic transformation, Testability, DFT, BIST

Abstract: In this paper, a new logic transformation method to achieve both low power consumption and high testability is proposed. Our method is based on the redundancy insertion approach. We describe the structure of redundant connections that operate as test points in the test mode. The results of experiments on MCNC benchmark circuits show that the transformed circuit consumes less power in the normal mode and has higher testability in the test mode than the original circuit

1. INTRODUCTION

In CMOS circuits, the dynamic power dissipation component goes up to about 90% of the total power [1]. This component is due to charging and discharging operations. Every time a capacitive node switches from ground to V_{dd} or from V_{dd} to ground, the power of $0.5CV_{dd}^2$ is consumed where C is the load capacitance. The average dynamic power consumption E_{avg} of a circuit is given by:

$$E_{avg} = \frac{1}{2}\sum_{i=1}^{n} C_{load}^{i} \cdot V_{dd}^{2} \cdot f \cdot P^{TR}(i) \qquad (1)$$

C^{i}_{load} and $P^{TR}(i)$ are the load capacitance and the expected number of transitions at a node i, respectively. And n is the number of all nodes in the circuit and f is the clock frequency. Under the assumption that the values app lied to each primary input are temporally independent, $P^{TR}(i)$ is given by:

$$P^{TR}(i) = 2 \cdot P(i) \cdot (1 - P(i)) \qquad (2)$$

$P(i)$ is the signal probability of node i which is the probability that the logic value of the node is 1. Many approaches were proposed to reduce the power consumption of circuits [2,3,4,5,6].

The ever-increasing complexity of the circuit makes the logic testing a harder problem, and the testing cost may be larger than 50% of the total cost for some applications. One approach to reduce the testing cost is DFT(Design for Testability) technique that considers the testability of a circuit as early as possible in the design cycle. Dynamic power dissipation component and testability of an internal node is dependent on the signal probability. For improving testability, this must be close to 0.5. However, low power design methods generally make the signal probabilities of internal nodes close to 0 or 1 and reduce the testability of circuits.

In this paper, we propose a logic design method to achieve both low power consumption and high testability. The rest of this paper is organized as follows. In section 2, the proposed logic design method is explained. In section 3, we will show that the low power design has some negative effects on the testability. A simple solution to reduce those effects is also presented. In section 4, the results of experiments on MCNC benchmark circuits are given.

2. THE PROPOSED LOGIC DESIGN METHOD

Preliminaries

A Boolean network N is represented by a directed graph $G_N(E, V)$. Let the number of vertices (nodes) in G_N be n and the number of edges (interconnections) in G_N be m, that is, $|E|$=m and $|V|$=n. A node is the output of a gate, that is, node i is the output of gate i. Adjacency and path relations are defined in terms of the digraph G_N.

A node i is an input or fanin of a node j if $(i,j) \in E$. The set of all fanins of node j is denoted by the set P_j. A node j is an output or fanout of a node i if $(i,j) \in E$. The set of all fanouts of node i is denoted by S_i. If there is a path from node i to node j in G_N, i is called transitive fanin of node j, and the set of all transitive fanins of node j is denoted by P^*_j. If there is a path from node i to node j in G_N, j is called transitive fanout of node i, and the set of all transitive fanouts of node i is denoted by S^*_i.

$l(i)$ represents the logic level of node i and can be determined by the following formula:

$$l(i) = \max(l(p)) + 1, \quad where \quad p \in P_i \tag{3}$$

When i is a primary input, $l(i)$=0. $v(i)$ is the logic value of node i and $c(l)$ is the control value of gate l. So, $c(l)$=0 means that l is an AND or NAND gate. Let O and S be the set of all primary outputs and the set of all fanout stems of a circuit, respectively.

[Definition 1] Subcircuit $N_i=\{k_1, k_2, \ldots, k_m\}$ is a fanout free region(FFR) if it is represented by a connected digraph and satisfies the following three conditions.

(1) $l(k_1) \geq l(k_2) \geq \cdots \geq l(k_m)$

(2) $k_i \in P_{k_1}$, where $i = 2,\cdots,m$

(3) $k_i \notin S \wedge k_i \notin O$, where $i = 1,2,\cdots,m$

[Definition 2] A gate is a dominator of i if all paths from i to all reachable primary outputs contain the gate. D_i is the set of dominators of i.

From definition 1 and 2, we can see that k_l is the output of the FFR. If k_l is an output of a FFR, dominators of k_l are also dominators of all other internal nodes in the FFR.

[Definition 3] A fanin of a node of D_i is called off-path input of i if it is not on the path from i to any reachable primary output. T_i denotes the set of all off-path inputs of i and can be determined by the following formula.

$$T_i = \bigcup_{j \in D_i} \{x \mid x \in P_j \ \wedge \ x \notin S^* i\} \qquad (4)$$

As every path from any node in a FFR to a primary output must go through the output of the FFR, we can write the following observation.

[Observation 1] Let k be the output of a FFR. The FFR is unobservable at all primary outputs under the following circuit condition:

$\exists i \in T_k$ such that $v(i) = c(m)$, where $m \in D_k$ and $i \in P_m$

The proposed logic transformation method

For an example, consider the circuit shown in Figure 1(a). The signal probability of each node is shown near the node name. $O=\{j,l\}$ and $F=\{b,e,f\}$. $N_l=\{k,h,g\}$ is a FFR and $D_k=\{l\}$ and $T_k=\{i\}$. When the logic value of node i is 0, the N_l is unobservable at the primary ouput l because i forces l to be 0.

A noninverted line from i to k is redundant because i=0 makes k unobservable and i=1 doesn't make a difference in function at the primary output l. An inverted line from i to h is also redundant. A node, which has very low transition probability and controls one of the dominators of a FFR with very high probability, must be selected as the source of the redundant connection. The sink of the redundant connection is a gate that is in the FFR and whose transition probability is very high.

In Figure 1(a), all nodes in N_l are considered as the candidates for the sink. Because $D_k=\{l\}$, $T_k=\{i\}$, and l is an AND gate that is a dominator of the FFR, i=0 means that the N_l is unobservable at the primary output. We denote the condition of i=0 as $(i,0)$. After performing the backward logic implication procedure to find out other conditions satisfying i=0, we can get $\{(i,0), (f,0), (c,0), (b,0), (a,0)\}$ as the set of all conditions of N_l being unobservable at the primary output. From the Figure 1(a), we can see that $(i,0)$ has the highest probability among the conditions.

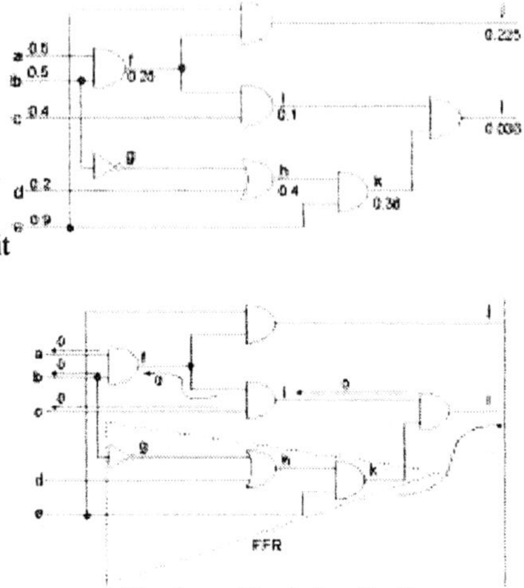

(a) example circuit

(b) backward logic implication

Figure 1. An example circuit

Table 1 shows types of redundant connections that can be added to the circuits. If a FFR is unobservable when the value at the source is 0, the non-inverted connections from the source to INV, BUF, AND, or NAND gates in the FFR are redundant. If the value of source is 0 and sinks are OR or NOR gates, the inverted connections are redundant.

Table 1. Types of redundant connection

Control value of source	Sink	Redundant Connection Type
0	INV or BUF	non-inverted
	AND or NAND	non-inverted
	OR or NOR	Inverted
1	INV or BUF	non-inverted
	AND or NAND	Inverted

If a circuit consists of simple gates and has no reconvergent fanout node, the exact signal probabilities can be computed using the following equation[12] :

$$\begin{aligned} P(o) &= 1 - P(i) \quad \textit{for INV gate} \\ P(o) &= \prod_{i \in FIN(o)} P(i) \quad \textit{for AND gate} \\ P(o) &= 1 - \prod_{i \in FIN(o)} (1 - P(i)) \quad \textit{for OR gate} \end{aligned} \tag{5}$$

where o is the ouput of the gate. In [13], the authors described a polynomial simulation method to calculate switching activities in a general-delay combinational logic circuit. They showed that ignoring spatial correlation for signals that reconverge in a few logic levels introduces negligible error. So, we may estimate signal probability of each node in the FFR by (5) without losing the proper precision.

Let *r* be the node that has the highest probability to make a FFR unobservable at all primary outputs. We may assume that *r*=0 makes the FFR unobservable and a non-inverted line from *r* to a node *k* of the FFR is redundant. If we let $P_{bef}(k)$ be the signal probability of *k* before redundancy insertion and $P_{aft}(k)$ that of *k* after redundancy insertion, $P_{aft}(k)$ is given by:

$$P_{aft}(k) = P_{bef}(k) \cdot P(r) \tag{6}$$

From (2), we can compute the transtion probability of *k* before and after redundancy insertion as follows.

$$\begin{aligned} P_{bef}^{TR}(k) &= 2 \cdot P_{bef}(k) \cdot (1 - P_{bef}(k)) \\ P_{aft}^{TR}(k) &= 2 \cdot P_{bef}(k) \cdot P(r) \cdot (1 - P_{bef}(k) \cdot P(r)) \end{aligned} \tag{7}$$

Then, the reduction factor Δ_k, which denotes the amount of reduction of transition probability resulted by redundancy insertion, is given by:

$$\Delta_k = P_{bef}^{TR}(k) - P_{aft}^{TR}(k) = 2 \cdot P^* \cdot (\alpha - \beta \cdot P^*) \tag{8}$$

where $P^* = P_{bef}(k)$, $\alpha = (1-P(r))$, and $\beta = (1-P(r)^2)$. P^*, α, and β for all possible cases are listed in Table 2. V(*s*) is the value of source *s* that makes sinks unobservable.

Table 2. P^*, α, and β according to v(*s*) and sink type

Sink	v(*s*)	P^*	α	β
AND	0	P(O)	1-P(r)	1-P(r)^2
	1		P(r)	1-(1- P(r))^2
NAND	0	1-P(O)	1- P(r)	1- P(r)^2
	1		P(r)	1-(1- P(r))^2
OR	0	1-P(O)	P(r)	1-(1- P(r))^2
	1		1- P(r)	1- P(r)^2
NOR	0	P(O)	P(r)	1-(1- P(r))^2
	1		1- P(r)	1- P(r)^2

Figure 2 shows the two redundant connections that are inserted from *i* to *g* and from i to *h*. Compared to Figure 1(a), inserting a non-inverted line from i to g makes the signal probabilities of g, h, and k to be 0.95, 0.04, and 0.036, respectively. Δ_g is 0.405.

If we let the load capacitance of the gate be the number of fanouts, the total swithced capacitance of the circuit reduces from 5.45 to 4.37. We may also insert an inverted line from i to h as shown in Figure 2(b). The total switched capacitance of the circuit, however, increases from 4.37 to 4.73 because the second redundant connection has little effect on the signal probabilities of all its transitive fanouts. Δ_h is negligible. On the other hand, the amount of switched capacitance due to the inserted INV and the inserted redundant fanout at *i* is larger than that reduced at the FFR. The second redundant connection must be removed from the circuit.

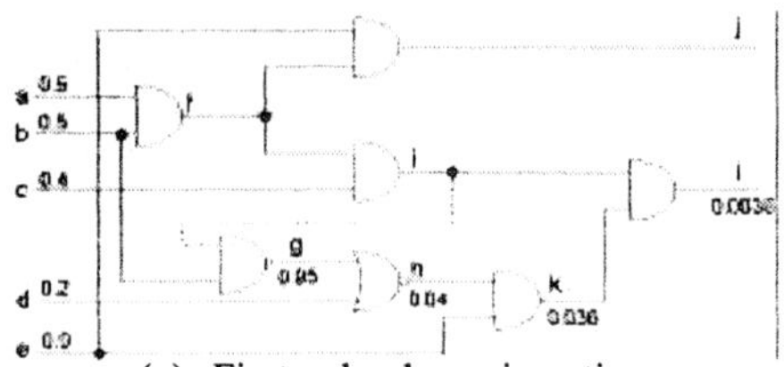

(a) First redundancy insertion

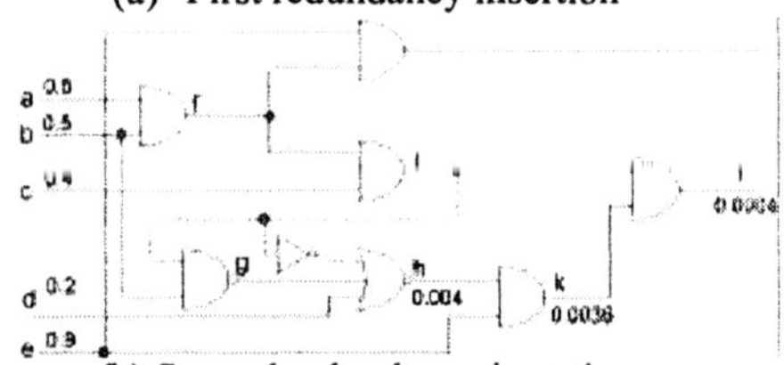

(b) Second redundancy insertion

Figure 2. The result of logic transformation

Proposed Logic Transformation Method

Figure 3 shows the proposed logic transformation algorithm. First of all, we find the FFRs in the circuit and save them in the set S_{FFR}. Then, we select a FFR from S_{FFR}, compute D_{FFR} which contains the dominators of the output of the selected FFR, and estimate the transition probability of the FFR by equation (2) after logic simulation. During the backward implication procedure, we can get $S_{candidate}$ that contains the conditions of making the selected FFR unobservable at all primary outputs. From $S_{candidate}$, we choose the condition that has the highest probability and we can get the best source S_{best}. The check_and_insert_redundancy routine estimates what effect the inserted redundancy has on the total switched capacitance of the circuit and determines whether redundancy is inserted or not. Repeat the above procedures for all FFRs.

3. CONSIDERING THE TESTABILITY

Signal probability and testability

BIST(Built-In Self-Test) provides a simple, low-cost test solution by building test circuitry, which is composed of TPG (test pattern generator), ORA (output response analyzer) and BIST controller, inside the chip. The commonly used TPG is LFSR (linear feedback shift register) that is used to generate pseudo-random sequences for CUT.

```
Power_optimize( ) {
  SFFR = find_all_FFR( )
  logic_simulation( )
  while (SFFR ≠ Ø) {
    R = select_a_FFR(SFFR)
    DFFR = find_all_dominator(R)
    Icontrol = find_control_fanin_of_dominator(R, DFFR)
    estimate_signal_activity(R)
    Scandidate = backward_implication(Icontrol, R)
    Sbest = select_best_source(Scandidate)
    check_and_insert_redundancy(Sbest, R)
  }
}
```

Figure 3. The proposed algorithm

The stuck-at fault model is the most widely accepted fault model in the industry. In this model it is assumed that the fault causes a line of the circuit to behave as if it is permanently at logic 0(stuck-at 0 fault) or logic 1(stuck-at 1 fault). Detection of stuck-at faults needs two conditions, that is, fault activation and fault effect propagation. Driving the faulty node to have the opposite of the faulty value is fault activation. The activated fault effect must be propagated to the primary outputs so as to be detected. This is fault effect propagation. If the pseudo-random sequences generated by LFSR can't drive a node to be 0 (1), the stuck-at-1 (0) fault at the node is a redundant one and cannot be detected. So, the signal probability of a node is deeply related to its testability. This is because the signal probability of a node indicates the ratio of the number of patterns that drive the node to be 1 to the number of all possible patterns. Lower the signal probability of a node, fewer the patterns that activate the stuck-at-0 fault at the node. Higher the signal probability, fewer the patterns that activate the stuck-at-1 fault at the node.

Consider stuck-at-1 (0) fault of a node k whose signal probability is $P(k)$. Let $L_0(k)$ denote the order of a test pattern that first drives k to be 0. $L_0(k)$=i means that the node can't be driven to be 0 by the first (i-1) test patterns and can be driven by the i-th test pattern. As the signal probability of k is $P(k)$, the probability of $L_0(k)$=i is $P(k)^{i-1}\cdot(1-P(k))$. Let $L_1(k)$ denote the order of a test pattern that first drive k to be 1. Under the assumption that the stuck-at 1 fault and the stuck-at 0 fault are equally likely at a node, we can derive the following theorem on the fault activation length that indicates the number of test patterns needed to activate a stuck-at fault.

[Theorem 1] Let $P(k)$ is the signal probability of a node k. If we let $L(k)$ denote the fault activation length at k, $L(k)$ is given by:

$$L(k) = \lceil \{2\,P(k)\,(1-P(k))\}^{-1} \rceil$$

(Proof) $L_0(k)$ is the number of patterns needed to drive the node to be 0. Table 3 shows the probability of $L_0(k)$=i. The probability of $L_1(k)$=i is also shown.

Table 3. The probability of fault activation

L_0	1	...	i	...
$P(L_0)$	(1-P(k))	...	$P(k)^{i-1}\cdot(1-P(k))$	...
L1	1	...	i	...
P(L1)	P(k)	...	$(1-P(k))^{i-1}\cdot P(k)$	...

$L_{0,exp}(k)$ and $L_{1,exp}(k)$, which are the expected values of $L_0(k)$ and $L_1(k)$ respectively, are given by:

$$L_{0.exp}(k) = 1 \cdot Prob(L_0 = 1) + 2 \cdot Prob(L_0 = 2) + \cdots$$

$$= (1 - P(k)) + 2 \cdot P(k) \cdot (1 - P(k)) + \cdots = (1 - P)^{-1}$$

As stuck-at-0 fault and stuck-at-1 fault are equally likely at k, the activation number L(k) is computed from $L_{0,exp}(k)$ and $L_{1,exp}(k)$ as follows.

$L = \lceil (0.5 \cdot L_{0,exp}(k) + 0.5 \cdot L_{1,exp}(k)) \rceil = \lceil \{2 P(k) (1-P(k))\}^{-1} \rceil$

[QED]

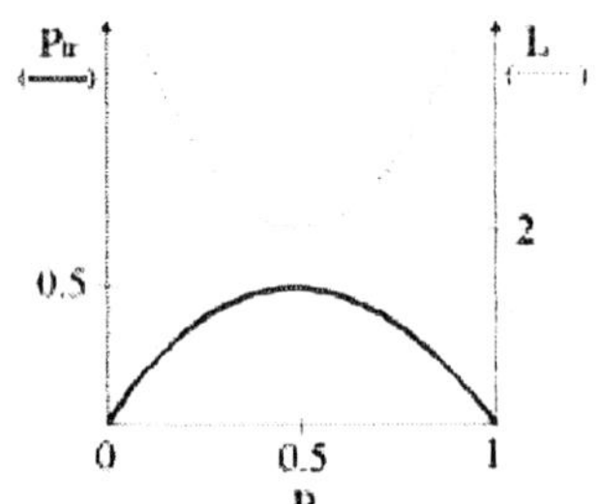

Figure 4. The relationship among L, P^{TR}, and P

Figure 4 shows the plots of L and P^{TR} versus the signal probability P at a node. In case of P=0.5, P^{TR} is the maximum and L is the minimum. As P is closer to 1 or 0, P^{TR} decreases to 0, and L increases rapidly.

When P=0 (P=1), the stuck-at-0 (1) fault at the node is an redundant fault. Logic transformation is the process that makes the signal probabilities of some nodes closer to 1 or 0 in order to reduce P^{TR}, so L increases for those nodes in FFRs.

After fault activation, the fault effect must be propagated to any primary output in order to detect the activated fault. If the fault effect is arrived at a fanin of a gate whose logic value is controlled by other fanins, it can't be propagated to the output of the gate. This is called fault masking.

The effect on the testability of the circuit that the inserted redundant connections have is well explained in Figure 5. To reduce the switched capacitance, a redundant connection k' is inserted from k to the FFR. D is one of the dominator of the FFR. When the node k controls D, the fault effect activated in the shaded region A is masked at D and not propagated to the region B. The redundancy k' also makes the activation of the faults of the black region C very difficult because the signal probabilities of the nodes in the region C are closer to 0 or 1 due to k'. The faults occurred at the region C are likely to be randompattern-resistant faults.

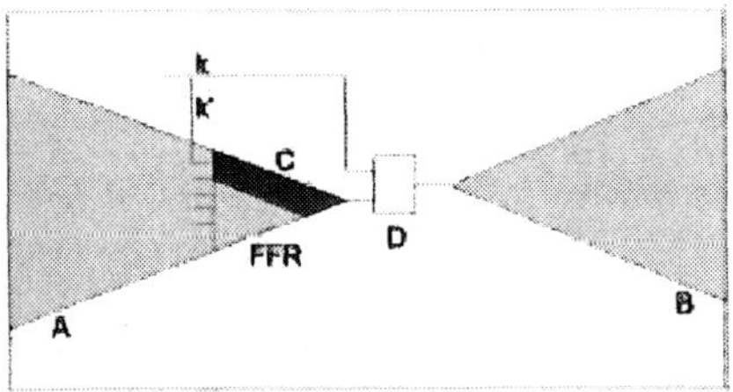

Figure 5. The effect of redundancy on the testability

We applied the logic transformation algorithm shown in Figure 3 to some MCNC benchmark circuits. Table 4 shows the fault simulation results. $FC_{Original}$ is the fault coverage of the original circuit and FC_{LT} is that of the logic transformed circuit. The used test set is composed of 2000 pseudo-randomly generated patterns.

Table 4. Fault simulation results

Circuit	$FC_{Original}$	FC_{LT}
cm85	99.25	95
cu	86.36	85.71
ttt2	83.65	83.63
frg1	42.58	40.19
rot	96.56	96.23
x2	100	94.22
cmb	61.19	57.85
cm163	94.30	91.66

Improving testability of circuits

Figure 6 shows the structure of redundancy to improve the testability of the circuit. MS(Mode Select) is a 2-input gate used as a test point and T_m is a primary input used for selecting the operation mode. In case of the normal mode(T_m=0), the circuit consumes less power. In case of the test mode(T_m=1), k and k' have the non-control values and the faults of the region A are easily propagated to region B without being masked at the dominator D.

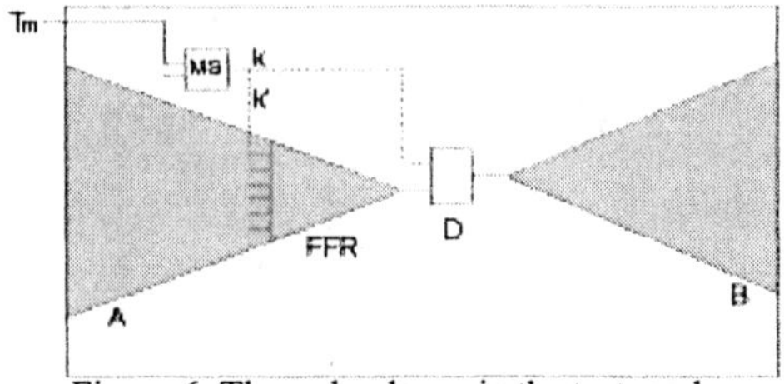

Figure 6. The redundancy in the test mode

Our testing strategy is consisted of two phases. In the first phase, a half of all test patterns are applied to the circuit with T_m=0, so fiults of all region but A are detected. In the second phase, the rest of test patterns are applied with T_m=1 and faults of region A can be easily propagated to region B through D.

4. EXPERIMENTAL RESULTS

We perform some experiments on MCNC '91 benchmark circuits. The procedure of experiments is explained in the following. Using SIS package, we optimize MCNC '91 benchmark circuits with 2-input gates. After extracting the netlist of the optimized circuits in SLIF format, we perform logic simulation with 2000 pseudo-random patterns and compute the signal probabilities of all nodes. We transform circuits using the proposed method explained in section 2 without considering the testability. Atalanta, which is the ATPG program developed at Virginia Tech., provides the information of removable redundancy. Total switched capacitances of the transformed circuits are computed and compared with those of the original circuits. Areas of all circuits are compared after technology mapping with mcu.genlib.

Table 5 shows the experimental results. #PI, #PO, and #lit columns contain the numbers of primary inputs, primary outputs, and literals of original benchmark circuits, respectively. Area column shows the ratio of the area of the transformed circuit to that of the original circuit. The ratio of switched capacitance of the transformed circuit to that of the original circuit. The proposed method can reduce the total power consumption of some circuits by 13% with low area overhead.

Figure 7 shows the change of fault coverage versus the number of test patterns for two benchmark circuits, that is, cmb and rot. We assume that the conventional test-per-clock BIST scheme is used. In case of cmb, 6 redundant connections are inserted and the number of sources is 3. Figure 7(a) shows the fault coverage of cmb. The graph **org** shows the fault coverage of the original circuit. **ndft** and **dft** show the fault coverage of the transformed circuit without considering the testability and that of the transformed circuit with considering the testability. In case of **dft**, we can get high fault coverage with a few test patterns.

Table 5. Experimental Results

Circuit	Initial			After	
	#PI	#PO	#lit	Area	SA
cm138	6	8	30	1.03	0.87
Decod	5	16	54	1.00	0.88
cm85	11	3	56	1.08	0.95
cu	14	11	72	1.04	0.96
f51m	8	8	320	0.99	0.97
ttt2	24	21	428	1.00	0.97
frg1	28	3	811	1.03	0.96
rot	135	107	1259	1.00	0.99
x2	10	7	62	0.99	0.93
cm42	4	10	27	1.00	0.96
cmb	16	4	65	1.09	0.90
cm163	16	5	57	1.06	0.96

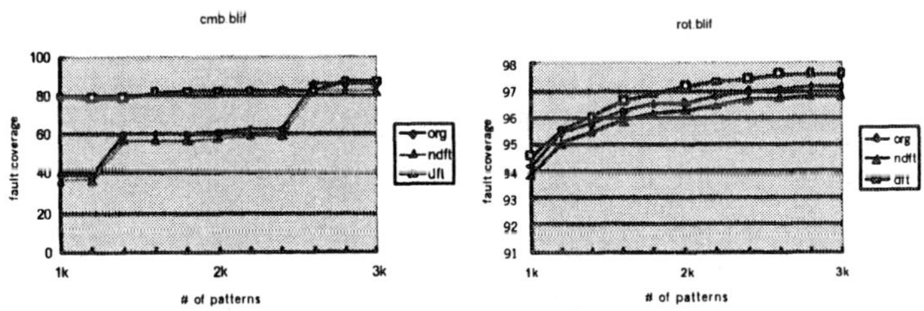

(a) cmb.blif (b) rot.blif

Figure 7. Fault Coverage

After logic transformation, Rot has 26 redundant connections, and only 3 among them are replaced with the proposed redundancy structure to avoid much area overhead. Figure 7(b) looks like showing little enhancement in the fault coverage. But, in case of **dft**, we can get 97% fault coverage with about 2000 test patterns. But **org** needs more than 2600 test patterns and **ndft** needs more than 3000 test patterns to achieve the same fault coverage. By considering the testability of the circuit in logic transformation, we can reduce the test application time by about 30% or more.

5. CONCLUSION

In this paper, a new logic transformation method targeting low power consumption and high random pattern testability simultaneously was proposed. The proposed method uses the redundancy insertion and removal approach. Under the condition that a FFR of the circuit is unobservable at all primary outputs, proposed method reduce the switching activities of the circuit with low area overhead. It is explained what effects the logic transformation method has on the testability. The structure of redundant connections is described to improve the testability of the circuit. From the experiments on the MCNC benchmark circuits, the results show that proposed method improves the power consumption and testability of circuits simultaneously.

REFERENCES

[1] J. M. Rabaey, M. Pedram, 1st ed., LOW POWER DESIGN METHODOLOGIES, Kluwer Academic Publishers, 1996.

[2] B. Rohfleisch, A. Kolbl, and B. Wruth, Reducing Power Dissipation after Technology Mapping by Structural Transformations, Proc. DAC, pp. 789-794, 1996.

[3] S. Chang, M. Marek-Sadowska, and K. Cheng, Perturb and Simplify: Multilevel Boolean Network Optimizer, IEEE Trans. CAD, vol. 15, no. 12, pp. 1494-1504, Dec. 1996.

[4] K. T. Cheng and L. A. Entrena, Multi-level Logic Optimization By Redundancy Addition and Removal, Proc. European DAC. pp. 373-37, 1993.

[5] Q. Wang and S. B. Vrudhula, Multi-level Logic Optimization for Low Power using Local Logic Transformations, Proc. ICCAD, pp. 270~277, 1996.

[6] Ki Sook Chung and C. L. Liu, Local Transformation Techniques for Multi-Level Logic Circuits Utilizing Circuit Symmetries for Power Reductions, Proc. ISLPED, pp. 215~220, 1998.

[7] Jacob Savir and Robert Berry, At-speed test is not necessarily an AC test, Proc. ITC, pp. 722-728, 1991.

[8] Mitrajit Chaterjee and Dhiraj K. Pradhan, A Novel Pattern Generator for Near-Perfect Fault-Coverage, Proc. VLSI Test Symp., pp. 417-425, 1995.

[9] J. Savir, Reducing the MISR size, IEEE Trans. Computers, vol. 45, pp. 930-938, 1996.

[10] M. F. AlShaibi and C. R. Kime, MFBIST : A BIST Method for Random Pattern Resistant Circuits, Proc. ITC, pp. 176-185, 1996.

[11] P. H. Bardell, W. H. McAnney, and J. Savir, 1st ed., Built-In Test for VLSI : Pseudorandom Techniques, John Wiley & Sons, 1987.

[12] H. Goldstein, Controllability/observability of digital circuits, IEEE Trans. Circuits and Systems, pp. 685-693, 1979.

[13] J. C. Costa, J. C. Monteiro, Srinivas Devadas, Switching Activity Estimation using Limited Depth Reconvergent Path Analysis, proc. ISLPED, pp. 184~189, 1997.

[14] H. K. Lee and D. S. Ha, Atalanta: an Efficient ATPG for Combinational Circuits, Technical Report, pp. 93-12, Dept. of Electrical Eng., Virginia Polytechnic Institute and State University, 1993.

[15] H. K. Lee and D. S. Ha, An efficient forward fault simulation algorithm based on the parallel pattern single fault propagation, Proc. ITC, pp. 946-955, 1991.

Section 8.2

System Synthesis for Optically-Connected, Multiprocessors On-Chip

Neal K. Bambha and Shuvra S. Bhattacharyya
Department of Electrical and Computer Engineering, and Institute for Advanced Computer Studies University of Maryland, College Park

Abstract: Optical interconnects are being considered as a possible solution to the well-known problems of scaling in VLSI interconnects. Along with enabling higher speed interconnects, optics allows the construction of highly connected and irregular networks that are streamlined for particular applications. Using these networks, it is possible to implement application mappings that allow flexible, single-hop communication patterns between processors. This has advantages for reduced system latency and power. Such optically connected multiprocessors are particularly promising for embedded digital signal processing (DSP) applications, which are highly parallel, and typically have tight constraints on latency and power consumption. This paper addresses novel trade-offs involving communication routing flexibility, power consumption, and performance that arise in the context of system synthesis of optically-interconnected multiprocessors. We report on experimental results that expose these trade-offs, and propose systematic techniques to address them efficiently. We demonstrate the performance of these techniques on several benchmark examples.

Key words: System-on-chip, low power, adder cells.

1 INTRODUCTION

In recent years, optics has played an increasing role in multiprocessor systems. Various studies have predicted that the energy consumed by data commu nication will ultimately limit the processing speed in electronic processors [8], [10]. Light signals do not suffer from effects such as electromagnetic interference and capacitive effects, which limit electrical interconnects. Many research groups have demonstrated optically-connected multiprocessor systems (e.g., see [4], [5], [7]). However, relatively little work has been done to develop synthesis techniques and automated mapping tools to take advantage of such systems. A key advantage of optics for a multiprocessor system is that it allows highly parallel data links and a large degree of connectivity between processors. This work addresses some fundamental issues related to mapping of applications onto optically-connected systems, particularly for generating low-latency, low-power schedules.

The problem of generation of communication topologies at the system level has been examined for bus-based implementations [3], [9]. These techniques try to optimize communication between a set of processes, and try to minimize the total communication costs. Many techniques have also been presented for scheduling application graphs onto multiprocessor networks with various fixed topologies (e.g., see [2]). Our work is different because we incorporate communication topology synthesis, limited hop communication, and optical interconnects which are based on unidirectional point-to-point links and give rise to new issues of schedule deadlock and communication flexibility.

2. OPTICALLY CONNECTED SYSTEMS

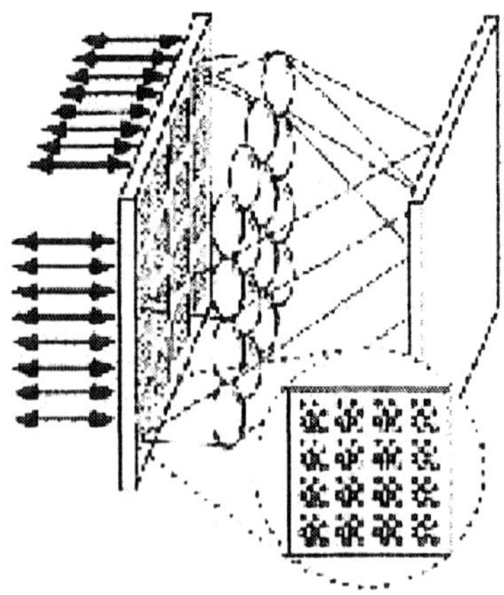

Figure 10. [FAST-Net prototype system]

One example of an optically connected system is the *FAST-Net* prototype [5], a high-throughput data-switching concept that uses a reflective optical system to globally interconnect a multi-chip array of processors. The three-dimensional optical system links each chip directly to every other with a dedicated bidirectional parallel data path. This is illustrated in *Figure 10* [5]. Initiatives such as the DARPA VLSI

Photonics Program [13] are pushing this technology from multi-chip arrays of processors towards single-chip arrays of processors.

3 TASK GRAPH SCHEDULING

The computational model used in this work is that of conventional acyclic *task graphs*, in which graph vertices (*tasks* or *nodes*) represent computations and each edge represents the communication of a packet of data from the source task to the sink task. Henceforth in this paper, we refer to a task graph representation of an application as an *application graph*. A vast range of scheduling techniques for task graphs has been developed (e.g., see [12] for a review of several representative approaches); however, these techniques typically assume a fixed communication network, and do not systematically take connectivity constraints into account. By connectivity constraints, we mean the inability of certain pairs of processors to communicate with each other. These constraints arise because it is advantageous to configure the schedules in such a way that multi-hop communication is minimized, and the relative abundance of communication links is used instead to achieve the required communication flexibility.

One contribution of this paper is to develop a general framework for extending arbitrary list scheduling approaches to avoid deadlock, and operate efficiently in the presence of connectivity constraints. We also apply this framework to jointly streamline the communication network and task graph mapping for a given application. This can be used both for minimum-cost dedicated implementations, and for reconfigurable networks, where the goal is to save power consumption by activating a minimal subset of available laser-detector pairs.

4 COMMUNICATION FLEXIBILITY

One major consequence of single-hop communication schedules is that each processor p is restricted to send data to a subset of the set of all processors Φ, and to receive data from a (possibly different) subset of Φ. If these constraints are not considered, deadlock can easily occur during the scheduling process. We define a *feasible set* of processors $F[v]$ for a task v as the largest subset of Φ on which v can be scheduled without deadlock. We have developed an efficient algorithm to determine the feasible set of processors $F[v]$ for all $v \in G$ at any point during the scheduling process.

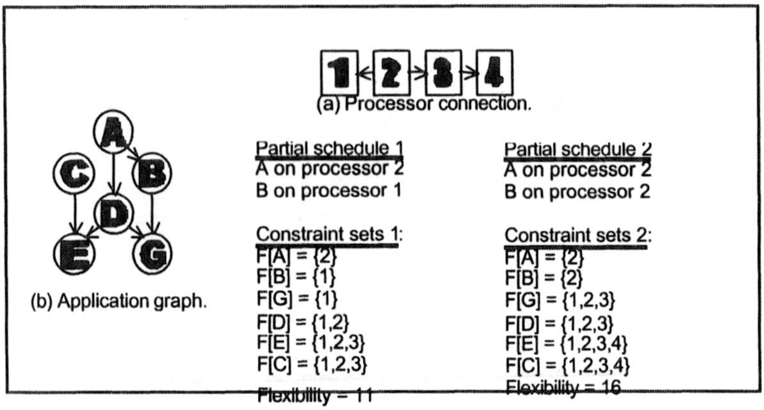

Figure 8.22. [Example demonstrating flexibility metric.]

We define the *communication flexibility* (or simply *flexibility* for short) of the system at any point during the scheduling process as the sum of the sizes of $F[v]$ for all $v \in G$. The flexibility gives some measure of the degree of constraint imposed on all tasks by a given scheduling move. Figure 2 depicts a simple example of an application graph with six tasks scheduled on four processors.

5 SINGLE-HOP SCHEDULING FOR LOW POWER

The general class of list scheduling algorithms can easily be adapted to produce single-hop (or n-hop) schedules by incorporating our constraint algorithm. This is advantageous because it allows us to leverage a large library of useful scheduling techniques.

In list scheduling, a priority list L of tasks is constructed. The priority list is a linear ordering $(v_1, v_2, \ldots, v_{|V|})$ of the tasks in the application graph $G = (V, E)$ such that for any pair of distinct tasks v_i and v_j, v_i is to be given higher scheduling priority than v_j if and only if $i < j$. Each task is mapped to an available processor as soon as it becomes the highest-priority task according to L among all tasks that are ready. This process is repeated until all tasks are scheduled.

The concepts of feasibility and flexibility can be incorporated into the general framework of list scheduling by restricting the set of candidate processors to include only those that are feasible at the given scheduling step, and by taking flexibility into account in designing the priority metric through which tasks are ordered.

Single-hop communication is desirable for low power, since every communication hop incurs an additional energy cost. However, restricting the communication to single-hop reduces the flexibility of the scheduler to fewer possible moves in each scheduling step. In general, this would not be optimal for producing the lowest makespan schedules. We show in Section 7 that restricting the schedule to single-hop communication can produce significant power savings when compared to multi-hop schedules, while not penalizing the makespan significantly.

If the communication is not constrained to be single hop, some routing algorithm must be employed to schedule the interprocessor communication. In our experiments, we used a greedy load-based minimal routing algorithm. Minimal routing algorithms consider only shortest paths between processors. Since each communication hop in an optical interconnect incurs an energy cost, shortest path routes are desirable. A weight is assigned to each edge on all shortest paths as a function of the change in the relative load that would occur if it were to be used by the route. A shortest path from source processor to destination processor is then computed with respect to those weights. Such algorithms are often employed in multiprocessor systems [1]. Although this routing algorithm runs in polynomial time [1], in the context of a fixed optimization budget, the time spent computing the routes in a multi-hop schedule reduces the number of possible scheduling moves that the scheduling algorithm can explore when compared to a single-hop schedule.

6 TPLA ALGORITHM

Realistic optical networks may incorporate relatively high, but not necessarily complete (fully connected), levels of connectivity. Even in fully-connected systems, such as *FAST-Net* [5], it is still desirable from the viewpoint of power and heat dissipation to have a minimal interconnect mapping, since for a given application, non-essential transmitters can be turned off. In this section, we illustrate how our single-hop scheduling strategies, and the underlying concept of communication flexibility, can be used to guide the synthesis of application-specific interconnect structures. The main idea here is that for embedded multiprocessors, the interconnect topologies should be driven by the specific application mappings that will execute across them, and jointly designing the two is advantageous.

Specifically, we have developed a greedy, heuristic algorithm, called the *two-phase link adjustment* (TPLA) algorithm, to synthesize an interconnect and an associated multiprocessor schedule for a given application. The TPLA algorithm starts with a fully connected network, and operates in *down* and *up* phases. Input to the algorithm is either a makespan constraint for the application, or a constraint on the total number of links.

Each step of the down phase in TPLA removes one link, while each step of the up phase adds one link. One step of the down phase consists of assigning each existing link a score based on the schedule makespan resulting from its removal, and removing the link with the lowest score. A history of scores is kept for each link. For the first pass through the down phase, ties between links are broken randomly. On subsequent passes, the link history is used to break ties. The down phase continues until all the links are removed.

Conversely, one step of the up phase in TPLA consists of assigning a score to each missing link based on the makespan resulting from its addition. The up phase continues until the network is fully connected. Repeated, alternating invocations of down and up phases are executed for some time limit (determined by the user), and the best result found is taken. Given a makespan constraint, this best result minimizes the number of links. Alternatively, given a constraint on the number of links, the best result minimizes the makespan.

7 EXPERIMENTS

Our techniques for scheduling and interconnection pattern synthesis operate in conjunction with a given list scheduling strategy. In these experiments, we employed the DLS algorithm [11] as the underlying list scheduling strategy, although, as described in Section 5, any list scheduling algorithm could have been used.

We examined a set of DSP application benchmarks and scheduled them using two different scheduling modes, one that incorporates only feasibility information (to avoid deadlock), and another that takes both feasibility and flexibility into account. We refer to these as the *feasibility-only* and *feasibility-flexibility* modes, respectively. To evaluate the performance across a range of connectivity levels, we scheduled the applications onto networks with varying degrees of connectivity. We also compared the makespan and power of single-hop schedules to that of multi-hop schedules. For the multi-hop scheduling, we used the greedy load-based minimal routing algorithm discussed earlier.

In the feasibility-only mode, the processor P considered for a given task v at each scheduling step was restricted to be in the feasible set $F[v]$ for v, as described earlier, and no modification was made to the task prioritization metric of the underlying list scheduling strategy (DLS).

In the feasibility-flexibility mode, the processor P considered at each scheduling step was again restricted to be in the feasible set for v; however, whenever two processor assignments for v resulted in equal priority levels $L(v)$, where L represents the priority metric of the original DLS algorithm, priority was given to the assignment that resulted in a higher value of flexibility. In other words, priority was given to assignments that offered greater flexibility for future scheduling decisions.

For each application, we chose a number N of processors, and then generated a fully connected network with $N(N-1)$ links. We scheduled the application using both feasibility-only and feasibility-flexibility modes onto this network. Next we removed one link from the network at random, and again scheduled the application using both scheduling modes. We continued this process of removing links until no links remained, resulting with all the tasks scheduled on a single processor. We define the relative improvement of the feasibility-flexibility mode over the feasibility-only mode by comparing the average makespan over all points.

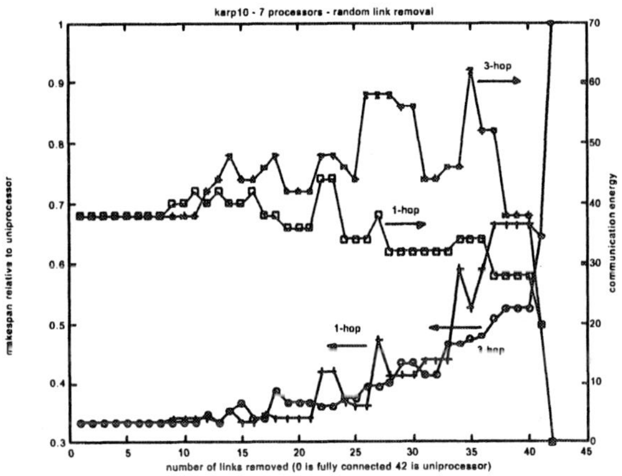

Figure 8.22. [Makespan and communication energy with single-hop and three hop maximum communication]

Figure 2 compares the makespan and communication energy for single-hop and three hop communication for the karp10 application. For this application, the single-hop schedule required 24% less communication energy while producing on average a 4% longer makespan than the three-hop schedule. For this application, the makespan obtained without using the flexibility metric was 26% longer than the 1-hop makespan. Table 1 summarizes this comparison for several other DSP application benchmarks.

We performed experiments with the following application graphs: fft1, irr, fft3, karp10, qmf4, Laplace, sum1, and RBFNN. The fft graphs are different implementations of the fast Fourier transform [6]. Karp10 refers to the Karplus-Strong music synthesis algorithm with 10 voices (21 nodes); qmf4 is a quadrature mirror filter bank (14 nodes). Laplace is a laplace transform. Irr is an adaptation of a physics algorithm, and sum1 is an upside down binary tree representing the sum of products computation. RBFNN is a neural network classifier algorithm.

Table 4. [Relative makespan improvement obtained by using flexibility information in the scheduling process.]

Application	N	1-hop makespan improvement using flexibility	Reduction in communication energy	Makespan penalty for 1-hop vs. 3-hop
Fft1	7	30	16	8
Karp10	6	26	24	4
Irr	8	17	16	(2)
Qmf4	7	19	32	3
Nn16-3-4	8	21	58	2
Sum1	6	8	1	4
Laplace	7	23	4	(3)
Fft2	7	15	12	2

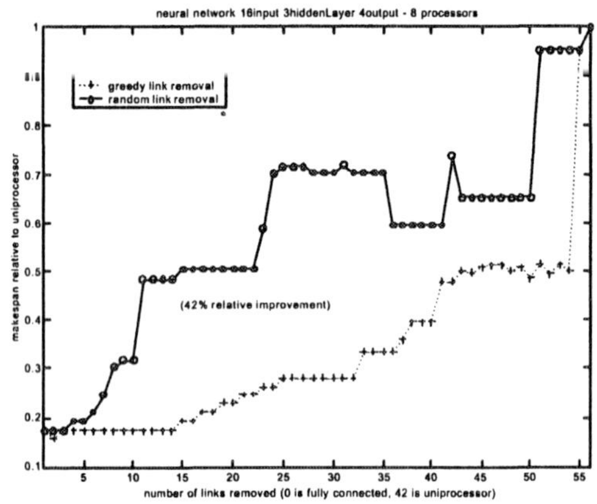

Figure 3. [Link synthesis using the TPLA algorithm.]

We evaluated the TPLA algorithm on the RFBNN benchmark described above. The bottom curve of Figure 3 shows the best makespan achieved for each level of connectivity between 0 and fully connected, after one "down" phase and one "up" phase. This gives a Pareto curve of the trade-off between number of links and makespan for the application. For purposes of comparison, the upper curve of Figure 3 shows the makespan achieved by starting with fully connected and randomly removing one link at a time. The TPLA algorithm shows a significant improvement (42% relative improvement) over random removal, and is thus a promising starting point for developing more sophisticated link synthesis methods. More broadly, it demonstrates the effectiveness of feas ibility- and flexibility-driven scheduling in iterative co-design of optical interconnection structures.

8 CONCLUSIONS

Optical interconnect technology is promising for global communication in embedded multiprocessors, since the interconnection patterns can flexibly be streamlined and reconfigured to match the target applications. However, due to the power consumption characteristics of optical links, it is useful to restrict communication across them to single-hop transfers. We have demonstrated an effective algorithm for determining the set of feasible processors that will avoid schedule deadlock in a single-hop schedule, and a useful metric, called communication flexibility, for the degree to which a given scheduling decision constrains future decisions (in the context of the given communication topology).

We used this algorithm and the flexibility metric in conjunction with the DLS algorithm to map several DSP applications across a wide range of interconnect topologies. These experiments demonstrated that incorporating the flexibility metric into existing scheduling algorithms improves the schedule makespan, and that single-hop schedules can significantly reduce power consumption of the system. We also demonstrated that these scheduling techniques could be used to effectively guide an algorithm for jointly synthesizing the interconnection network together with the mapping of an application onto the network.

ACKNOWLEDGEMENTS

This work was supported by the DARPA funded Optoelectronics Center for Innovative Photonic Chipscale Technologies (Contract number MDA972-00-1-0023). We would like to thank Vida Kianzad for her compilation of the benchmark examples.

REFERENCES

[1] J. Aspnes, Y. Azar, A. Fiat, S. Plotkin, O. Waarts, "On-Line Routing of Virtual Circuits with Applications to Load Balancing and Machine Scheduling," *Journal of the ACM*, vol. 44, no. 3, pages 486-504, May 1997.

[2] S. Banerjee, D. Picker, R. Fellman, and P. Chau, "Improved Scheduling of Signal Flow Graphs onto Multiprocessor Systems Through an Accurate Network Modeling Technique," *Proceedings of the International Workshop on VLSI Signal Processing*, pages 157-167, 1994

[3] M. Gasteier and M. Glesner, "Bus-Based Communication Synthesis on System-Level," *Proceedings of the 9th International Symposium on System Synthesis*, pages 65-70.

[4] P. S. Guilfoyle, "Digital Optical Computing Architectures for Compute Intensive Applications," *Proceedings 1994 International Conference on Optical Computing*, 1994.

[5] M. W. Haney, M. P. Christensen, P. Milojkovic, "Description and Evaluation of the FAST-NET Smart Pixel-Based Optical Interconnection Prototype," *Proceedings of the IEEE*, vol. 88, no. 6, June 2000.

[6] A. Kahn, C. McCreary, J. Thompson and M. McArdle, "A Comparison of Multiprocessor Scheduling Heuristics," *Proceedings of 1994 International Conference on Parallel Processing*, vol. II, pages 243-250, 1994.

[7] N. McArdle, M. Naruse, A. Okuto, and M. Ishikawa, "Implementation of a Pipelined Optoelectronic Processor: OCULAR II," *Technical Digest of Optics in Computing '99.*

[8] J. D. Meindl, "Low Power Microelectronics: Retrospect and Prospect," *Proceedings IEEE*, vol. 83, no. 4, 1995.

[9] S. Narayan and D. Gajski, "Synthesis of System Level Bus Interfaces," *Proceedings of European Design and Test Conference '94*, pages 395-399, Feb. 1994.

[10] P.W. Smith, "On the Physical Limits of Digital Optical Switching and Logic Elements," *Bell System Technical Journal*, vol. 61, no. 8, 1982.

[11] G. C. Sih, *Multiprocessor Scheduling to Account for Interprocessor Communication, Ph.D. Thesis*, Department of Electrical Engineering and Computer Sciences, University of California at Berkeley, April 1991.

[12] S. Sriram and S. S. Bhattacharyya. *Embedded Multiprocessors: Scheduling and Synchronization*. Marcel Dekker, Inc., 2000

[13] http://www.darpa.mil/MTO/VLSI.

Section 8.3

Low Power System-on-Chip Platform Architecture for High Performance Applications

W.-C. Lo[1], A.T. Erdogan[2] and T. Arslan[1,2]

[1]Institute for System Level Integration, The ALBA Campus, Livingston, EH6 4 7BH, Scotland, United Kingdom.

[2]University of Edinburgh, Department of Electronics & Electrical Engineering, Edinburgh EH9 3JL, Scotland, United Kingdom

Abstract This paper describes work on the development of a scheme for implementation of low power high performance Digital Signal Processing intensive AMBA based System-On-Chip platforms. The scheme is based on a novel interfacing scheme which utilises the bus hierarchy within AMBA in order to allow single and multiple high performance DSP Intellectual Property cores to be integrated into the SoC platform. The paper describes the overall AMBA SoC architecture, the integration scheme, and an SoC generic wrapper which allows multiple low power high performance DSPs to be connected to the platform. Initial results are provided with a low power FIR filter core demonstrating impact of power on various sections of the platform.

Key words : Low power, System on a Chip, DSP

1. INTRODUCTION

Today's System-On-Chip (SoC) devices are targeting complex multi-media applications where there is a need for significant amount of Digital Signal Processor (DSP) computing power in order to achieve the various data processing tasks which could include both video and audio. In order to complete these tasks various DSP Intellectual Property (IP) cores, which are optimised for the application in hand, are utilised. In order to satisfy the constraint of fast time to market effective platform architectures are required. High performance IP cores can be connected to the above platforms on a plug-in basis in order to maintain performance constraints as well as that of the fast time to market [1][2].

Although numerous researchers in the literature have considered the development of interfaces and wrappers [3][4], only a few discuss implementation issues and impact at the SoC level. This paper will discuss the implementation of an efficient interface strategy and its effect on the overall system. The paper describes the development of an SoC platform for low power wireless applications. The platform utilises the LEON Processor, which is a general SPARC compatible processor [5].

In order to integrate a number of DSP IP cores, an interface is required. This interface connects the DSP IPs directly to the Advance Microcontroller Bus Architecture (AMBA) [6] of the LEON Processor. In comparison to the bus interface design that contains a Virtual Component Interface (VCI) [7], the AMBA interface described in this paper eliminates the VCI overhead. As discussed in [8], it is less efficient to design extra VCI for both the bus interface and DSP IPs. Therefore, the AMBA interface discussed in this paper connects the DSP IPs directly to the processor without any additional control signals. This is achieved through gated clock and look ahead features. At the same time, these features have made the interface consumed less power.

Gated clock has been employed normally to selectively stop the clock in portions of circuits where active computation is not being performed [9] [10]. In [9], the Finite State Machine (FSM) is partitioned into several sub-FSMs and these sub-FSMs are clocked when necessary. However, the AMBA interface discussed in this paper is partitioned into Bus Address and Control Decoders, a AHB Data FSM and a Clock Controller. The AHB Data FSM is clocked when there is data available from the processor and the selected DSP IP is clocked when the data is available from the interface.

The DSP IP core used in this paper is an FIR filter that has a transpose direct form structure [11][12]. This structure reduces the switching activity at data inputs of the multiplier. The data input of the multiplier remains unchanged until it is multiplied by all the coefficients. In addition, the transpose direct form structure can be exploited by different numerical ordering algorithms for manipulating filter coefficients in order to reduce the switching activity at the coefficient input of the multiplier.

2. PLATFORM ARCHITECTURE

The overall SoC platform architecture is demonstrated in Figure 1. The architecture is built around the LEON processor, and includes a number of high performance DSP IP cores.

The LEON processor is connected to the AMBA interface through the AHB (Advance High Performance Bus) and the APB (Advance Peripheral Bus). Within the LEON processor, the AHB is intended for high clock frequency system modules and enables access to high bandwidth memory devices, while the APB is optimised for minimal power consumption and suits low bandwidth peripheral modules. Any data that the LEON processor sends or receives (from the AMBA interface and to the DSP IP core) will be through these buses. The AMBA interface functions as a middleman that monitors the AHB and APB buses for any possible transaction being directed to the DSP Core. If a read or write operation is required, then the AMBA interface will load or store data (from the AHB or APB buses) to the DSP IP respectively. The DSP IP only accepts input data, processes this data and outputs the resulting data. Therefore, the AMBA communication protocol with the LEON processor is transparent to the DSP IP core as these protocols are handled by the AMBA interface. Likewise, the DSP IP's clock is actively controlled by the AMBA interface to save power.

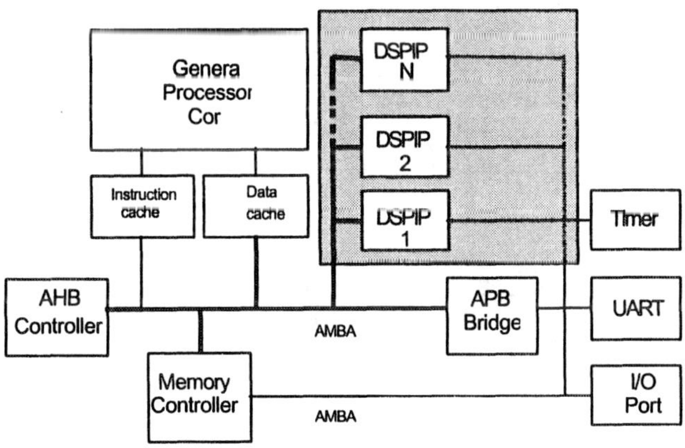

Figure 1. The Low Power Wireless Processor

3. THE AMBA INTERFACE

As illustrated in Figure 1, the scheme in this paper can allow multiple high performance IP cores to be attached to the platform. In this paper only a single DSP IP core is used to demonstrate the above scheme. The DSP IP core is an FIR filter. An interface is required to handle the AMBA communication protocols on behalf of this core. This interface is implemented as both an AHB slave as well as an APB device. Inputs into the interface must go through the AHB, while outputs from the interface can go through AHB or APB, depending on the urgency of output data required by the LEON processor.

An AMBA-based low power clocking scheme is devised for the interface to reduce the switching activity in the AMBA interface and the connecting DSP IP Cores. This clocking scheme is based on a 3 level hierarchy of clocks, where the relevant module is clocked when required.

The AMBA interface is first partitioned into 4 modules, consisting of the AHB Address and Control Decoder, AHB Data FSM, APB Address and Control Decoder, and the Clock Controller. This is shown in Figure 2.

The AHB address and control decoder will decode all the address and control signals from the LEON processor and send a signal to the AHB FSM to indicate its next state operation for the next clock cycle. This is possible due to the pipelined nature of the bus where the address and data are overlapped to allow high performance operation. The decoded address also enables the decoder to decide whether to clock the FSM and the connecting cores, by sending signals to the clock controller.

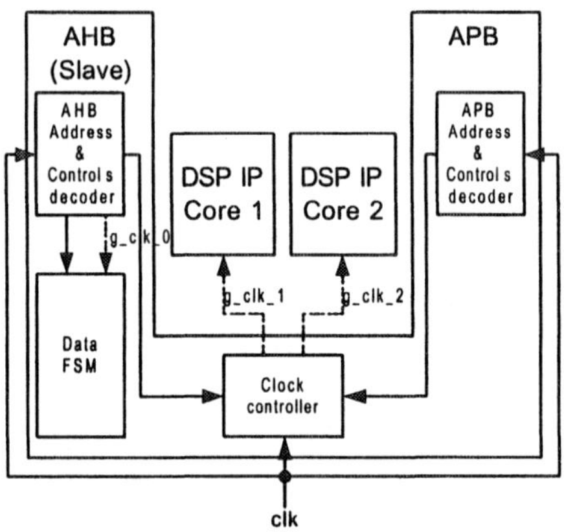

Figure 2. Partitions of the AMBA interface with 2 DSP IPs

The AHB FSM reacts to the signals send by the AHB address and control decoder, and routes the AHB write data bus to the input of the selected DSP IP core, or routes the data from the output of the selected core back to the AHB read data bus. This AHB FSM is clocked based on the gated clock signal from the clock controller.

The APB address and control decoder decodes the address and control signals send by the APB Bridge. It will only respond to read operations while write operations are ignored. During a read operation, the output data from the selected core will be routed back to the APB read data bus, and transferred back to the APB Bridge. Since this decoder only handles output data from the cores, a FSM is therefore not required.

The Clock controller will determine whether to clock the cores based on the signals from the AHB or APB address and control decoders. An AHB read or write operation will trigger the AHB address and control decoder to send signals to the Clock controller, indicating which DSP IP core has to be clocked. Meanwhile, an APB read operation will trigger the APB address and control decoder to send signals to the Clock controller to clock the selected core.

4. RESULTS AND DISCUSSION

The whole system was simulated at RTL level with Mentor Graphics' ModelSim simulator. A section of the timing diagram from the system is shown in Figure 3. Due to the pipelined nature of the AHB operations, the data arrives on the next clock cycle following the address. Therefore, it gives the AHB Address and Control decoder 1 clock cycle time to anticipate and decide the next clock operation as well as knowing which relevant DSP IP Core to clock.

Using Figure 3 as an example, the address of 0xA3000000 indicates that on the next clock cycle, the data of 0x000042DE has to be transfer from the AHB write data bus to the DSP IP input. The AHB address and control decoder clocks AHB FSM twice. First time is to load the FSM's next state, while the second time is to transfer the next state to current state. At the same time, the AHB or APB address and control decoder sends signal to the Clock controller to request for 1 clock cycle for the selected DSP IP core to load the data to its input.

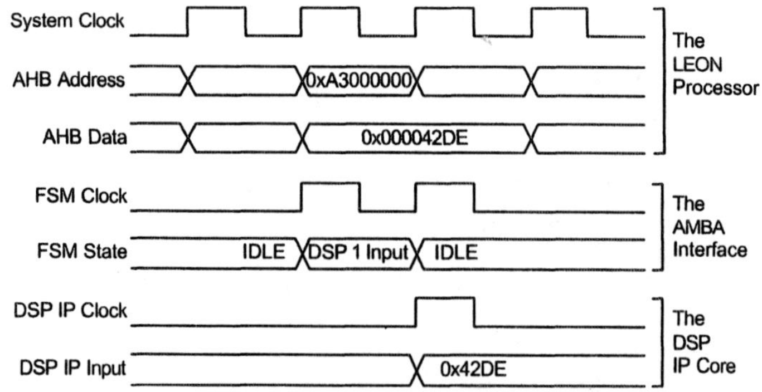

Figure 3. Timing diagram for data transfer from the LEON Processor to the DSP IP via the AMBA interface

The power analysis was then performed with Synopsys Power tools. The power distribution of the main blocks of the system is shown in Figure 4. Clearly, the most power is consumed by the LEON processor, followed by memory controller and the timers.

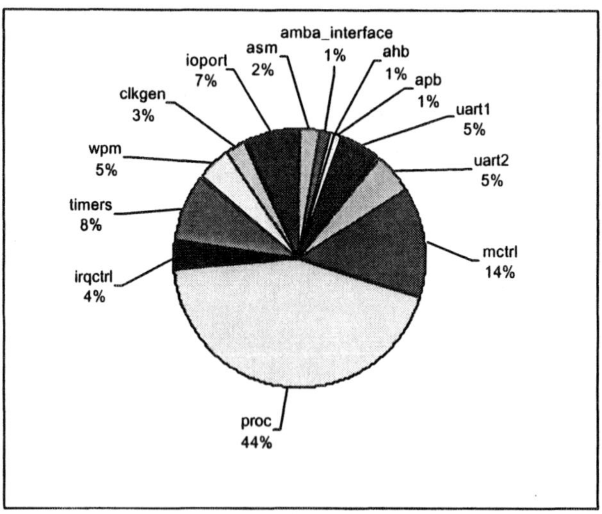

Figure 4. The Power Distribution of the Main Blocks

proc: *Processor and cache sub-system*

mctrl:	*Memory controller*
ahb:	*AHB arbiter/decoder*
apb:	*AHB/APB bridge*
uart:	*Universal receiver/transmitter*
asm:	*AHB status register*
wpm:	*AHB write protection register*
irqctrl:	*Interrupt controller*

5. CONCLUSION

The authors have presented a low power platform for the implementation of high performance DSP intensive tasks. In addition, a generic wrapper architecture was demonstrated to integrate DSP IP cores to the platform. Simulations were carried out with an FIR filtering IP core integrated on to the platform. Results were provided for timing and power consumption of the whole SoC platform. The work is subject of on going research. Current work involves synthesis and simulation of complete systems with multiple high performance DSP IP cores which include blocks for audio and image manipulation together with IP cores for transmitting these over a range of channels.

REFERENCES

[1] HUANG, J.R., IYER, M.K., CHENG, K.T.: 'A self-test methodology for IP cores in bus-based programmable SoCs', VLSI Test Symposium, 19th IEEE Proceedings on VTS 2001, 2001 pp. 198 –203.

[2] WILTON, S.J.E., SALEH, R.: 'Programmable logic IP cores in SoC design: opportunities and challenges', *Custom Integrated Circuits, 2001, IEEE Conference on 2001* pp. 63–66.

[3] KIM, K.W., KWANG, H.B., SHANBHAG, N., LIU, C.L., KANG S.M.: 'Coupling-driven signal encoding scheme for low-power interface design', *Computer Aided Design, 2000. ICCAD-2000. IEEE/ACM International Conference on*, 2000, pp. 318 -321.

[4] YOO, S.J.; NICOLESCU, G., LYONNARD, D., BAGHDADI, A., JERRAYA, A.A.: 'A generic wrapper architecture for multi-processor SoC cosimulation and design', *Hardware/Software Codesign, 2001. CODES 2001. Proceedings of the Ninth International Symposium on*, 2001 pp. 195 -200.

[5] GAISLER J.: 'The LEON Processor User's Manual', Version 2.3.7, August 2001, Gaisler Research, www.gaisler.com.

[6] AMBA™ Specification, Revision 2.0, May 1999, © ARM Ltd., www.arm.com/Pro+Peripherals/AMBA

[7] HELLMICH, H.H.H., ERDOGAN, A.T., ARSLAN, T. 'Re-Usable Low Power DSP IP embedded in an ARM based SoC Architecture', *IEE Coloquim on Intelectual Property,* Edinburgh, UK, July 2000, pp. 10/1 - 10/5.

[8] YEUNG, C., MATTHEWS, G., MORRIS, J., HAVERINEN, A., ZAIDI, J.: 'Standard bus vs. bus wrapper: what is the best solution for future SoC integration?', Design, Automation and Test in Europe, 2001. Conference and Exhibition 2001. Proceedings, 2001 pp. 776 - 776.

[9] OELMANN, B., O'NILS, M.: 'Asynchronous control of low-power gated-clock finite-state-machines', *Electronics, Circuits and Systems, 1999. Proceedings of ICECS '99. The 6th IEEE International Conference on,* Volume: 2, pp. 915 -918.

[10] WU, Q., PEDRAM, M., WU, X.W.:'Clock-gating and its application to low power design of sequential circuits', Circuits and Systems I: Fundamental Theory and Applications, IEEE Transactions on, Volume: 47 Issue: 3, March 2000, pp. 415 -420.

[11] ERDOGAN, A.T., ARSLAN, T.: 'Low power multiplication scheme for FIR filter implementation on single multiplier CMOS DSP processors', *Electronics Letters,* Volume: 32 Issue: 21, 10 Oct. 1996, pp. 1959 -1960.

[12] ERDOGAN, A.T., ARSLAN, T., HORROCKS, D.H.: 'Low power multiplication schemes for single multiplier CMOS based FIR digital filter implementations', *Circuits and Systems, 1997. ISCAS '97., Proceedings of 1997 IEEE International Symposium on ,* Volume: 3 , 1997 pp. 1940 -1943 vol.3.

9

INTERCONNECT TECHNOLOGY

The advanced technologies that enable System-on-Chip device densities, also bring along their own problems. A major concern, as we move to 100nm and smaller minimum geometries, are the limitations imposed by interconnect. Interconnect covers both external connections to the chip as well as the internal connections between blocks. The very high signal speeds that accompany these advanced fabrication techn ologies also impose their own problems with respect to system and cross-talk noise and propagation time limitations. Interconnect is also responsible for a large fraction of the total power that is dissipated by an SOC chip. With these factors in mind, this chapter presents three sections that deal with various aspects of the interconnect problem.

In section 9.1, "SoC Interconnect in Deep Submicron technologies", by M. Bayoumi of the Centre for Advanced Computer Studies at the University of Louisiana at Lafayette, the author introduces issues associated with the migration to Ultra Deep-submicron (UDSM) processes of 0.25 µm and below. These SoCs technologies introduce various challenges in terms of design flows, CAD tools and virtual components. A design methodology that allows component reuse and intellectual property is necessary for achieving the required functionality, performance and testability while minimizing the cost and time to market. This design methodology relies on the use of a standardized connection interface, such as a shared bus, which presents increasing difficulties in SoC design. This section discusses current research directions and various levels of design abstraction to increase the performance of interconnect. The author's survey includes buffer insertion, wire sizing and spacing, network ordering, and minimized coupled switchings through signal encoding. He also discusses approaches to adopt new analytical models for interconnects, physical design levels and finally ways to face these challenges early in the design flow. To maximize the benefits of this section, an extensive set of references is provided.

In section 9.2, "Optimizing Inductive Interconnect for Low Power", by Madgy A. El-Moursy and Eby G. Friedman, of the University of Rochester, the authors consider the power dissipation problem associated with interconnect. The width of an interconnect line affects the total power consumed by a circuit. The short-circuit power of an overdriven interconnect line decreases with the line width, while the dynamic power increases, so a trade-off exists between the dynamic power and the short-circuit power dissipated in inductive interconnect. In this work the optimum line width that minimizes the total transient power dissipation is determined, and a closed form solution for the optimum width, with an error less than 5%, is presented. The optimum line width is shown to be more effective in reducing the total transient power as the line becomes longer. With aluminum interconnect, a reduction in power of about 80% and 37% is obtained as compared to thin and wide wires. For copper interconnect, a reduction in power of 68% and 42% is obtained for the same conditions. Greater power reduction is achieved for optimally sized lines with higher resistivity as compared to minimum width lines. The optimum interconnect width depends upon the driver size and the number of gates being driven. With this solution, the optimum driver and wire size can be simultaneously determined.

The "Skin Effect" causes interconnect resistance and inductance to be frequency dependent. In the final section, "Skin Effects in System on a Chip Interconnects", by Shizhong Mei and Yehea I. Ismail of Northwestern University, the authors address the problem of efficiently estimating the signal characteristic of any RLC network when skin effect is significant. A new type of moments is defined that simplifies the interconnect simulation, namely, the square root moments. The time to calculate the square root moments is similar to the time to calculate the traditional moments, and the new moments preserve the recursive properties of the traditional moments. The method introduced in this section can handle the more complex problem of interconnect simulation with skin effect at almost no overhead compared to constant element interconnect simulation. Using these new square root moments, higher order approximations can be used compared to traditional moments. Finally, this method is applied to a sample circuit, and the effectiveness of the technique demonstrated by the high accuracy of the results.

Section 9.1

SoC Interconnect in Deep Submicron

Magdy Bayoumi
Center for Advanced Computer Studies (CACS)
University of Louisiana at Lafayette

Abstract: The migration to using Ultra Deep-submicron (UDSM) process, 0.25 μm or below has enabled the integration of complete electronic systems onto one single chip. These Systems-on-Chip (SoCs) introduce various challenges in terms of design flows, CAD tools and virtual components. A design methodology that allows component reuse and intellectual property is necessary for achieving the required functionality, performance and testability while minimizing the cost and time to market. This design methodology relies on the use of a standardized connection interface like a shared bus, which presents increasing difficulties in (SoC). This paper describes research directions and various levels of design abstraction to increase the performance of interconnect. These directions include approaches to adopt new analytical models for interconnects, physical design levels and finally ways to face these challenges early in the design flow. To maximize the benefits of this paper, an extensive set of references is given.

Key words : SoC, interconnect, noise, deep submicron, routing

1. INTRODUCTION

SoC is defined as the integration of several functional elements of a complete end product into a single chip, (e.g., a single chip radio or a single chip camera). These elements include programmable processors, on-chip memory, DRAM, accelerating functional units, analog components, microelectronic mechanical systems (MEMS), embedded software, reconfigurable logic, RF, and optical input/output. This complex integration has been motivated by the vast advance in silicon process technology. *Figure 11* illustrates a typical structure of such a SoC.

SoC has evolved the reuse of complex components like Intellectual Property blocks in a platform-based design approach. A platform is defined as a coordinated family of hardware-software architectures that satisfy a set of architectural constraints [44]. The design and development of SoCs encounters several challenges [1], mainly: the silicon complexity (reliability, power, clock, interconnect), the system complexity (logic, memories, RF, FPGA, MEMS), the design procedures (300 to 800 people for the design of a single chip, IP reuse, design levels), verification and testing at different levels. It is very clear that these challenges are getting worse especially in submicron technology due to interconnect delays and noise, many processors and very high frequencies. The design methodologies for SoC rely on the use of a standardized connection interface like a shared bus. This kind of interface has limitations to satisfy the higher communication costs. The increases in wire resistance and capacitance mean that wire delays are becoming more significant compared to gate delays as process feature sizes shrink. This paper focuses on research directions and various levels of design abstraction to increase the performance of interconnect.

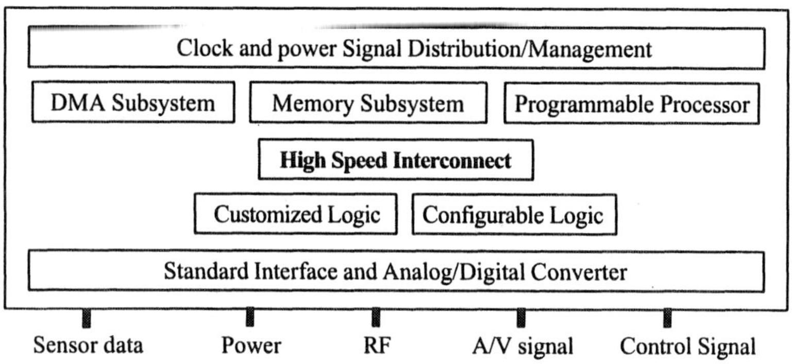

Figure 11. A typical SoC Architecture

Recent development in the area of SoC, as stated in [45], lies in several areas. First, moving from the digital-dominated mixed-signal Soc to one where analogue, digital, and other processing domains (such as RF, MEMS, optical, etc.) are all considered equal partners for design and integration of complex systems. Second, bringing the analogue and digital design teams together and offering a co-design and co-verification environment. This requires the support of both top-down and bottom-up methodologies, new variants of the standard languages such as Verilog-AMS and VHDL-AMS, an understanding of capacity, speed and various accuracy tradeoffs, and equal attention paid both to cell library models and semiconductor device models. Finally, is to treat on-chip communications between IP blocks or islands of functional processing not as simple point-to-point connections or through the use of standard board-style buses, but as the design of an on-chip communications network closer to that in telecom/datacom networks today.

The function of interconnects or wiring system is to distribute the clock and other signals and to provide power/ground to and among the various circuits/systems on the chip. Current leading-edge logic processors have 6–7 levels of high-density interconnect, and current leading-edge memory has three levels [2]. Distribution of the clock and signal functions is accomplished on three types of wiring: local, intermediate, and global. *Local wiring*, consisting of very thin lines, connects gates and transistors within an execution unit or a functional block (such as embedded logic, cache memory, address adder) on the chip. Local wires usually span a few gates and occupy first and sometimes second metal layers in a multi-level system. The lengths of these wires tend to scale down with technology. *Intermediate wiring* provides clock and signal distribution within a functional block with typical lengths up to 3–4 mm. Intermediate wires are wider and taller than local wires to provide lower resistance signal/clock paths. *Global wiring* provides clock and signal distribution between the functional blocks, and delivers power/ground to all functions on a chip. Global wires, which occupy the top one or two layers, are longer than 4 mm and can be as long as half of the chip perimeter.

Figure 12 shows the delay of local and global wiring in future generations [2]. Repeaters can be incorporated to mitigate the delay in global wiring but they consume power and chip area. Also techniques like inserting pipelined latches into the datapath, using a delay insensitive signaling scheme, and splitting the datapath into a number of independent, narrower datapaths, can be used.

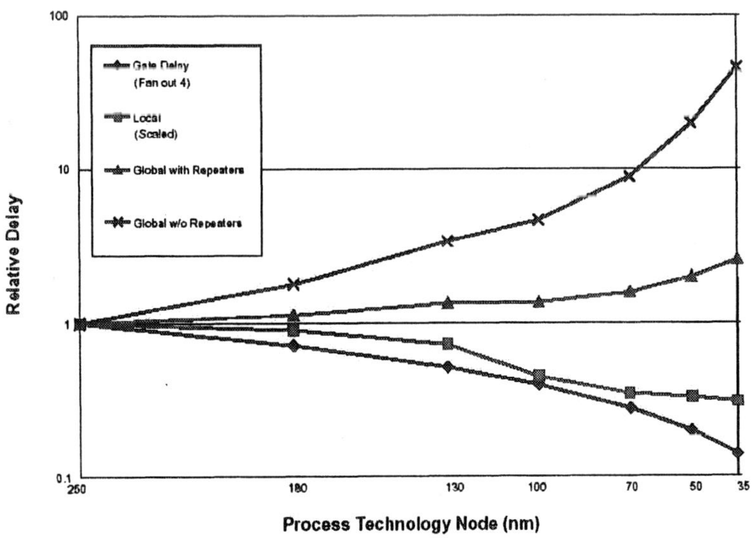

Figure 12. Delay for Local and Global Wiring versus Feature Size [2]

In the long term, new design or technology solutions (such as coplanar wave guides, free space RF, optical interconnect) will be needed to overcome the performance limitations of traditional interconnect. Inductive effects will also become increasingly important as frequency of operation increases, and additional metal patterns or ground planes may be required for inductive shielding. As supply voltage is scaled or reduced, crosstalk has become an issue for all clock and signal wiring levels; the near term solution adopted by the industry is the use of thinner metallization to lower line-to-line capacitance. This approach is more effective for the lower resistivity copper metallization, where reduced aspect ratios (A/R) can be achieved with less sacrifice in resistance as compared with aluminum metallization. The 1999 Roadmap [2] reflects this design trend by featuring reduced aspect ratios (as an alternative means of reducing capacitance) and less aggressive scaling of dielectric constant in comparison with the 1997 Interconnect Roadmap. The latter change expands the development window to address the difficulty in integrating low *k* dielectrics into a damascene architecture.

In this paper we survey new trends for interconnect synthesis and optimization techniques. We emphasize interconnect delay and noise models, techniques and algorithms to optimize them, and comment on advantages and drawbacks of each. In order to achieve this goal, we give a detailed explanation of what noise in interconnects is, how to deal with the different types of noise either separately or all together and how each type affects the delay or a function failure.

2. INTERCONNECT NOISE SOURCES

Noise in digital circuits is defined as any deviation of a signal from its stable value in those subintervals of time when it should be otherwise stable. In this section we describe interconnect noise sources and their scaling.

Interconnect Crosscapacitance Noise

Interconnect crosscapacitance noise refers to charge injected in quiet nets (victims) by switching on neighboring nets (aggressors) through the capacitance between them (crosscapacitance). This is perceived to be the most significant source of noise in current processes (

Figure 13). Crosscapacitance can also affect delay and slew depending on whether the aggressor signals are switching in the same or the opposing direction to the victim net.

Figure 13. Crosscapacitance noise in digital circuits

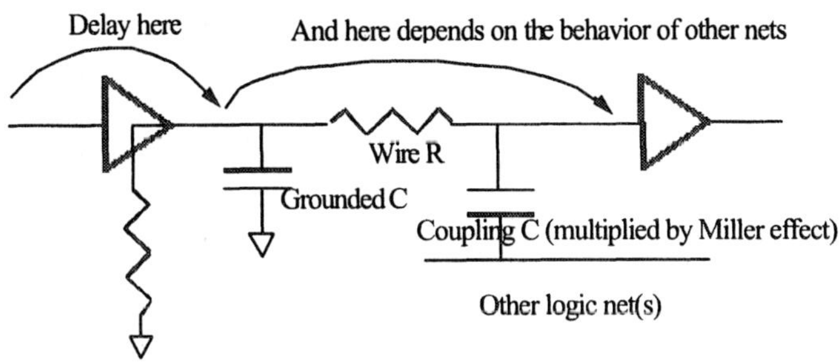

Power Supply Noise

Power supply noise is the difference between the local voltage references of the driver and receiver, which can appear as a spurious signal to the receiver and cause circuit failure. There are two components of power supply noise. The low-frequency component, known as IR drop, which is a variation in the dc power supply and ground levels. This component is managed well by flip-chip C4 packaging, which provides a very low resistance current path. The simultaneous switching of off-chip drivers and internal circuits produces the high frequency component known as delta-noise. This sudden demand for current leads to large powe r supply bounce since the current must be supplied through inductance of the chip power grid and chip-package. Power supply noise is a dominant factor in the design of wide domino circuits and in circuits using contention where the AC logic level is shifted with respect to power supply rails.

Mutual Inductance Noise

When signal switching causes transient current to flow through the loop formed by the signal wire and current return path, a changing magnetic field is created and the mutual inductance noise occurs [43] (

Figure 14). A voltage on a quiet line, which is in or near this loop, is induced. For several signals in a bus switching simultaneously, these noise sources can be cumulative. Crosscapacitance is a short-range phenomenon; however, mutual inductance can be a long-range phenomenon, which is worse in the presence of wide busses. This noise is very significant in current technologies due to faster switching speeds and wider, synchronous bus structures. Combined with capacitive noise, inductive noise can cause even worse noise than shown in

Figure 14. The analysis of inductive effects is quite complex and highly dependent on layout. Therefore, the trend is to design the problem through rules rather than to analyze arbitrary configurations.

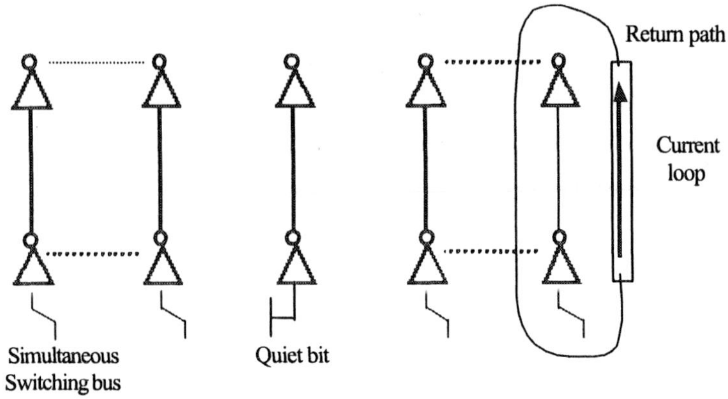

Figure 14. Mutual inductance noise from simultaneous switching on a wide bus [43]

Thermal Noise

Ebctrical power distribution and signal transmission through interconnects are always accompanied by thermal noise due to self-heating (or Joule heating) caused by the current flow. Thermal noise affects both interconnect design and its reliability. First, to limit the temperature increase in interconnects, the maximum allowable RMS current density is limited (since the RMS value of the current density is responsible for heat generation). Second, thermally induced open-circuit metal failure, under short-time high peak currents (including ESD), is a reliability concern [5], and can introduce latent electromigration damage, (transport of mass in metals under an applied current density) [6]. Third, interconnect lifetime (reliability), which is limited by (EM), has an exponential dependence on the inverse metal temperature. Hence, temperature rise of metal interconnects due to self-heating phenomena can also limit the maximum allowed average current density, since EM capability is dependent on the average current density [4]. It has been shown that thermal effects will dominate interconnect design rules that determine maximum current densities for circuit designers [7]. Recently, Hunter [8] has followed up on this issue by solving the EM lifetime equation for Al-Cu, and the 1D heat equation in a self-consistent manner. In this approach, both EM and self-heating can be comprehended simultaneously.

3. INTERCONNECT OPTIMIZATION

The main goal of interconnect optimization is to satisfy the predefined conditions on area, delay, and noise in interconnects. In order to analyse delay and noise induced in interconnects and to be able to include the model in the inner loop of an optimization technique, an accurate and sufficiently fast interconnect model is needed. In this section, we survey some of the interconnect models and show what parameters should be taken into account for that model. Then we show what techniques are used to alleviate the delay and noise in interconnects. This includes buffer and shield insertion, wire sizing and spacing, and net ordering. All of these techniques affect interconnect topology and the layout.

Analytical Models

The authors in [42] introduce a linear driver model that accurately models the noise pulse induced on a switching signal net due to cross coupling capacitance. The proposed model effectively captures the non-linear behavior of the victim driver gate during the transition and has an average error below 8% whereas the traditional approach using a Thevenin model incurs an average error of 48%. They also discuss the worst case alignment of the aggressor net transitions with respect to the victim net transition, emphasizing the need to maximize not merely the delay of the interconnect alone but the combined delay of the interconnect and receiver gate. They show that the worst case alignment of an aggressor net transition is a function of the receiver gate output loading, victim transition edge rate, and the noise pulse width and height and hence propose a pre-characterization approach to efficiently predict the worst-case alignment. The proposed methods were implemented in an industrial noise analysis tool called ClariNet. Results on industrial designs are presented to demonstrate the effectiveness of their approach. In that paper, they propose a more accurate model by replacing the standard Thevenin resistance R_{th} with a *transient holding resistance* R_{tr}. They determine this transient holding resistance such that it produces a matching noise waveform with noise injected on the nonlinear victim driver. Their approach is outlined as follows. First, they obtain the aggressor noise on the victim net by performing a linear simulation using the standard Thevenin resistance for the shorted victim driver. From the simulation result, they calcu late the associated noise current that is injected into the victim driver output. Then, they simulate the non-linear victim driver switching with an effective load, both with and without this injected noise current. Since the victim driver is switching whe n the aggressor driver injects noise on the victim, they cannot directly observe the noise injected on the victim line. They can only construct it from the difference of the driver responses with and without injected noise. Thus, they subtract the two driver output waveforms to obtain the noise waveform at the victim driver output. Then they calculate a transient holding resistance that yields a matching noise pulse (this is shown in

Figure 15).

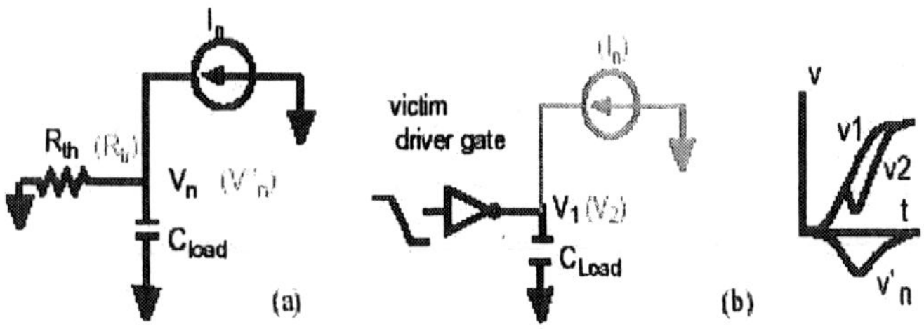

Figure 15. Transient holding resistance (R_{tr}) calculation [42]

The optimization problem chosen in [28] is to reduce the amount of crosstalk in a routed channel. The method takes an initial routing solution that attempted to minimize the number of tracks, and modifies the solution to reduce the crosstalk-induced delay while leaving the number of tracks in the channel unchanged. It permits the direct incorporation of the delay in the objective function, and instead of permuting full tracks; this approach allows segments of tracks to be permuted. A simulated annealing approach is used to perform the permutations. The procedure has a polynomial time complexity in the worst case. In that work the authors emphasize that the exact functional form that is used to estimate the capacitance and the delay are not important. The only requirement that the delay model must satisfy is that an increase [decrease] in the coupling capacitance s hould be translated into an increase [reduction] in the delay of a net. The author also suggests that it may be possible to use the same technique to simultaneously satisfy requirements on delay and noise.

The authors in [32] address the problem of crosstalk computation and reduction using circuit and layout techniques. They provide expressions for crosstalk amplitude and pulse width in resistive, coactively coupled lines. They claim that the expressions hold for nets with arbitrary number of pins and of arbitrary topology under any specified input excitation. The expressions are used to motivate circuit techniques such as transistor sizing, and layout techniques, such as wire ordering and wire width optimization, to reduce crosstalk. They do not deal with coupling due to magnetic field interactions. They claim that the magnitude of inductive coupling is small in the presence of good ground return paths close to signal lines. They used their expression in formulations aimed at reducing crosstalk. These methods include transistor sizing, wire ordering, wire width optimization and wire spacing. They failed to get success in some of these formulations. In wire ordering which is the work reported in [3], they propose a new heuristic since they did not report results for this work. Further, reduced order modeling of interconnect [24][26] has allowed the analysis of large interconnect networks in a computationally efficient manner.

Buffer Insertion

The dependence of the propagation delay on the length of an RC line is quadratic. For long interconnects, wire sizing or spacing, explained in next subsections, are not sufficient to limit the interconnect delay. So, buffer insertion is used to trade-off the active device area for reduction of interconnect delays. The shorter the sections of the line, the greater the interconnect delay is reduced and that it also reduces coupling noise. However, the additional inserted buffers increase the overall gate delay. Thus, there is an optimal number of repeaters that should be inserted into an RC line to minimize the overall propagation delay. This trend is qualitatively illustrated in *Figure 16* from [50].

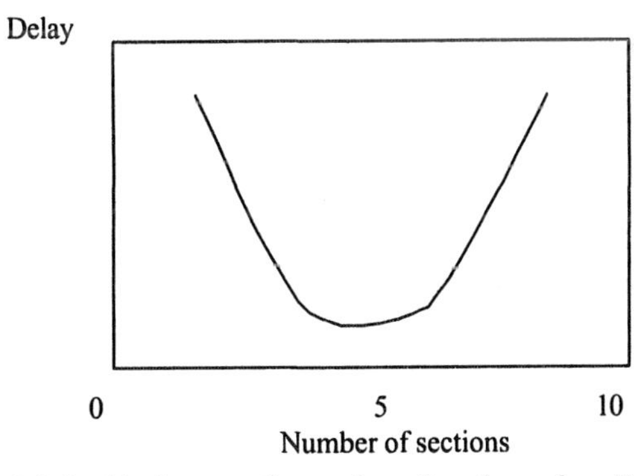

Figure 16. Relationship between the number of sections of an RC line and the total propagation delay

During the routing of global interconnects, macroblocks form useful routing regions which allow wires to go through but forbid buffers to be inserted. They give restrictions on buffer locations. In [31], the authors take these buffer location restrictions into consideration and solve the simultaneous maze routing and buffer insertion problem. Given a block placement defining buffer location restrictions and a pair of pins (a source and a sink), they give a polynomial time exact algorithm to find a buffer route from the source to the sink with minimum Elmore delay.

Wire Sizing

It is known that proper wire sizing can effectively reduce the interconnect delay especially in deep submicron or nanometer designs when the wire resistance becomes significant.

An optimal wire-sizing algorithm was developed in [48],[49] for a single source RC interconnect tree to minimize the sum of weighted delays from the source to timing-critical sinks under the Elmore delay model.

The work in [39] presents an efficient approach to perform global interconnects sizing and spacing (GISS) for multiple nets to minimize interconnect delays with consideration of coupling capacitance, in addition to area and fringing capacitances.

Symmetric wire sizing, *Figure 17*, may be too restrictive for interconnect sizing and spacing, especially when coupling capacitance is considered. In that paper, they propose an asymmetric wire-sizing scheme where they may widen or narrow above and below the centre line of the original wire segment asymmetrically.

In the global interconnect sizing and spacing problem for multiple nets; they assume that an initial layout of n routing trees Ti's is given. With consideration of the area, fringing and coupling capacitances, the GISS problem for multiple nets is to find a symmetric wire assignment or an asymmetric wire assignment for all Ti's such that, the summation of the weighted performance measure of all nets is minimized. The optimal wire sizing and spacing problem for a single net with fixed surrounding wire segments can be solved by adapting the bottom-up dynamic programming (DP)-based buffer insertion and wire-sizing algorithm proposed by [40].

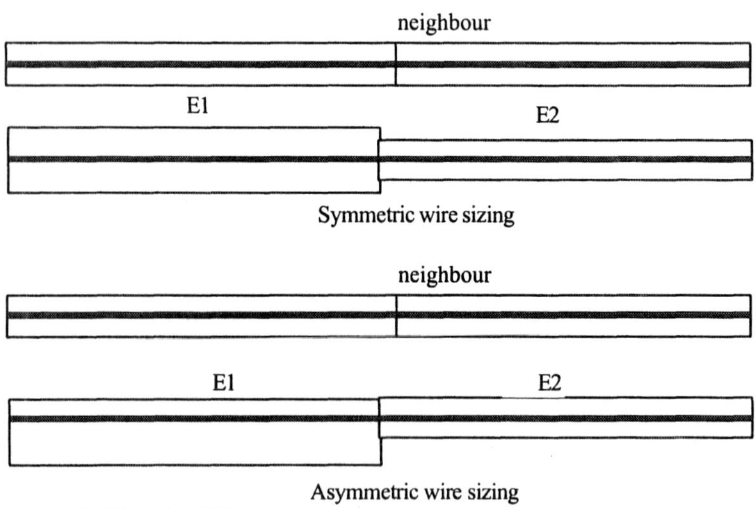

Figure 17. Wire sizing [39]

The difficulty of the GISS problem for multiple nets arises from the fact that the lateral coupling capacitance coefficient for each wire segment changes with the width and spacing. Moreover, there is no closed form representation for lateral coupling capacitance. The key to solving the multiple-net wire sizing and spacing problem is the effective-fringing property of wire-sizing solutions under different spacing conditions. The beauty of this property is that it is possible to reduce the GISS problem with variable lateral coupling capacitance to an optimal wire-sizing problem with consideration of only constant area capacitance coefficient and different effective-fringing capacitance coefficients for different wire segments. Such a reduction allows us to compute global upper and lower bounds of the optimal wire sizing and spacing solution for all nets.

Wire Spacing

As we move to deep submicron era, we have a higher aspect ratio for interconnects. This leads to higher cross capacitance. In order to decrease the value of this capacitance, either we increase the spacing between wires or insert shields. This subsection describes how wire spacing can decrease the crosscapacitance value. For example, in [41], the graph-based optimizer preroutes wires on the global routing grids incrementally in two stages - net order assignment and space relaxation. The timing delay of each critical path is calculated taking into account interconnect coupling capacitance. The objective is to reduce the delays of critical nets with negative timing slack values by tuning net ordering and adding extra wire spacing.

In a graph base framework for the crosspoint assignment problem, there could be multiple critical nodes to which they need to allocate routing resources [41]. However, there may be multiple directed paths connecting two critical nodes in a channel graph, the space allocation for one node may degrade the possible space allocation of other critical nodes on the same directed path. To allevia te the ordering problem in space allocation, they allocate only partial space resources around a node at each iteration. At each iteration, the critical node *nc* (the longest segment) of the most critical net *Nc* (*timing_slackNc* = *min*(*timing_slack*)) is selec ted for space relaxation. The available spaces around the selected node are allocated by two stages of relaxation optimizations - *self relaxation* and *neighbor relaxation*. The former has higher priority than the latter. Node *nc* is said to be *self-relaxable* if it's movable in the graph. In the self relaxation stage, one of the upper and lower sides of *nc* is selected, based on the least cost, and all neighbors on that side are moved outward by a predetermined incremental spacing (e.g. 0.5 track). Its cost function is equal to the increase in the total capacitances of interesting nets due to this operation. In the neighbor relaxation stage, the neighboring node is selected, based on the least cost, to add space to *nc*. The cost function is the same as that of self relaxation.

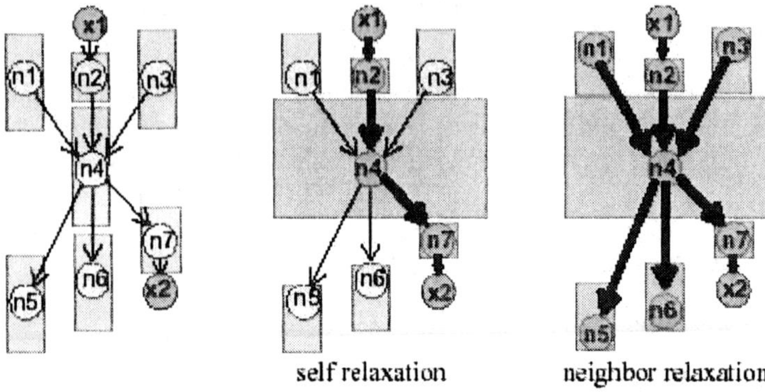

Figure 16. Self relaxation of node 4 and its neighbor relaxation [41]

An example of self relaxation and neighbor relaxation on node 4 is shown in Figure 18. The light shaded region of each node in (a) stands for the movable region, and the dark nodes in (b) are non-relaxable because extra spaces (dark shaded areas) are added. The relaxable neighbors *n1, n3, n5, n6* are further relaxed outward by adding extra spacing until none of them is movable.

SHIELD INSERTION

As described in the previous subsection, shield insertion is another technique for decreasing crosstalk. In the work presented in [38], the authors show that existing net ordering formulations to minimize noise are no longer valid with presence of inductive noise, and shield insertion is needed to minimize inductive noise. They formulate two simultaneous shield insertion and net ordering (SINO) problems: the optimal SINO/NF problem to find a min-area SINO solution that is free of capacitive and inductive noise, and the optimal SINO/NB problem to find a min-area SINO solution that is free of capacitive noise and is under the given inductive noise bound. They claim that this work is the first to present an in depth study on the simultaneous shield insertion and net ordering problem to minimize both capacitive and inductive noise. Throughout this work they consider only parallel coplanar interconnect structure with all wires having the same length. They also assume that an appropriate sensitivity matrix is given a priori. This matrix stores the sensitivity relation between every pair of wires. They defined the sensitivity graph and proved a theorem, which states that the maximum clique size in the sensitivity graph is a lower bound of the number of blocks required in all optimal SINO/NF solutions.

They developed four approximate algorithms for solving the SINO/NB problem: greedy-based shield insertion (SI) algorithm, net ordering for minimizing C_x noise followed by SI algorithm (NO+SI algorithm), graph -coloring based SINO algorithm (SINO/GC algorithm), and simulated annealing based SINO algorithm (SINO/SA algorithm). They solve the SINO/NF problem by using the SINO algorithms and setting the noise bound to zero. The SINO/SA always performs the best for any given setting.

NETWORK ORDERING

It has been shown that signal ordering can effectively reduce the crosstalk noise on interconnects. In other words, arranging signals in an order that does not let a wire fights against its neighbour, making a transition in the same directions, can reduce the crosstalk noise. For example, [37], they propose two interconnect layout design methodologies for minimizing the "cross-coupling effect" in the design of full-custom datapath. These are the control signal ordering scheme and the track assignment algorithm.

The high-performance datapath is usually done by full custom layout design. In the datapath design, most control signals are related to the multiplexers. To implement low-power and high speed multiplexer, pass transistor logic gates are commonly used. In the multiplexer design, the control signals are generated from the standard cell block, globally routed to the datapath, and then drive the multi-bit datapath multiplexer cells.

Figure 17. (a) A multiplexer example showing the signal coupling effects and (b) three different orderings [37]

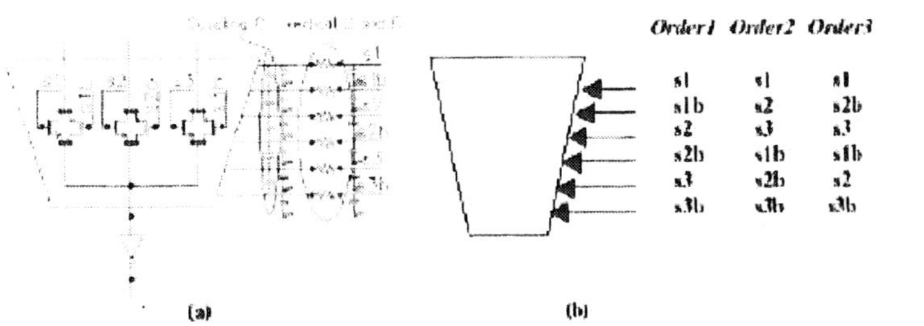

A conventional scheme for the multiplexer signals ordering is shown in Figure 9 (b). In the layout, the signals are very sensitive to the cross-coupling effect. The situation is not improved with the "order 2" layout. To minimizing the opposite direction transitions of neighbor wire, we can use the following rules.

. ***Rule 1:*** Do not place si/sib pair in neighborhood.

. ***Rule 2:*** Do not place si/sj pair in neighborhood for all i, j.

. ***Rule 3:*** Do not place sib/sjb pair in neighborhood for all i, j.

Following these rules, we can get the "Order 3" scheme in

Figure 17 (b). Table 5 summarizes the comparison of the frequency of opposite transitions among three different orderings. We can see that the transitions of "Order 3" are reduced to zero.

The cost of "Order 3" is its larger cell area. However, in modern technology, the wiring is dominant of the whole area, the cell area of a 3-1 MUX using "Order 3" is only 2.1% larger than that of "Order 2" using 0.18um technology. At the same time, the wiring complexity outside the cell can be kept the same as that of "Order 2".

Another solution is to offset the switching time of adjacent wires to reduce the transitions at the same time between the wires. Another useful way is to arrange the buses to travel in opposite directions, or to insert inverters along the bus in a staggering location way. In practice, however, the nonzero delay of the inverter results in a time offset and hence coupling capacitance of the wire.

The datapath track assignment based on evolutionary programming is also presented in this work. The goal of track assignment is to minimize the total number of tracks used. It is known that the "*Left-Edge algorithm*" gives optimal results. To mitigate the coupling capacitance between long interconnects, they should minimize the common runlength between adjacent wires. They propose an "*evolutionary programming-based track assignment (EPTA)*" algorithm considering both length and switching activity. It was shown that the solution quality of EP is quite good and the converging speed is very fast compared to SA and GA. More than anything else, EP does not require an annoying genetic operator design, which makes the implementation of EP very easy.

Table 5 Switching Directions of selection signals for 3 odering cases (a) order1 (b) order 2 (c) order 3

		Selection signal changes from a to b (a->b)						Total number of opposite transition
		1->2	1->3	2->1	2->3	3->1	3->2	
(a)	s1	↓	↓	↑	0	↑	0	12
	s1b	↑	↑	↓	1	↓	1	
	s2	↑	0	↓	↓	0	↑	
	s2b	↓	1	↑	↑	1	↓	
	s3	0	↑	0	↑	↓	↓	
	s3b	1	↓	1	↓	↑	↑	
(b)	s1	↓	↓	↑	0	↑	0	8
	s2	↑	0	↓	↓	0	↑	
	s3	0	↑	0	↑	↓	↓	
	s1b	↑	↑	↓	1	↓	1	
	s2b	↓	1	↑	↑	1	↓	
	s3b	1	↓	1	↓	↑	↑	
(c)	s1	↓	↓	↑	0	↑	0	0
	s2b	↓	1	↑	↑	1	↓	
	s3	0	↑	0	↑	↓	↓	
	s1b	↑	↑	↓	1	↓	1	
	s2	↑	0	↓	↓	0	↑	
	s3b	1	↓	1	↓	↑	↑	

4. INTERCONNECT-DRIVEN FLOORPLANNING

A big pressure on floorplanning has been placed because of the deep submicron era. Quick floorplanning tools are needed in the domain. These tools should account for the prediction of interconnects delay and noise early in the floorplanning phase. So, we mentioned some references about interconnect driven floorplanning.

In the current design flow, interconnect optimizations are mainly used in post layout stages [27]. As the global interconnect are largely determined by floorplanning, it becomes critical for floorplanning engines to be able to handle efficient interconnect planning and optimizations, so that the overall timing and design convergence can be achieved.

Due to the inherent complexity of the IDFP problem, they use multi-stage, adaptive cost functions within IDFP to gradually consider more interconnect optimization, planning, and/or global routing features. Using the adaptive cost functions gives them flexibility to tune for different design objectives, and to tradeoff between performance and run time. They use simulated annealing algorithm to achieve this flow. They divide the simulated annealing temperature region into four different stages from high to low temperatures. From Stage 1 to Stage 4, they use progressively more accurate interconnect performance and routability measurements as the temperature goes down. A, TL, ML, D, CJ1, CJ2 are the parameters used in the weighted cost function and they mean area, total wire length, maximum length, maximum net delay, congestion estimation, the average tile boundary congestion respectively. These terms are having adaptive relative importance in the cost function. This work adopted a net-based delay model, which is not practical.

5. ROUTING

Traditionally, the objective of the routing problem is dependent on the nature of the chip. For general purpose chips, it is sufficient to minimize the total wire length, while completing all the connections. For high performance chips, it is important to route each net such that it meets its timing budget. Recently, an important parameter, crosstalk, appeared that should be planed during routing. In this section we survey global and detailed routing techniques that consider crosstalk noise.

Global Routing

The crosstalk noise is usually considered in the detailed routing stage because of the need to know the wiring track adjacency information. However, the flexibility to change the wiring topology is limited at this stage. In [46] -[47], the authors consider crosstalk avoidance in global routing with the idea of exploiting the greater flexibility that is available at the global routing stage to alleviate the crosstalk problem. The primary goal of the work in [33] is to examine global interconnect effec ts to determine if there are any significant roadbacks which will prevent National Technology Roadmap for Semiconductors (NTRS) performance expectations from being met. Their analysis indicates that due to global RC delays as well as TOF considerations, th e global clock will necessarily be slower than the achievable local clock frequency. Also they have found that crosstalk at the global level will not be as significant as the local level due to the use of large repeaters whose capacitance will dampen the effects of coupling capacitance. They propose a wiring hierarchy which complements the modular design methodology proposed in [34]. This modular methodology proposes the use of 50 000 to 100 000 gate modules of bgic to eliminate the impact of interconnect at the local level. These modules are arranged together in isochronous (or locally-clocked) regions, which run at a higher clock speed than the global clock. These isochronous come together to form tho ontiro design.

Detailed Routing

The authors in [29] present a multilayer gridless detailed routing system for deep submicrometer physical designs. Their detailed routing system uses a hybrid approach consisting of two parts: 1) an efficient variable-width variable-spacing detailed routing engine and 2) a wire-planning algorithm providing high-level guidance as well as ripup and reroute capabilities. They use a nonuniform grid graph which has been proved to guarantee a gridless connection of the minimum cost in multilayer variable-width and variable-spacing routing problem. They suggested further improving their wire-planning algorithm and fine tuning of ripup and rerouting algorithm.

The authors in [30] developed an integer linear programming (ILP) formulation that yields tight bounds on the quality of the achievable solution. The input to crosstalk-aware wire-packing comprises a set of wires, a set of variable tracks (per layer), and a crosstalk graph XG that determines forbidden wire adjacencies. The output is a legal assignment such that no forbidden adjacencies and no electrical shorts exist. The key technical insight is to model all constraints (both geometric and crosstalk conflicts) as cliques in an appropriate conflict graph; these cliques can be extracted quickly from the interval structure of the slice.

The authors in [36] developed methods to reduce interconnect delay and nois e caused by coupling. The coupling free routing (CFR) takes a set of nets and tries to find a one-bend couple-free routing for a subset of nets. A routed net must not couple with any other routed net. They define coupling as a Boolean variable, which is true when the coupling is greater than some threshold. They develop an exact algorithm for the CFR decision problem via a transformation to 2-satisfiability. This algorithm runs in linear time. They also present the implication graph which models the dependencies associated with CFR. They also develop an algorithm for the maximum coupling-free layout (MAX-CFL) problem. Given a set of nets, the MAX-CFL is defined as finding a subset of nets that are coupling-free routable. The subset should have maximum size and/or criticality. They call this algorithm implication algorithm. They present the coupling capacitance between two wires i and j as follows:

$$C_c(i,j) = \frac{f_{ij} \cdot l_{ij}}{d_{ij}} \frac{1}{1 - \frac{w_i + w_j}{2d_{ij}}}$$

where w_i and w_j are the sizes of wires i and j (w_i, $w_j > 0$), f_{ij} is the unit length fringing capacitance between wires i and j, l_{ij} is the overlap length of wires i and j and d_{ij} is the distance from the center line of wire i to the center of wire j.

Since the nets are routed at most one bend, they have minimum wire length. In addition, coupling-free routing minimizes the coupling of the routed nets. Compiling these two factors, they have a routing of the critical nets with minimal interconnect delay.

It has been shown that the delay of a wire of length l increases at the rate of $O(l^2)$ without wire sizing, $O(l\sqrt{})$ with optimal wire sizing and linearly with proper buffer insertion [35]. The criticality function can easily be changed to incorporate some other functions.

6. MINIMIZING COUPLED SWITCHINGS

Increased coupling effect between interconnects in ultra deep sub-micron technology not only aggravates the power-delay metrics but also deteriorates the signal integrity due to capacitive and inductive crosstalk noises. Conventional approaches to interconnect synthesis aim at optimal interconnect structures in terms of interconnect topology, wire width and spacing, and buffer location and sizes [9].

Signal encoding schemes have been proposed to minimize transition activities on buses while ignoring cross-coupled capacitances. When statistical properties are unknown a priori, the bus-invert method [10] and the on-line adaptive scheme [11] can be applied to encode randomly distributed signals. On the other hand, highly correlated access patterns exhibit a spatio-temporal locality, which can be exploited for energy reduction [12] in Gray code [13],[14], the TO method [15], the working-zone encoding [16], the combined bus-invert/TO [22], and the coupling-driven method [23]. Lower bounds for minimum achievable transition activity have been derived for noiseless buses in [17] and for noisy buses in [18]. In [19], a segmentation method was introduced to reduce power consumption. In [20], specification transformation approaches were used to reduce the number of memory accesses at the behavioral level. The effectiveness of various encoding schemes was compared at the system level in [21].

7. CONCLUSIONS

In this paper, we present a global overview for Interconnects in SoC, explain the obstacles to be tackled, and review the recent developments in that area. We concentrated on research directions to increase the performance of interconnects. We survey areas like buffer insertion, wire sizing and spacing, network ordering, and minimized coupled switchings through signal encoding. Also interconnect optimization should be tackled at higher levels and during some earlier phases of the design flow. For example, we mentioned interconnect driven floorplanning. The conventional approach for interconnect, which is a tristate bus, is no longer valid for the deep submicron era. An alternative approach for SoC interconnect is a delay-insensitive, asynchronous approach to interconnect over long paths using 1-of-4 encoded channels. This is the new trend for bus interconnects. We also listed some references that deal with minimized coupled switching thru signal encoding without changing the interconnect topology.

REFERENCES

[1] Piguet, C.; Renaudin, M.; Omnes, T.J.-F., "Special session on low-power systems on chips (SOCs)," *Proceedings Design, Automation and Test in Europe, Conference and Exhibition*, 2001, PP. 488-494.

[2] http://public.itrs.net/files/1999_SIA_Roadmap/Int.pdf

[3] A. vital and M. Marek-Sadowska, "Crosstalk reduction for VLSI," *IEEE Trans. Computer-Aided Design*, Vol. 16, Mar. 1997.

[4] B. K. Liew, N. W. Cheung, and C. Hu, "Projecting Interconnect Electromigration Lifetime for Arbitrary Current Waveforms," *IEEE Trans. Electron Devices*, Vol. 37, No. 5, May 1990.

[5] K. Banerjee, A. Amerasekera, N. Cheung, and C. Hu, "High-Current Failure Model for VLSI Interconnects under Short-Pulse Stress Conditions," *IEEE Electron Device Lett.*, Vol. 18, No. 9, September 1997.

[6] K. Banerjee, A. Amerasekera, and C. Hu, "Characterization of VLSI Circuit Interconnect Heating and Failure under ESD Conditions," *Int. Reliability Physics Symp.*, Dallas, TX, April-May 1996.

[7] P. K. Chatterjee, W. R. Hunter, A. Amerasekera, S. Aur, C. Duvvury, P. E. Nicollian, L. M. Yang, and P. Yang, "Trends for Deep Submicron VLSI and Their Implications for Reliability," Int. Reliability Physics Symp., Las Vegas, NV, April 1995.

[8] W. R. Hunter, "Self-Consistent Solutions for Allowed Interconnect Current Density–Part I: Implications for Technology Evolution," *IEEE Trans. Electron Devices*, Vol. 44, No. 2, February 1997.

[9] J. Cong, "An Interconnect-centric design flow for nanometer technologies," Int. Symp. VLSI Technology, Systems, and Applications, pp. 54–57, June 1999.

[10] M. R. Stan and W. P, "Burleson. Bus-invert coding for low-power I/O," *IEEE Trans. on VLSI Systems*, pp. 49–58, Mar. 1995.

[11] L. Benini, A. Macii, E. Macii, M. Poncino, and R. Scarsi, "Synthesis of low-overhead interfaces for power-efficient communication over wide buses," *ACM/IEEE Design Automation Conf.*, pp. 128–133, 1999.

[12] P. R. Panda and N. D. Dutt, "Reducing address bus transitions for low power memory mapping," *European Design and Test Conf.*," pp. 63–37, Mar. 1996.

[13] H. Mehta, R. M. Owens, and M. J. Irwin, "Some issues in Gray code addressing," *the Great Lakes Symp. VLSI*, pp. 178–180, Mar. 1996.

[14] C. L. Su, C. Y. Tsui, and A. M. Despain, "Saving power in the control path of embedded processors," *IEEE Design and Test of Computers*, 11(4):24–30, 1994.

[15] L. Benini, G. De Micheli, E. Macii, D. Sciuto and C. Silvano, "Asymptotic zero - transition activity encoding for address busses in low-power microprocessor-based systems," *the Great Lakes Symp. VLSI*, pp. 77–82, 1997.

[16] E. Musoll, T. Lang, and J. Cortadella, "Working -zone encoding for reducing the energy in microprocessor address buses," *IEEE Trans. on VLSI Systems*, 6(4):568–572, Dec. 1998.

[17] S. Ramprasad, N. R. Shanbhag, and I. N. Hajj, " Information-theoretic bounds on average signal transition activity," *IEEE Trans. on VLSI*, Sept. 1999.

[18] R. Hegde and N. R. Shanbhag, "Energy-efficiency in presence of deep submicron noise," *IEEE/ACM Int. Conf. Computer Aided De-sign*, pp. 228–234, 1998.

[19] Y. Zhang, W. Ye, and M. J. Irwin, "An alternative architecture for on-chip global interconnect: segmented bus power modeling," *Asilomar Conf. on Signals, Systems, and Computers*, pp. 1062–1065, 1998.

[20] F. Catthoor, F. Franssen, S. Wuytack, L. Nachtergaele, and H. D. Man, "Global communication and memory optimizing transformations for low power signal processing systems," *VLSI Signal Processing VII*, pp. 178–187, 1994.

[21] W. Fornaciari, D. Sciuto, and C. Silvano, "Power estimation for architectural exploration of HW/SW communication on system-level buses," *Int. Workshop on Hardware/Software Codesign*, pp. 152– 156, 1999.

[22] L. Benini, G. De Micheli, E. Macii, D. Sciuto, and C. Silvano, "Address bus encoding techniques for system-level power optimization," *Design, Automation and Test in Europe*, Feb. 1998, pp. 861–866.

[23] K.W. Kim, K.H. Baek, N. R. Shanbhag, C. L. Liu, and S. Kang, "Coupling-driven signal encoding scheme for low-power interface-design," 2000 Intl Conf. on Computer-Aided Design.

[24] L. T. Pillage and R. A. Rohrer. Asymptotic waveform evaluation for timing analysis. IEEE Transactions on Computer-Aided Design of ICs and Systems, 9:352-366, April 1990.

[25] P. Feldmann and R. W. Freund. E_cient linear circuit analysis by Pade approximation via the Lanczos process. IEEE Transactions on Computer-Aided Design of ICs and Systems, 14(5): 639-649, May 1995.

[26] L. M. Silveira, M. Kamon, I. Elfadel, and J. White. A coordinate-transformed Arnoldi algorithm for generating guaranteed stable reduced-order models of RLC circuits. IEEE International Conference on Computer-Aided Design, pages 288-294, November 1996.

[27] C. -C. Chang, J. Cong, D. Zhigang, and X. Yuan, "Interconnect-Driven Floorplanning with Fast Global Wiring Planning and Optimization," *Proc. SRC Techcon Conference*, September 21-3, 2000, Phoenix.

[28] Sachin S. Sapatnekar, "A Timing Model Incorporating the Effect of Crosstalk on Delay and its Application to Optimal Channel," *IEEE Trans. Computer-Aided Design*, Vol. 19, No. 5, May 2000.

[29] Jason Cong et al, "DUNE-A Multilayer Gridless Routing System," *IEEE Trans. Computer-Aided Design*, Vol. 20, No. 5, May 2001.

[30] Rony Kay and Rob A. Rutenbar, "Wire Packing-A strong Formulation of Crosstalk-Aware Chip-Level Track/Layer Assignment with an Efficient Integer Programming," *IEEE Trans. Computer-Aided Design*, Vol. 20, No. 5, May 2001.

[31] Hai Zhou, D. F. Wong, I-Min Liu, and Adnan Aziz, "Simultaneous Routing and Buffer Insertion with restrictions on Buffer Locations," *IEEE Trans. Computer-Aided Design*, Vol. 19, No. 7, July 2000.

[32] Ashok Vittal and marek-Sadowska, "Crosstalk in VLSI Interconnections," *IEEE Trans. Computer-Aided Design*, Vol. 18, No. 12, December 1999.

[33] Dennis Sylvester and Kurt Keutzer, "A Global Wiring Paradigm for Deep Submicron Design," *IEEE Trans. Computer-Aided Design*, Vol. 19, No. 2, February 2000.

[34] D. Sylvester and K. Keutzer, "Getting to the Bottom of Deep Submicron," in *proc. ICCAD*, 1998, pp. 203-211.

[35] J.Cong and D.Z. Pan "Interconnect Delay Estimation Models for synthesis and design planning". In *proc. Asia and South Pacific Design Automation Conference*, January 1999.

[36] Ryan Kastner and Majid Sarrafzadeh, "An Exact Algorithm for Coupling-Free Routing," *International Symposium on Physical Design (ISPD)*, April 2001.

[37] J.S.Yim and C.M.Kyung, "Reducing Cross-Coupling among Interconnect Wires in Deep-Submicron Datapath Design," 36th Design Automation Conference (DAC), New Orleans, USA, PP.485-490, June 1999.

[38] Lei He and Kevin M. Lepak, "Simultaneous Shield Insertion and Net Ordering," proceedings of the *international symposium on Physical design*, April 9 - 12, 2000, San Diego, CA USA

[39] J. Cong, L. He and C.-K. Koh, "Global Interconnect Sizing and Spacing with Consideration of Coupling Capacitance ," *Proc. ACM/IEEE Int. Conf. on Computer-Aided Design*, November 1997.

[40] J. Lillis, C. K. Cheng, and T. T. Y. Lin, "Simultaneous routing and buffer insertion for high performance interconnect," in *Proc. the Sixth Great Lakes Symp. on VLSI*, 1996.

[41] Hsiao-Ping Tseng, Louis Scheffer, and Carl Sechen, "Timing and Crosstalk Driven Area Routing," Design Automation Conference, pp. 378-381, 1998.

[42] Supamas Sirichotiyakul, David Blaauw, Chanhee Oh, Rafi Levy, Vladimir Zolotov, Jingyan Zuo "Driver Modeling and Alignment for Worst-Case Delay Noise," *DAC 2001*, June 18-22, 2001, Las Vegas, Nevada, USA.

[43] http://developer.intel.com/technology/itj/q12001/articles/art_5.htm

[44] A. Sanfiovanni-Vincentelli, and A. Ferrari, "System Design – Traditional Concepts and New Paradigms," *Proc. of ICCD 99*, October 1999, pp. 2, Austin, USA.

[45] Martin, G.; Chang, H., "Tutorial 2 System-on-chip design," Proceedings. 4th International Conference on ASIC, 2001, PP. 12 –17.

[46] H. Azhou, D.F. Wong, "Global Routing with Crosstalk Constraints," *proceedings of the ACM/IEEE Design Automation Conference*, 1998, PP. 374-377.

[47] H. Azhou, D.F. Wong, "Global Routing with Crosstalk Constraints," IEEE Trans. Computer Aided Design 18(11), 1999, PP. 1683-1688.

[48] J. Cong, K.S. Leung, and D. Zhou, "Performance-Driven Interconnect Design based on Distributed RC Model," *Proc. Design Automation Conference*, pp. 606-611, 1993.

[49] J. Cong and K.S. Leung, "Optimal Wire Sizing Under the Distributed Elmore Delay Model," *Proc. Int. Conference on Computer Aided Design*, pp. 634-339, 1993.

[50] Yehea I. Ismail and Eby G. Friedma n, *On-Chip Inductance in High Speed Integrated Circuits*, Kluwer Academic Publishers, 2001.

Section 9.2

Optimizing Inductive Interconnect for Low Power

Magdy A. El-Moursy and Eby G. Friedman
Department of Electrical and Computer Engineering University of Rochester Rochester, New York 14627-0231

Abstract: The width of an interconnect line affects the total power consumed by a circuit. A trade off exists between the dynamic power and the short-circuit power dissipated in inductive interconnect. The optimum line width that minimizes the total transient power dissipation is determined in this paper. A closed form solution for the optimum width with an error less than 5% is presented. For a specific set of line parameters and resistivities, a reduction in power approaching 78% is achieved as compared to the minimum wire width. Considering the driver size in the design process, the optimum wire and driver size that minimizes the total transient power is also determined.

Keywords: transient power dissipation,inductive interconnect, underdamped systems, short- circuit power, characteristic impedance, dynamic power

1 INTRODUCTION

As the feature size of CMOS circuits and wiring has decreased, interconnect design has become an important issue in high speed, high complexity integrated circuits (IC). With the increase in signal frequencies and the corresponding decrease in signal transition times, the interconnect impedance can behave inductively [1], increasing the on-chip noise. Furthermore, considering inductance within the design process increases the computational complexity of IC synthesis and analysis tools. However, inductive behaviour can also be useful. As shown in [3], a properly designed inductive line can reduce the total power dissipated by high-speed clock distribution networks. Clock networks can dissipate a large portion of the total power dissipated within a synchronous IC, ranging from 25% to 75 %[4,5]. The technique proposed here can be used to reduce the overall power being dissipated by long interconnect such as a high-speed clock distribution network or data buss.

Many algorithms have been proposed to determine the optimum wire size that minimizes a cost function such as delay [6] or power. Some of these algorithms address reliability issues by reducing clock skew. The work described in [7] considers simultaneous driver and wire sizing using the Elmore delay model with simple capacitance, resistance, and power models. As the inductance becomes important, certain algorithms have been enhanced to consider an RLC model [8].

In this paper, the tradeoff between short-circuit and dynamic power in inductive interconnect is introduced. The optimum line width that minimizes the total power dissipation is determined. As the line driver has an important effect on the signal and power dissipation characteristics, a closed form solution for the simultaneous driver and wire-sizing problem that minimizes the total transient power dissipation is presented. An analytical solution for the transition time at the far end of a long inductive interconnect is also provided. These results are used to determine a closed form solution for the width of an inductive interconnects line that minimizes the power dissipated by that line.

The paper is organized as follows. In section 2, the transient power characteristics of inductive interconnect line are discussed. The signal behaviour at the end of an inductive line is described in section 3. In section 4, a power optimization criterion is formulated. The effect of line material and length on the optimum interconnect width is demonstrated in section 5. Some conclusions are discussed in section 6.

2. POWER CHARACTERISTICS OF INDUCTIVE INTERCONNECT

The transient powers characteristics of inductive interconnect are presented in this section. The research described [3] uses wire sizing techniques to reduce the total transient power dissipated by a clock distribution network, but does not provide a closed form solution to determine the optimum interconnect width. Their model also ignores the change in the circuit behavior that occurs when the width of the line is increased. The matching response between the line and the driver plays an important role in the transient power dissipation as discussed in section 3. In [3], the driver size is also not considered as a design parameter.

Issues that affect wire sizing are discussed in this section. In subsection 2.1, the trade off between dynamic and short-circuit power dissipation in inductive interconnect is described. A lossy transmission line model used in the development of a closed form solution for the interconnect width that minimizes the total transient power is presented in subsection 2.2. Simultaneous wire and driver sizing criteria are described in subsection 2.3.

Transient Power in Inductive Lines

A trade-off exists between dynamic and short-circuit power in sizing inductive interconnect. As shown in Fig1, an optimum interconnect width at

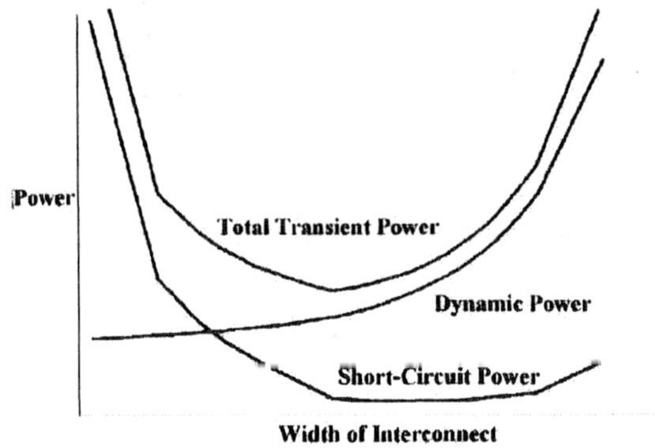

Figure.1: Dynamic, short-circuit, and total transient power as a function of the connect line width

which the total transient power is a minimum exists for overdriven lines. This trade off does not occur if the line is under driven, as described in section 3.

Transient power dissipation is composed of dynamic power and short-circuit power [2]. For the circuit shown in Fig. 2, both Inv1 and Inv2 dissipate transient power during switching. The change in the line width primarily affects the dynamic power of Inv1.P1d and the short-circuit power of Inv2 P2sc Closed form expressions for the dynamic and short-circuit power are given by (1) and (2), respectively [9]. The dynamic power of Inv2 depends on the load capacitance, and is not affected by the wire size. The change in the short-circuit power of Inv1 is negligible, assuming a fixed signal transition time at the input of Inv1

$$P_{id} = f V_{dd}^2 C, \quad (1)$$

$$P_{2sc} = f(1/12) k_2 \ \tau_o (V_{dd} - 2 V_{t2})^3 \quad (2)$$

Where f is the operating frequency, C1 is the total capacitance driven by Inv1, K_2 and V_{t2} are the trans conductance and threshold voltage of Inv_2, respectively, and τ_o is the transition time of the signal at the input of Inv_2.

Figure.2: CMOS inverter driving another inverter through a long interconnect line

Only C_l and τ_0 are affected by a change in the line width. Changing the line width has a significant effect on the transition time, as described in section 3. To determine the optimum width, closed form expressions for both C_l and τ_0 are provided in subsections 2.2 and 3.1, respectively.

Transmission Line Model

To obtain a closed form solution for the optimum width, expressions for the line impedances which model the interconnect are presented. Neglecting the line dielectric losses, a lossy transmission line is represented by the line resistance R, inductance L, and capacitance C, all per unit length. R, L, and C are expressed in terms of the line dimensions in (30-(5) respectively.

$$R=\rho / W_{int\,T} \tag{3}$$

where ρ,T and W_{int} are the line resistivity, thickness, and width, respectively. Assuming a single line, C is given by [16]

$$\frac{C}{\in_{ox}} = 1.13\frac{W_{INT}}{H} + 1.44(\frac{W_{INT}}{n})^{0.11} - 1.47(\frac{T}{H})^{0.42} \tag{4}$$

Note that C increases super linearly with the line width, which increases the dynamic power P_{ld}. L is the self-inductance of the line and is given by [12]

$$L = 200(\ln(\frac{2l}{W_{INT}+T}) + 0.5 + 0.22\frac{W_{INT}+T}{l}) \tag{5}$$

The self-inductance decreases with increasing line width. Note that the line inductance may increase if the total area of the interconnect is fixed, since the mutual inductance can increase with smaller line-to-line spacing. For a single line, however, the interconnect inductance decreases with increasing width.

The inductive time constant of a line can be characterized by $\sqrt{l/R}$ [10].

This time constant increases with increasing line width, as the reduction in resistance is much greater than the reduction in the inductance. An increase in the inductive time constant and line capacitance affects the transition time as described in section 3.

Simultaneous Wire and Driver Size

The geometric width of the driver also plays an important role in the total transient power dissipation. Two complementary effects occur. As the driver size increases, the transition time of the output signal decreases and, consequently, the short-circuit power of the load gate decreases. At the same time, the driver gate input capacitance increases as the width of the driver becomes larger, increasing the power required to charge the gate capacitance. A general expression for the total transient power dissipation including the power dissipation due to the driver input gate capacitance is

$$P_{tdrive} = P_{id} + P_{2sc} + P_{drive} \qquad (6)$$

Where $P_{drive} = f\, V_{dd}^2\, C_{drive}$ and C_{drive} is the driver gate input capacitance.

Equation (6) is therefore a function of both the wire width and the driver size. To achieve the global minimum for the transient power dissipation, both of these design variables need to be simultaneously determined.

3 TRANSITION TIME FOR A SIGNAL AT THE FAR END OF AN INTERCONNECT LINE

Wire sizing techniques do not consider line-matching characteristics as the line width changes. For inductive interconnect, the matching response plays an important role in the signal characteristics. It is shown in this section that, for an under driven line, the transition time increases as the line becomes wider. A closed form expression of the signal transition time at the far end of a line is presented in subsection 3.1.The effect of wire sizing on the line matching characteristics and the transition time is described in subsection 3.2.

Closed Form Expression for the Transition Time

To determine a closed form expression for the high-to-low signal transition time at the far end of an inductive interconnect line, a lumped RLC equivalent circuit is used to model the interconnect impedance [2]. Assuming the PMOS transistor of the driving inverter is off, an analytical expression for the signal at the far end of the line is

$$V(t) = V_c(\tau_p OFF)\, e^{-\alpha n(t-\tau p\ off)} \qquad (7)$$

Where τ_pOFF is the time at which the PMOS transistor of the driver turns off, and α_n is a constant that depends upon R, C, L, and the driver transistor characteristics such as the transconductance, mobility, and threshold voltage. V_c (τ_pOFF) is the voltage at the load capacitance at τ_pOFF. The transition time is expressed by $\tau_0 = t_{10\%} - t_{90\%}$ where $t_{10\%}$ and $t_{90\%}$ are the times at which the signal reached 10 % and 90% of the final value, respectively.The transition time based on this analytical expression is compared to SPICE in Fig. 3

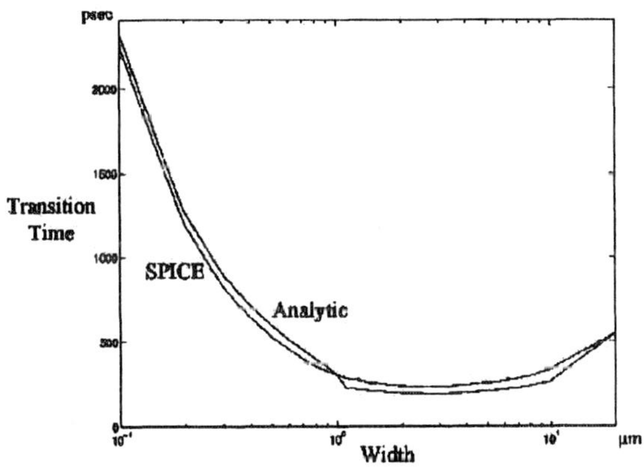

Figur.3: Analytical solution of the trasistion time as compared with SPICE for different line widths

The values of R, C and L are determined from (3) –(5) based on H = 2 µm, T =2µm , ρ= 2.5 µΩcm and l= 5 mm. A 0.24µm CMOS inverter with W_n= 15 µm and W_p = 30 µm is assumed. The maximum error in the analytical expression as compared to SPICE is less than 20% and is typically around 14%.

As shown in Fig3, as the line width increases, the signal transition time decreases until a minimum transition time is reached. The signal transition time increases again after exceeding a certain line width. The behavior of the transition time in terms of the line impedances is described in the following section.

3.1 EFFECTS OF CHANGING THE INTERCONNECT WIDTH ON THE LINE CHARACTERISTICS

To describe the signal behavior in terms of the interconnect wid th, an equivalent circuit of an inverter driving an inductive interconnect line is used. The characteristic impedance of a lossy line is given by the well-known formula $Z_{lossy} = \sqrt{R + jwL/ jwC}$. The input signal driving the line is assumed to be a ramp signal. Different approximations have been made to estimate Z_{lossy} in terms of the per unit length parameters [14,15]. A general form for Z_{lossy} is $Z_0 + g R$ where g is a constant which depends on the line parameters.

For the high-to-low input transition, the NMOS transistor is assumed to be off. The inverter can be replaced with an ideal voltage source with a variable output resistance R_{tr} as shown in Fig 4.

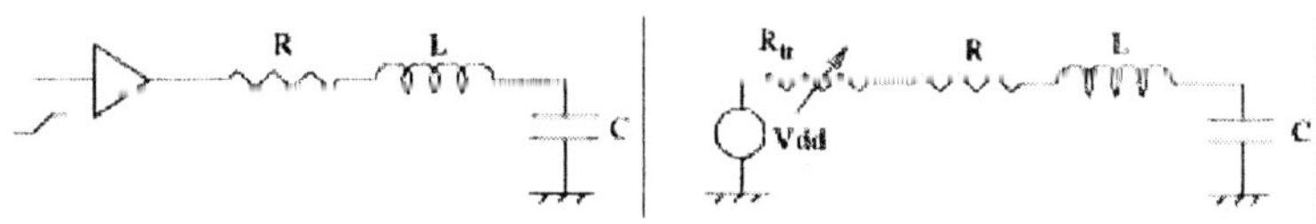

Figure.4: Equivalent circuit of an inverter driving an *RLC* interconnect line

At small interconnect widths; the characteristic line impedance is large as compared to the equivalent output resistance of the transistor. Thus, the line is overdriven and the waveform at the far end of the line contains overshoots and undershoots (the under damped condition).Z_{lossy} decreases with increasing line width, and the line remains under damped until Z_{lossy} equals R_{tr}.. A further increase in the line width under drives the line, as Z_{lossy} becomes less than R_{tr}[13]. As the line width is increased, the line driving condition changes from overdriven to matched to under driven.

For an overdriven line, increasing the line width makes the line more inductive as the resistance decreases linearly while the inductance has a logarithmic dependence on the width (for a single line). As described in [3] , the line impedance approaches becoming lossless, and the attenuation constant approaches zero at large line widths. This effect reduces the signal transition time. A further increase in the line width decreases Z_{lossy} and matches the line impedance with the driver impedance. At this width, the transition time reaches a minimum, as shown in Fig 3.Increasing the width further under drives the line. At this width, the capacitance begins to dominate the line impedance. The line becomes highly capacitive which further increases the transition time, increasing the short-circuit power dissipation in the load inverter. For an overdriven line, the short-circuit power dissipation changes with the line width as shown in Fig. 1. However, for an under driven line, an increase in the line width increases the short-circuit power. If the line is under driven, the line should be made as thin as possible to minimize the total transient power.

4. INTERCONNECT WIDTH OPTIMIZATION FOR MINIMUM POWER

Using the closed form expressions for C and τ_o, the transient power components can be expressed in terms of the line width. A criterion for determining the optimum interconnect width is applied in this section to a simple example circuit and compared with SPICE.

The dynamic power of a driver increases with the interconnect width as the line capacitance increases. The short-circuit power within the load gate decreases as the line becomes wider (more inductive). When the width reaches the matched condition, the short-circuit power is minimum. An increase in the interconnect width beyond the matching condition increases the short-circuit power, since the transition time increases (see Fig. 3). For an inverter driving N gates, as shown in Fig. 5a, the total transient power is given by

$$P_{tDrive} = P_{tDrive} + N P_{2sc} + P_{drive}$$

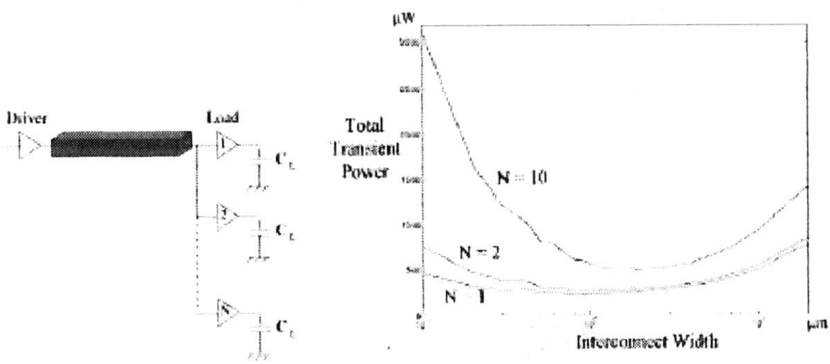

(a) (b)

Figure.5: Inverter driving N logic gates, a0 circuit structure, b0 total transient power dissipation as determined by SPICE

For a specific driver size, the total transient power dissipation is a function of the line width. Given a set of line parameters, the optimum line width can be obtained by determining the interconnect width which minimizes P_{tDrive}. For circuit specifications as described in subsection 3.1, the total simulated power dissipation for different loads, N = 1,2,and 10, is shown in Fig. 5b. A comparison between the analytic solution and simulation is listed in Table 1. The error between the analytic solution and SPICE for the chosen range of values is less than 5%.

Number of Loads N	$W_{INT\ Optimum}$ (µm)		Error (%)
	Analytical	SPICE	
1	1.08	1.10	-1.9
2	1.13	1.15	-1.8
10	1.99	1.90	+4.8

For N = 10, the total transient power dissipation of a symmetric driver is shown in Fig6. Considering the driver size as a design variable, a different local minimum for the transient power dissipation exists for each driver size. Furthermore, for each line width, a minimum transient power dissipation also exists for each driver size. A global minimum for the transient power is obtained by determining the optimum value of each design variable. Considering the driver size as a design parameter, (8) is a function of two variables, permitting the global minimum for the power dissipation to be obtained. For the example circuit shown in Fig. 5a, the minimum power is achieved at W_{INT}= 2.8 µmand W = 57 µm

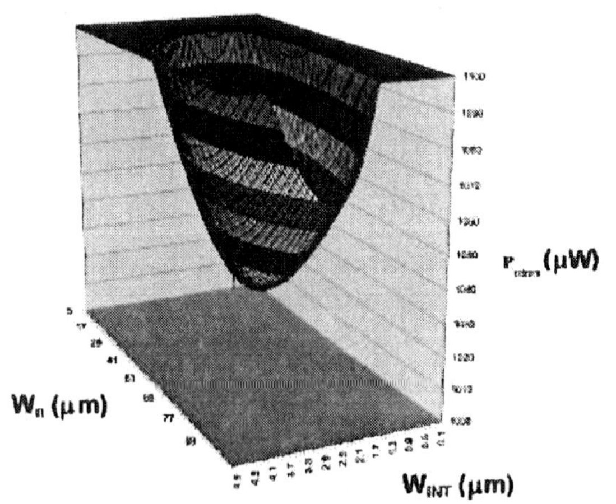

Figure.6: Total power dissipation with different wire and driver sizes where N=10.

5. EFFECTS OF INTERCONNECT RESISTIVITY AND LENGTH ON POWER

To evaluate the effectiveness of the optimum solution, different interconnect line parameters are considered. To demonstrate the importance of the optimum solution on inductive lines, the optimum width is obtained for two line lengths, l = 1mm and 5 mm. The transient power dissipation is determined for three line widths, thin, optimum, and wide. As listed in Table 2, the optimum width of a copper line reduces the total transient power by 68.5% for l = 5 mm as compared to 28.6% for l = 1 mm. For aluminium, a reduction of 77.9% is achieved as compared to 37.8%. As the line become longer (i.e. more inductive), the power decreases further. The more inductive the interconnect, the more sensitive the power dissipation is to a change in the line width. Wire width optimisations is, therefore, more effective for longer inductive lines.

Also listed in Table 2 is that the power reduction in aluminium is greater than that in copper. The reduction in power in these inductive lines increases further as the lines become more resistive (see, for example, aluminium lines versus copper lines in Table 2.) Comparing the optimum width with the minimum width, the reduction in power is greater in more highly resistive lines.

Resistivity ρ μΩcm	Total Transient Power Dissipation (μW)				
	Resistive line(l=1 mm)				
	Optimum	Thin	Reduction	Wide	Reduction
1.7 (copper)	583	817	28.6%	88 0	33.8%
2.5(Aluminium)	606	976	37.8%	81 3	25.4%
	Inductive Line(l=5 mm)				
1.7 (copper)	1121	3563	68.5%	1931	41.9%
2.5(Aluminium)	1236	5592	77.9%	1973	37.4%

6 CONCLUSION

It is shown in this paper that a trade off exists between the dynamic and short-circuit power in inductive inter connect. The short-circuit power of an overdriven interconnect line decreases with the line width, while the dynamic power increases. When the line exceeds the matched condition, the short-circuit power also increases with increasing line width. For a long inductive interconnect line, an optimum interconnect width exists that minimizes the total trans ient power dissipation. A closed form solution is presented for determining this optimum width. This solution has high accuracy, producing an error of less than 5%. The optimum line width is shown to be more effective in reducing the total transient po wer as the line becomes longer. With aluminum interconnect, a reduction in power of about 80% and 37% is obtained as compared to thin and wide wires, respectively. For copper interconnect, a reduction in power of 68% and 42% is obtained for the same conditions. Greater power reduction is achieved for optimally sized lines with higher resistively as compared to minimum width lines. The optimum interconnect width depends upon both the driver size and the number of gates being driven. With this solution, the optimum driver and wire size can be simultaneously determined.

REFERENCES

[1] Y. I. Ismail, E. G. Friedman, and J. L. Neves, "Figure of Merit to Characterize the Importance of On-Chip Inductance," IEEE Transactions on Very Large Scale Integration (VLSI) Systems}, Vol. 7, No. 4, pp. 442-449, December 1999.

[2]K. T. Tang and E. G. Friedman, "Delay and Power Expressions Characterizing a CMOS Inverter Driving an RLC Load," Proceedings of the IEEE International Symposium on Circuits and Systems}, Vol. III, pp. 283-286, May 2000

[3]Y. I. Ismail, E. G. Friedman, and J. L. Nerves, "Exploiting On-Chip Inductance in High Speed Clock Distribution Networks," IEEE Transactions on Very Large Scale Integration (VLSI) Systems}, Vol. 9, No. 6, pp. 963 - 973, December 2001.

[4]P. E. Gronowski et al., "High-Performance Microprocessor Design," IEEE Journal of Solid-State Circuits}, Vol. 33, No. 5, pp. 676-686, May 1998.

[5]C. J. Anderson, et al., "Physical Design of a Fourth-Generation POWER GHz Microprocessor," Proceedings of the IEEE International Solid-State Circuits Conference, pp. 232-233, February 2001.

[6]J. J. Cong, K. Leung, and D. Zhou, "Performance -Driven Interconnect Design Based on Distributed RC Delay Model, "IEEE Transactions on Computer-Aided Design of Integrated Circuits and Systems, Vol. 14, No. 3, pp. 321-336, March 1995.

[7]J. J. Cong and C.-K. Koh, "Simultaneous Driver and Wire Sizing for Performance and Power Optimization," IEEE Transactions on Very Large Scale Integration (VLSI) Systems}, Vol. 2, No. 4, pp. 408-425, December 1994.

[8]Q. Zhu and W. M. Dai, "High-Speed Clock Network Sizing Optimization Based on Distributed RC and Lossy RLC Interconnect Models,"IEEE Transactions on Computer-Aided Design of Integrated Circuits and Systems, Vol. 15, No. 9, pp. 1106-1118, September 1996.

[9]H. J. M. Veendrick, "Short-Circuit Power Dissipation of Static CMOS Circuitry and its Impact on the Design of Buffer Circuits," IEEE Journal of Solid-State Circuits}, Vol. SC-19, No. 4, pp. 468-473, August 1984.

[10]Y. I. Ismail and E. G. Friedman, "Effects of Inductance on the Propagation Delay and Repeater Insertion in VLSI Circuits, "IEEE Transactions on Very Large Scale Integration (VLSI) Systems, Vol. 8, No. 2, pp. 195 - 206, April 2000.

[11] X. Qi, G. Wang, Z. Yu, R. W. Dutton, T. Young, and N. Chang, "On-Chip Inductance Modelling and RLC Extraction of VLSI Interconnects for Circuit Simulation," Proceedings of the IEEE Custom Integrated Circuits Conference}, pp. 487-490, May 2000.

[12]E. B. Rosa, "The Self and Mutual Inductances of Linear Conductors, "Bulletin of the National Bureau of Standards, Vol. 4, No. 2, pp. 301-344. Government Printing Office, Washington, January 1908.
[13]J. Wang and W. Dai, "Optimal Design of Self-Damped Lossy Transmission Lines for Multichip Modules," Proceedings of the IEEE International Conference on Computer Design, pp. 594-598, October 1994.
[14]Cappuccino and G. Cocorullo, "A Time-Domain Model Power Dissipation of CMOS Buffer Driving Lossy Lines,"Electronics Letters, Vol. 35, No. 12, pp. 959-960, June 1999.
[15]P.Heydari and M. Pedram, "Calculation of Ramp Response of Lossy Transmission Lines Using Two-Port Network Functions," Proceedings of the ACM International Symposium on Physical Design, pp. 152-157, April 1998.
[16]N. Delorme, M. Belleville, and J. Chilo, "Inductance and Capacitance Analytic Formulas for VLSI Interconnects," Electronics Letters, Vol. 32, No. 11, pp. 996-997, May 1996.

Section 9.3

Skin Effects in System on a Chip Interconnects

Shizhong Mei and Yehea I. Ismail
Electrical and Computer Engineering Northwestern University, Evanston, IL 60208
meisz@ece.nwu.edu, ismail@ece.nwu.edu

Abstract: Skin effect makes interconnect resistance and inductance frequency dependent. This paper addresses the problem of efficiently estimating the signal haracteristic of any RLC network when skin effect is significant, which complicates interconnect simulation. In this paper, a new type of moments is defined that simplifies the interconnect simulation,namely, the square root ents. The time to calculate the square root moments is similar to the time to calculate the traditional moments, and the new moments preserve the recursive roperties of the traditional moments. Hence, the method introduced here can handle the more complex problem of interconnect simulation with skin effect at almost no overhead compared to constant element interconnect simulation. Using the square root moments, higher order approximations can be reached as compared to traditional moments. This method is applied to a sample circuit with high accuracy

2 Introduction

In deep sub-micrometer technologies, the intrinsic gate delay tends to decrease dramatically. However, since the average length of global interconnect lines increases, interconnect delay dominates the overall gate delay in current deep sub-micrometer VLSI circuits [1], [2]. To achieve high quality circuit design in as short a period of time as possible, it is necessary to accurately model interconnect lines and to find efficient algorithms for quick delay es timation and circuit simulation. The increasing on-chip clock frequencies make skin effect on some nets, *e.g.*, clock and power distribution nets, increasingly prominent and non negligible. Consequently, model order reduction techniques for constant eleme nt circuits (no frequency dependence) become inaccurate when skin effect is encountered. Existing work dealing with skin effect focuses on building models for SPICE or generates too complex models for quick delay estimation for *RLC* network, e.g., [3]-[5]. However, the standard tools for large VLSI circuits use model order reduction techniques. Hence, model order reduction techniques are required which can handle frequency dependent elements. Many papers, *e.g.*, [6]-[7] have shown that the skin effect impedance can be described by $A+B\ S$ with high accuracy. Thus this approximation is used in this paper. To deal with the square root term S , a new type of moments is defined that simplifies the interconnect simulation, namely, the square root moments. Similar to the moments in case of constant elements, the square root moments obtained in this paper have the same recursive properties and are easy to calculate. The rest of this paper is organized as follows. A background for calculating delay in *RLC* circuits with constant elements is presented in section II. The square root moments of general *RLC* trees/grids with skin effect are derived in section III. The computation of time domain signals from the square root moments is given in section IV. The simulated node voltages of a sample circuit are offered in section V. Finally, some conclusions are given in section VI.

3 Background

In the S domain, the transfer function of any constant *RLC* circuit takes the following general form.

$$g(s) = \frac{1 + a_1 s + a_2 s^2 + \cdots + a_n s^{n}}{1 + b_1 s + b_2 s^2 + \cdots + b_{l_2} s^{l_2}} \qquad (1)$$

where a_i 's and bi 's are real positive coefficients and $l_2 > l_1$ The Taylor expansion of $g(s)$ exists and is given by

$$g(s) = 1 + g^{(1)} s + \cdots + \frac{g^{(k)}(0)}{k!} s^k + \cdots \qquad (2)$$

In order to efficiently evaluate the transient response,a popular method [8] is to neglect high order terms in the above expansion and use the truncated $g(s)$ to obtain the inverse Laplace transform by finding the dominant poles and their residues. Arbitrary accuracy can be achieved by matching higher number of additional moments. Elmore and Wyatt used only two terms of the expansion in equation (2) to find a closed form solution of the delay [9], [10]. The Wyatt approximation was extended in [11] to include the inductance of the interconnect. The capability of efficiently estimating the transient response makes the approximate methods mentioned above widely used in industry. Since the derived results do not apply to circuits with skin effect, it is desirable to find a new method that aims at calculating the transient response of circuits with skin effect with similar efficiency.

4 Square Root Moments

For a general tree (*e.g.*, a tree of 3 stages as in Fig. 1), the voltage at node *i* is given by

$$V_i(s) = V_{in}(s) - \sum_k C_k V_k(s) s[R_{ki}(s) + L_{ki}s] \tag{3}$$

where *Vin* (*s*) represents the input voltage, *Rki(s)* represents the total frequency dependent resistance shared by paths from the input port to node *i* and from the input port to node *k* , and *L ki* represents the total inductance shared by paths from the input port to node *i* and from the input port to node *k* .

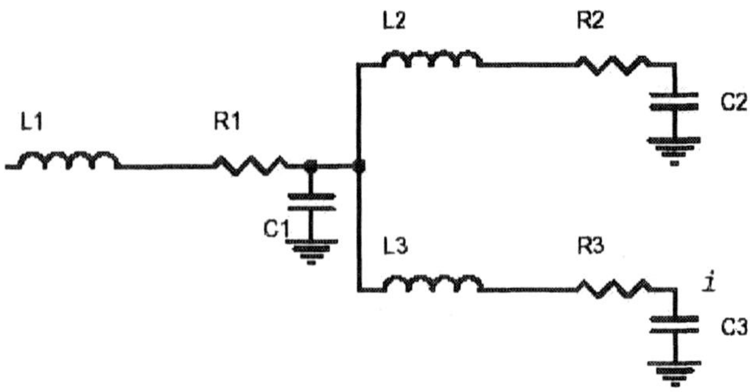

Figure 1. General *RLC* tree

The skin effect makes *Rki* (*s)* frequency dependent, which, according to the expression of the skin effect impedance $A+B\sqrt{S}$ [6], [7], consists of two parts, defined as *R dki* and *Raki* $\sqrt{s}$ for the dc and ac terms of the resistance, respectively. Therefore *Vi* (*s)* can be re-expressed as

$$V_i(s) = V_{in}(s) - \sum_k C_k V_k(s) s[R_{dki} + R_{aki}\sqrt{s} + L_{ki}s] \tag{4}$$

The normalized transfer function *gi(s)* at node *i* is the voltage *Vi*(*s)* when a unit impulse is applied to the tree (*Vin* (*s*)=*1*

$$g_i(s) = 1 - \sum_k C_k g_k(s) s [R_{dki} + R_{aki}\sqrt{s} + L_{ki} s] \tag{5}$$

As the $\sqrt{s}$ term represents singularity at $0 = s$,) (s gi can't be expanded in the power of s . To overcome this problem, a symbol y is introduced to replace $\sqrt{s}$ and $gi\ (s)$ becomes a function of y

$$\begin{aligned} g_i(y) &= g_i(\sqrt{s}) \\ &= 1 - \sum_k C_k g_k(y) y^2 [R_{dki} + R_{aki} y + L_{ki} y^2] \\ &= m_0^i + m_1^i y + m_2^i y^2 + \cdots + m_n^i y^n + \cdots \end{aligned} \tag{6}$$

The relation between $gi\ (y)$ and the square root moments at node i , m^i_n is

$$m_n^i = \left. \frac{d^{(n)} g(y)}{n!\, dy^n} \right|_{y=0} \tag{7}$$

Using this relation and equation (6), the moments are obtained as

$$m_0^i = 1 \tag{8}$$

$$m_1^i = 0 \tag{9}$$

$$m_2^i = -\sum_k C_k R_{dki} \tag{10}$$

$$m_3^i = -\sum_k C_k R_{aki} \tag{11}$$

$$m_n^i = -\sum_k C_k (m_{n-2}^k R_{dki} - m_{n-3}^k R_{aki} - m_{n-4}^k L_{ki}) \quad \text{for } n \geq 4 \tag{12}$$

Note that these moments fall back to the S moments when $R\ aki = 0$ since the odd square root moments become zeros. The closed form solutions above are not valid for circuits with resistive loops, *e.g.*, grids as shown in Figure 2.

Currents through the resistors differ from currents through the terminating Capacitors. Finding node voltages requires solving a complete set of independent equations that relate all the node voltages and inductive currents. For a circuit of n 1 capacitors and inductors, the application of Kirchoff's law leads to the complete set of independent equations given by

$$sX(s) = A_1 X(s) + \sqrt{s} A_2 X(s) + A_3 U(s) \qquad (13)$$

where $U(s)$ represents an $n2$ inputs to the circuit and $X(s)$, $A1$, $A2$, and $A3$ represent 1 1 · n , 1 1 $n\ n$ ·, 1 1 $n\ n$ ·, and 2 1 $n\ n$ · matrices, respectively. Each element of vector $X(s)$ is either a node voltage or an inductor current.

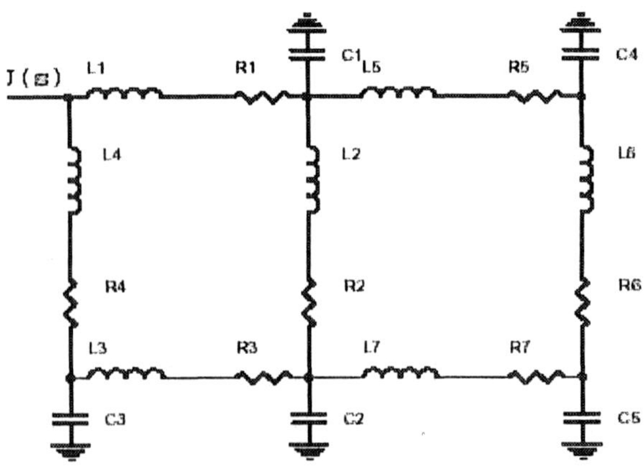

Figure 2. General RLC grid

A simple example of forming equation (13) is given for a single level RLC circuit as shown in Figure 3. A voltage $U(s)$ is applied to the input port and a current $I(s)$ is generated to flow through the resistance $R\ d + R_a \sqrt{s}$, the inductance L , and the capacitance C to ground.

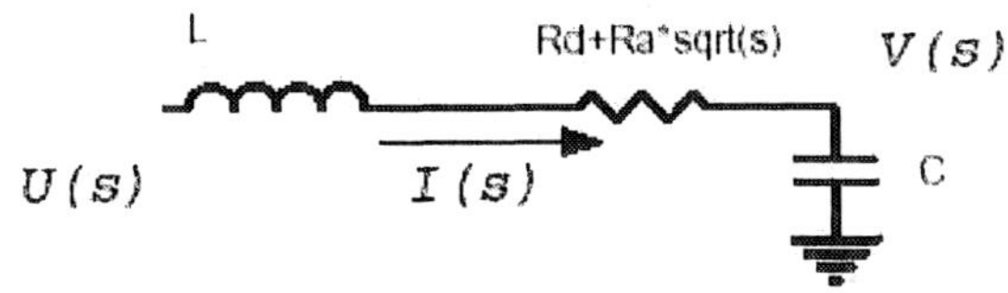

Figure 3. Single level *RLC* circuit

Denoting the voltage drop on the capacitor as $V(s)$, the following two equations are obtained

$$U(s)-V(s)=(R_d+R_a\sqrt{s})I(s)+sLI(s) \qquad (14)$$

$$I(s)=s\,C\,V(s) \qquad (15)$$

Rewriting the above equations in the form of equation (13) results in

$$s\begin{bmatrix} I(s) \\ V(s) \end{bmatrix} = \begin{bmatrix} -\frac{R_d}{L} & -\frac{1}{L} \\ \frac{1}{C} & 0 \end{bmatrix} \begin{bmatrix} I(s) \\ V(s) \end{bmatrix} + \sqrt{s}\begin{bmatrix} -\frac{R_a}{L} & 0 \\ 0 & 0 \end{bmatrix} \begin{bmatrix} I(s) \\ V(s) \end{bmatrix} + \begin{bmatrix} \frac{1}{L} \\ 0 \end{bmatrix} U(s) \qquad (16)$$

where $\begin{bmatrix} I(s) \\ V(s) \end{bmatrix}$, $\begin{bmatrix} -\frac{R_d}{L} & -\frac{1}{L} \\ \frac{1}{C} & 0 \end{bmatrix}$, $\begin{bmatrix} -\frac{R_a}{L} & 0 \\ 0 & 0 \end{bmatrix}$, and

$\begin{bmatrix} \frac{1}{L} \\ 0 \end{bmatrix}$ correspond to $X(s)$, A_1, A_2, and A_3 in the

in the equation (13), respectively.

The elements of the input $U(s)$ are set to 1.0 to obtain the moments of $X(s)$. Again, as the right hand side of equation (13) contains the term $\sqrt{s}$, which represents singularity at $s = 0$, $X(s)$ can't be expanded in the powers of s. This can be overcome by using y to replace s and expanding $X(y)$ in the powers of y

$$y^2 X(y) = A_1 X(y) + yA_2 X(y) + A_3 \quad (17)$$

$$X(y) = M_0 + M_1 y + \cdots + M_n y^n + \cdots \quad (18)$$

Substituting (18) into (17) and comparing the coefficients of y^n on both sides yields

$$M_0 = -A_1^{-1} A_3 \quad (19)$$

$$M_1 = -A_1^{-1} A_2 M_0 \quad (20)$$

$$M_n = A_1^{-1} M_{n-2} - A_1^{-1} A_2 M_{n-1}, \text{ for } n \geq 2 \quad (21)$$

These M_n 's are $n_1 .1$ matrices and fall back to the S moments if $A2 = 0$ since odd moments become zeros. The j *th* element of M *n*'s, namely Mn (j),represents the n *th* square root moments of the j *th* element of $X(y)$, which is either a capacitor voltage or an inductor current.

5 RESPONSES IN TIME DOMAIN

A reduced order transfer function of the form

$$g'_i(y) = \frac{1 + a_1 y + a_2 y^2 + \cdots + a_{q-1} y^{q-1}}{b_0 + b_1 y + b_2 y^2 + \cdots + b_q y^q} \quad (22)$$

is required to approximate the exact transfer function at node i, namely gi (y).This transfer function can match 2 q square root moments since it has $2q$ variables (a_1 ,...,a_{q-1} ,b_0, ...b_q) By matching the first 2 q moments of $gi(y)$,the transfer function $g'i(y)$ represents a low frequency approximation of $gi(y)$ since higher order terms in the Taylor expansion are negligible at high frequencies. The more the number of moments matched, the larger the range of frequency where $g'_i(y)$ is valid.

The coefficient b_0 equals $1/M_0(i)$ while the coefficients $b_1 b_2 \ldots b_q$ can be expressed in terms of $M_0(i), M_1(i), \ldots, M_{2q-1}(i)$ as [12]

$$\begin{bmatrix} b_q \\ b_{q-1} \\ \vdots \\ b_1 \end{bmatrix} = -\begin{bmatrix} M_0(i) & M_1(i) & \cdots & M_{q-1}(i) \\ M_1(i) & M_2(i) & \cdots & M_q(i) \\ \vdots & \vdots & \vdots & \vdots \\ M_{q-1}(i) & M_q(i) & \cdots & M_{2q-2}(i) \end{bmatrix}^{-1} \times \begin{bmatrix} M_q(i) \\ M_{q+1}(i) \\ \vdots \\ M_{2q-1}(i) \end{bmatrix} \times \frac{1}{M_0(i)} \tag{23}$$

Denoting the poles of $g'_i(y)$ as p_l ($q\ l = 1, \ldots, q$), equation (22) can be rewritten as

$$\begin{aligned} g_i(y) &= \sum_{l=1}^{q} \frac{k_l}{y - p_l} \\ &= m_0^i + m_1^i y + \cdots + m_{2q-1}^i y^{2q-1} \end{aligned} \tag{24}$$

where k_l's satisfy [12]

$$\begin{bmatrix} k_1 \\ k_2 \\ \vdots \\ k_q \end{bmatrix} = -\begin{bmatrix} \frac{1}{p_1} & 0 & \cdots & 0 \\ 0 & \frac{1}{p_2} & \cdots & 0 \\ \vdots & \vdots & \vdots & \vdots \\ 0 & 0 & \cdots & \frac{1}{p_q} \end{bmatrix}^{-1} \times \begin{bmatrix} \frac{1}{p_1} & \frac{1}{p_2} & \cdots & \frac{1}{p_q} \\ \vdots & \vdots & \vdots & \vdots \\ \frac{1}{p_1^{q-1}} & \frac{1}{p_2^{q-1}} & \cdots & \frac{1}{p_q^{q-1}} \end{bmatrix}^{-1} \begin{bmatrix} m'_0 \\ m'_1 \\ \vdots \\ m'_{q-1} \end{bmatrix} \tag{25}$$

If a unit step voltage is applied at the input port, then node voltage $Vi(y)$ becomes $g'i(y)/y^2$. The inverse Laplace transform of $g'i(y)/y^2$ defines the node voltage in time domain, $Vi(t)$, that is calculated to be

$$V_i(t) = -\sum_{l=1}^{q} \frac{k_l}{p_l}[1-(1-erf(-p_i\sqrt{t}))e^{p_i^2 t}] \tag{26}$$

where $erf()$ stands for error function.

6 SIMULATED DATA

Three simulated step responses of a node voltage in the time domain in a three branches tree are shown in Figure 4. The exact result is obtained by numerically solving the differential equations describing the step input response. The analytical expression of skin effect impedance [13] was used in getting the exact result. According to Figure 4, the approximate result based on constant *RLC* model underestimates the propagation delay. The other approximate result based on the skin effect impedance $A+B\sqrt{S}$ and the square root moments matches the exact result very well.

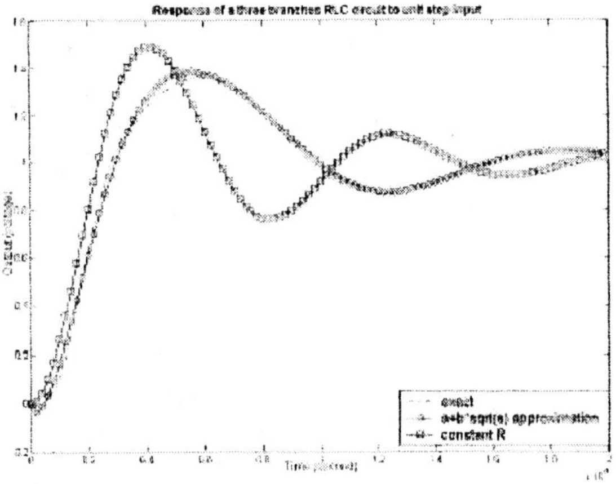

Figure 4. Simulated results

CONCLUSIONS

As can be seen from equation (26), using the square root moments, an Explicit expression of the time domain response can be obtained for *RLC* circuits with skin effect. In comparison, exact time domain responses for *RLC* circuits with skin effect are integrals and in most cases cannot reduce to closed-form expressions. The square root mome nt's technique introduced in this paper provides a much less expensive method than the exact solution in computing time domain responses of *RLC* circuits with skin effect. The accuracy of equation (26) depends on the size of circuits and number of Moments u sed. Theoretically, there's no limit on the number of moments to be used. However in practice, due to the limit number of significant digits applied in numerical computation, only a certain number of moments can be used in order to get negligible computational errors.

REFERENCES

[1] D. A. Priore, "Inductance on silicon for sub-micron CMOS VLSI," in *Proc. IEEE symp . VLSI circuits*, May 1993, pp. 17-18.

[2] "FD-TD modeling of digital signal propagation in 3-D circuits with passive and active loads," *IEEE Trans. Microwave Theory Tech.*, vol.42, pp. 1514-1523, Aug. 1994.M. P. May, A. Taflove, and J. Baron

[3] C.-S. Yen, Z. Fazarinc, and R. L. Wheeler, "Time-domain skin-effect model for transient analysis of lossy transmission lines," *Proceedings of the IEEE*, vol. 70, pp. 750-757, 1982.

[4] T. V. Dinh, B. Cabon, and J. Chilo, "New skin-effect equivalent circuit," *Electronic Letters*, vol. 26, pp. 1582-1584, 1990.

[5] A. R. Djordevic, T. K. Sarkar, "Closed-from formulas for frequency dependent resistance and inductance per unit length of microstrip and strip transmission lines," *IEEE Transactions on Microwave Theory and Techniques*, Vol. 42, pp. 241-248.

[6] N. S. Nahman and D. R. Holt, "Transient analysis of coaxial cables using the skin effect approximation $A+B\,S$," *IEEE Transactions on Circuit Theory,* Vol. 19, pp. 443-451, 1972.

[7] L. C. Calvey and J. L. Bihan, "Coefficient algorithm for time-domainresponse of skin effect lossy coaxial cables with arbitrary resistive terminations," *IEEE Trans. on CT*, vol. 3, pp. 1394-1415, Nov. 1990.

[8] L. T. Pillage, R. A. Rohrer, "Asymptotic waveform evaluation for timing analysis," *IEEE Trans. Computer-Aided Design*, vol. 9, pp. 352-366, Apr. 1990.

[9]W. C. Elmore, "The transient response of damped linear networks," *J.Appl. Phys.*, Vol. 19, pp. 55-63, 1948.

[10] J. L. Wyatt, Circuit analysis, simulation and design. North-Holland, The Netherlands: Elsevier Science, 1987.

[11]Y. I. Ismail, E. G. Friedman, and J. L. Neves, "Equivalent elmore delay for *RLC* trees," *IEEE Transactions on Computer Aided Design of Integrated Circuits and Systems,* Vol. 19, pp. 83-97, 2000.

[12] L. T. Pillage, R. A. Rohrer, and C. Visweswariah, Electronic circuit and system simulation methods, McGraw-Hill, 1995.

[13] S. Ramo, J. R. Whinnery, and T. V. Duzer, Fields and waves in communication electronics, John Wiley & Sons, 1984.

10

Micro-Electro-Mechanical Systems

System-on-Chip is not simply about building high density digital/analog circuits on a single silicon die, but building complete Microsystems. This more complete definition also encompasses the integration of disparate technologies into complete systems. Disparate technologies include MEMS devices, and this chapter presents four different aspects of MEMS technologies.

The first section 10.1, "A Road Map Towards Designing MEMS Devices with High-Reliability", by W. A. Moussa of the University of Alberta, a brief introduction to MEMS technology is provided. This section explains the basic concepts behind Micro-Electro-Mechanical Systems (MEMS) and how they are being applied to meet current technological challenges in many high technology industries such as aerospace, medicine and communication. The rapid expansion of MEMS application in these industries, particularly in space applications and drug delivery, introduces more cases of mechanical failure that can jeopardize the evident potential of MEMS. A reliability assurance process is therefore needed to be in place during the design and manufacturing processes of MEMS devices. This section discusses a potential road map that could be applied towards qualifying high-reliability MEMS devices.

In section 10.2, "A MEMS Socket Interface for SOC Connectivity", by Sazzadur Chowdhury, M. Ahmadi, G. A. Jullien, and W. C. Miller of the Universities of Windsor and Calgary, a solution is presented for the interconnection of disparate technologies to form microsystems. This is the design of a system of MEMS sockets that can provide temporary or semi-permanent connectivity to a CMOS System-on-Chip (SoC) environment. The system is comprised of a fixed MEMS socket that is thermosonically bonded to a SoC die and a carrier MEMS socket that acts as a removable insert for the fixed socket to provide connectivity between the SoC die and a CMOS die, a MEMS device or a photonics device. A key enabling element of the design is a low-contact resistance cantilevered bridge type spring contact structure. The system is expected to enable reliable, cost-effective testing of high density interconnection CMOS dies or to connect disparate technology devices, such as MEMS or photonics devices, to a SoC environment.

In section, 10.3, "On the Application of Finite Element Analysis to Investigate the Reliability of Electrostatic Comb-Drive Actuators Utilized in Microfluidic and Space systems, by Hesham Ahmed and Walied Moussa of the University of Alberta, and Wael Badawy of the University of Calgary, investigates the design parameters of lateral Electrostatic Comb -Drive actuators. The effect of these parameters on the actuation performance is explored using Finite Element Analysis (FEA), which is essential to the design process of system-on-a-chip MEMS applications. This is an important application since a Comb -Drive can represent the main source of actuation within many micro-systems. Of particular interest to this study is the application of the electrostatic Comb -Drive motor as a fluidic pump in on-chip systems used for drug delivery or in cooling of microprocessors used in space vehicles. The commercial FE package, ANSYS, is utilized to construct a robust model and solve this multi-physics interaction problem using Direct Coupled-Field Analysis. In this model, the thickness, gap and overlap between the Comb fingers are varied, and the method accounts for the change in the number of Comb fingers and the applied driving voltage. The calculated Comb displacement and generated electrostatic force are shown to be directly proportional to the number of Comb fingers. Moreover, the generated electrostatic force is found to be inversely proportional to the gap between the Comb fingers. As the thickness of these fingers increases, the displacement is found to increase for a given value of the driving voltage. The electrostatic force is also shown to be proportional with the offset (i.e. overlap) value between the fingers.

In section 10.4, "An HDL Model for a Vacuum-Sealed Micromachined Pressure Sensor", by Rafik S. Guindi of Cairo University in Egypt, we look at an aspect of MEMS devices that provide an interface to measuring the environment external to the SOC device; in this case the development of behavioural models for MEMS devices that measure pressure. The development of behavioural models for MEMS devices allows system-level simulation, and here the author describes a behavioural model for a vacuum-sealed micro machined pressure sensor that can be integrated with electrical components within an electrical simulator using an analog hardware description language. The model can be used within top-down methodologies, in system design, system level simulation, and optimization.

In Section 10.5,the final section, "Performance Analysis of MEMS-based Inertial Sensors for Positioning Applications", by Walid Abdel-Hamid, Aboelmagd Noureldin, Naser El-Sheimy and Gerard Lachapelle, of the University of Calgary, provides an analysis of commercial MEMS-based inertial sensors for the development of low cost and small size navigation and guidance systems. The aim of this section is to investigate the performance of these sensors and to analyse their impact on short-term positioning accuracy. The results show that modeling these components as second-order Gauss-Markov processes provides significant improvement in the estimation accuracy. This section also suggests some real-time digital signal processing methods, based on wavelet techniques, to improve the performance of these inertial sensors.

Section 10.1

Road Map Towards Designing MEMS Devices With High-Reliability

Walied A. Moussa
Department of Mechanical Engineeing, University of Alberta, Alberta, Canada, T6G 2G8

Abstract This section offers a brief description of the MEMS industry as it is currently standing. With the assumption of no background of the subject matter, it explains the basic concepts behind Micro-Electro-Mechanical Systems (MEMS) and how they are being applied to meet current technological challenges in the global market of many high tech industries such as, aerospace, medicine and communication. The rabid expansion of MEMS application in these industries, particularly in space and drug delivery, introduces more cases of mechanical failure that can jeopardize the evident potential of MEMS. A reliability assurance process is therefore needed to during the design and manufacturing processes of MEMS devices. This section introduces a road map that could be applied towards qualifying high-reliability MEMS devices.

Key words MEMS-bases inertial sensors, inertial navigation, noise modelling, and wavelets multi-resolution analysis.

1. INTRODUCTION

MEMS is a broad term that encompasses a fairly unformulated group of products. Essentially, MEMS are any product, ranging in size from a micron to a centimetre that combines mechanical and electrical structures. The scope of MEMS is fairly limitless. However, for the sake of conventions and the need for conciseness, the current discussion will only address the more common MEMS technologies.

Primarily, MEMS developed from technologies used in the semiconductor industry for the production of electronic circuits. Within the same decade that hosted the invention of the integrated circuit, H. C. Nathanson used microelectronic fabrication techniques to make the world's first micro-mechanical device [1]. Due to enormous improvements in processing technologies, micro-mechanical devices grew in popularity by the early 1980s. In the consequent years, a new industry was brought to light, where electromechanical systems could be realized on micrometer scales. The result was a whole new class of sensors and actuators that performed common tasks on smaller scales. This new class of micro-mechanical devices were ideally suited for mass production.

In its most conventional sense, MEMS refers to a class of batch-fabricated devices that employ both mechanical and electrical components to simulate macroscopic devices on a microscopic scale. This section focuses upon the conventional definition of MEMS. The essence of MEMS is that they are small devices that perform mechanical tasks in ways and, more importantly, in ranges that conventional devices cannot reach.

2. THE PROSPECTIVE OF MEMS

In the wake of the outburst of the microprocessor in the early eighties, the semiconductor industry revealed its unchangeable law that smaller is better. With economies of scale turning tiny firms into industrial giants, it became evident that mass miniaturization, along with mass distribution, could produce huge revenues and substantively change in the daily lives of average citizens. Given the unmitigated success of the microcircuit, it became almost certain that technologies would emerge that could bring machines to the microscopic world and produce similar results. With MEMS poised to do for machines what the transistor did for computers, there has been a vast outburst of interest, and thus funding, in MEMS research.

MEMS are used to perform the tasks of macroscopic devices at a fraction of the cost and with, occasionally, improved functionality and performance. By using simple mechanical structures and tailoring integrated circuits to suit specific tasks, designers have seen a drastic reduction in device scales and the implementations of functions that were previously unrealised. Their size alone makes them attractive for limited mass applications, with the biomedical, aerospace, communications, data storage, and automotive industries taking a keen interest in MEMS developments. Far more promising, though, is the possible reduction in costs offered by MEMS. By combining increasing throughput with fixed cost structures, manufacturers can linearly reduce prices by a comparable production increase. In summary, MEMS have the potential to revolutionize the industrial age by offering economies unique to the semiconductor industry,

The effects of MEMS could cause sweeping reforms within many high tech industries such as communications, medicine and space. This space industry, for examp le, hopes to eventually phase out the large satellites that it employs to reach the farthest points in the solar system. With every kilogram sent to outer space costing upwards of one million dollars, the potential of sending a fully integrated spacecraft weighing a few kilograms instead of the thousands of kilos offers significant monetary benefits. With MEMS capable of performing certain functions of macroscopic devices, the benefit of cutting the cost of research missions cannot be understated given this era of shrinking budgets. Space applications of MEMS are only a small part of their full potential. MEMS are also capable of revolutionizing the information age by changing the daily fabric of our global existence.

3. AVAILABLE MEMS TECHNOLOGIES

Understanding the stated advantages of MEMS, designers have started developing a range of products to suit their needs. The first major MEMS to hit markets were pressure sensors for engine control in cars. This development was followed by the introduction of micro-accelerometers, which were pioneered to provide zero-fault air bag deployment systems. Integrating a diagnostic circuit into a sensor, engineers were able to produce a device that could not only sense acceleration but that could also detect internal failures. Replacing a faulty system based on ball bearings and plastic tubing that was prone to misfire, these devices swept through the automotive industry. Building from the technological, as well as commercial, success of these initial designs, engineers have developed MEMS to act as a wide variety of motion sensors. Recently intense research has been conducted into producing micro-gyroscopes as part of a fully integrated inertial reference unit. Development has also commenced, seismometers, anemometers, tempera ture sensors, pressure sensors, and hygrometers which, when incorporated with accelerometers, could provide miniaturized weather stations.

MEMS have also shown promise for aerospace applications. Research into magnetometers shows that it may be possible to build devices that far outperform traditional solid-state sensors, which could provide cost saving reductions in the weight of spacecraft. Furthermore, the bulky propulsion systems in modern satellites will be phased out by advances in micro-propulsion coming from new generations of ion drives and micro-thrusters. Recent developments at universities have shown that MEMS micro-actuators, when placed upon the leading edge of aircraft, can offer significant drag reduction and thus increase fuel efficiency [2]. Some even more interesting research has led to the design of a MEMS controlled aircraft, where control surfaces are replaced by micro-machines, which could offer unprecedented control and diagnostic capabilities. In damage diagnostics of composite aerospace Micro Ultrasonic sensors is proven to be very helpful in detecting damage occurring in fuel tanks during charging process, which is an important feature toward its certification.

One of the more promising fields within MEMS is the concept of optical MEMS. Using micro-mirrors placed on top of memory arrays, researchers have developed a television projection unit on a semiconductor wafer that has all the functionality of a cathode ray tube [3]. Another promising development is in the field of optical switches. Conventional optical switching networks are costly and, with the forecasted growth in optical communications systems, cheaper alternatives are at a premium. Multiple groups have developed MEMS-based optical switches that can be produced at a fraction of the cost of conventional systems.

With the digital age largely upon the global market, MEMS are expected to offer greater improvements in computer technology. Given that power dissipation of the average microprocessor increases with every generation of microchip, microtubules research has been initiated to attempt to find better ways to conduct heat away from integrated circuits. This will require the introduction of micro fluidic systems driven by micro actuators such as comb drives or piezoelectric material. MEMS structures have also been developed as microprobes for integrated circuits [4]. Using MEMS, it may be possible to take point contact voltage and current measurements on microprocessors. Another exciting development has been the pioneering of nanometer scale data storage. With miniaturized tunneling tips now possible, engineers have developed systems that could eventually store information at commercially competitive speeds in an area twenty nanometers on a side.

Another field that shows promise is the development of biological sensors. MEMS provides an opportunity for the development of new sensors to monitor the human environment. Researchers at University of Alberta have begun to develop MEMS-based particles that can provide information about the digestive system. Another interesting application of MEMS has been in the development of new biological instruments. This technology is sometime referred to as the Lab-on-a-chip technology, which uses semiconductor-like micro-fabrication techniques to translate experimental and analytical protocols, developed in software, into chip architectures consisting of interconnected fluid reservoirs and pathways. The miniaturized fluid pathways shorten run times and the strong electrokinetic driving forces, which can be imposed in the microfluidics environment, improve analyte resolution. In this way, it is possible to create the functional equivalent of valves and pumps capable of performing manipulations such as reagent dispensing and mixing, incubation/reaction, sample partition and analyte detection. Among other developments, Researchers have also produced probes to measure the strength of the human heart cell [5].

While the potentials of MEMS are almost limitless, to this point production of commercial parts has been relatively limited. MEMS, as products of a young industry, remain largely prototypical. While their potential have been demonstrated their actual implementation has been relatively scarce, with commercial successes still the exception rather than the rule. In order for this rapid growth to be realized, the field of MEMS reliability will need to rapidly mature.

4. THE NEED FOR, AND ROLE OF, MEMS RELIABILITY

With MEMS still in their early years, the question has been posed as to the need for reliability assurance process in the MEMS design and manufacturing cycle. The answer to this question is not just focused on providing reliability information for the current designers but to set the standard of MEMS reliability for the foreseeable future. Given the unstoppable commercialisation of MEMS, reliability issues that have previously been ignored are destined to become of paramount importance. Researchers involved in the micro-systems field feel that these issues must be raised in parallel to the development of MEMS in order to assure their rapid insertion into industrial and space applications. Understanding the future of the MEMS industry, it would be shortsighted to ignore the importance of reliability.

In confronting the issues of MEMS reliability assurance, users will certainly have different requirements. Undoubtedly, a comb drive driven micro-pump will have a different set of requirements and specifications than a communications satellite, but there will be similar methodologies for assessing qualification for both. The answer to any MEMS reliability guidelines should be focused on utilizing the basic similarities in design requirements to provide a means of developing high-reliability MEMS parts. In order to produce a high reliability part one must not only examine the device itself, but one must also examine the entire process surrounding the part, from conception to finish. This means that the process must be qualified, with the supplier fully investigated, the design verified, and the packaging certified. With the MEMS industry as diverse as it is today, it seemed presumptuous to lay out specific tests for every MEMS device inserted into space as that kind of map would be unduly constrictive to some applications. Rather, any road map should be intended as a MEMS educational guide, offering descriptions of the most common devices and technologies and the steps required to meet the demands of the environment for the designed micro-device.

If we examined the meaning of reliability, we will easily notice that it is the conjugate of failure probability. Therefore, it is vital to start any reliability assessment process by discussing the different failure sources of a MEMS device. The next step in the design process should focus on assessing the effect of specific failure mechanisms that are likely to occur within the designed micro-device lifetime. It must be understood that each failure mechanism will have a different level of predominance on different devices. To examine this effect it is essential to adopt a technique that allows us to iterate the construction process of the designed micro device. And that is where Finite Element (FE) analysis can be very practical as a simulation tool that can emulate the same boundary conditions and geometries of various MEMS devices. The iterative FE analysis process will also investigate the effect of the manufacturing process on the resulting MEMS device, in particular, cases of distortion due to excessive heat and resulting residual stress that can significantly affect the durability of MEMS devices. Other variables that should be considered in the FE analysis include the variation of the material properties and its effect on the stability of the performance of the designed MEMS device. The Next section will demonstrate the application of FE analysis in investigating the performance of Electro-static Comb-Drive Actuators utilized in micro fluidic and space systems.

REFERENCES

[1] H. C. Nathanson, W. E. Newell, R. A. Wickstrom, and J. R. Davis, Jr., "The Resonant Gate Transistor", IEEE Transactions on Electron Devices, March, 1967.

[2] T. Tsao, F. Jiang, R. A. Miller, Y. C. Tai, B. Gupta, R. Goodman, S. Tung, and C. M. Ho, "An Integrated MEMS System for Turbulent Boundary Layer Control, " Technical Digest, 1997 International Conference on Solid-State Sensors and Actuators (Transducers '97), Chicago, IL, Vol. 1, pp. 315-318, Jun. 16-19 1997.

[3] Helvajian, H. ed. Microengineering Technology for Space Systems, The Aerospace Corporation Report Number ATR-95(8168)-2. El Segundo, CA (September 30, 1995).

[4] Handouts From EE 493: Microelectromechanical Systems, Professor Noel MacDonald, Cornell University, January 18- May 2, 1997.

[5] G. Lin, K.S.J. Pister and K. P. Roos, "Measurement of Contractile Characteristics from Isolated Heart Cells Using a Miniaturized Force Transducer System, " IEEE Engineering in Medicine and Biology Society 16th Annual International Conference, Baltimore, Maryland, November 3-6, 1994, pp. 1017 – 1018.

Section 10.2

A MEMS SOCKET INTERFACE FOR SOC CONNECTIVITY

Sazzadur Chowdhury, M. Ahmadi, G. A. Jullien, W. C. Miller
Electrical and Computer Engineering
University of Windsor
Windsor, Ontario, N9B 3P4
Canada

Abstract: The design of a system of microelectromechanical (MEMS) sockets that can provide temporary or semi-permanent connectivity to a CMOS System-on-Chip (SoC) environment is presented. The system is comprised of a fixed MEMS socket that is thermosonically bonded to a SoC die and a carrier MEMS socket that acts as a removable insert for the fixed Socket to provide connectivity between the SoC die and a CMOS die, a MEMS device or a photonics device. Pressure dependent, solderless, high density (150 μm pitch), temporary or semi-permanent connectivity is achieved using a cantilevered bridge-type platinum coated copper spring contact mechanism. The proposed 100 x 100 μm^2 area cantilevered bridge -type spring contact has excellent elastic properties and features a contact resistance of 19.3 mΩ at a contact force of 1.35 mN. The system is expected to enable reliable, cost-effective testing of high density interconnection CMOS dies or to connect disruptive technology devices, like MEMS or photonics devices to a SoC environment.

Key words: MEMS, SoC, Socket, Spring Contact, Connectivity, Interface

1. INTRODUCTION

A System-on-Chip (SoC) implementation may require IP cores of diverse technological origins, such as, analog and/or digital CMOS, MEMS, photonics, electro-biological or electrochemical devices. A trend of decreasing feature size in CMOS technology drives requirements for fine pitch, high I/O density interconnects [1]. Consequently, there is a requirement for connectivity to a high I/O density SoC implementation that can accommodate a variety of interconnection needs. Temporary or semi-permanent connectivity is also required to interconnect disruptive technology devices that are designed for one time use, e.g., a disposable chemical sensor or some bio-specimen analyzer. The MEMS socket system proposed in this paper is designed to meet flexible connectivity requirements. The proposed socket interface is comprised of a fixed MEMS socket that is thermosonically bonded to a SoC die and a removable carrier MEMS socket that has cantilevered bridge-type spring contacts fabricated on both sides. This interface is capable of providing a highly reliable cost-effective temporary or semi-permanent connectivity between a SoC die and a CMOS die or a MEMS or a photonics device. The 100 x 100 μm^2 area cantilevered bridge type spring contact ensures a pressure dependent, low-force, low-contact resistance, and high-density interconnection mechanism. The platinum coated copper spring contacts have excellent elastic properties and features a contact resistance of 19.3 mΩ at a contact force of 1.35 mN. The MEMS socket system has been designed and simulated using IntelliSuite CAD tools with satisfactory results and fabrication arrangements are being pursued.

2. MEMS SOCKET CONCEPT

The conceptual geometry of the MEMS socket system for a SoC implementation of a die tester is illustrated in Figure 1. A fixed MEMS socket having geometry as shown in the Figure 1, is surface mounted to a CMOS Tester-on-Chip (ToC) die using an enabling gold-to-gold thermosonic bonding technology [2]. The fixed MEMS socket is fabricated using deep reactive ion etching (DRIE) of a silicon wafer to create vertical columnar structures on the top. The fixed socket and the CMOS ToC die are packaged together and the hermetically sealed interior of the package is filled with dry nitrogen. Platinum coated aluminum socket pin electrodes, fabricated on the vertical columnar structures, connect to the thermosonically bonded ToC metal I/O pad by a wafer-through-hole aluminum-via-interconnect. A removable MEMS carrier socket acts as a die-specific carrier for the CMOS die under test. The removable carrier socket, is also a silicon wafer pre-fabricated to have a specified geometry (Figure 1) and has cantilevered bridge-type platinum coated copper spring contacts on both sides. The size of the removable carrier is adaptable to any die size, thus adding flexibility to the interconnection environment. The device under test (DUT) will be placed in the removable carrier socket by a robotic manipulator. A load mass will then be positioned on the DUT to provide the necessary downward force to keep the DUT tightly pressed against the carrier. Due to the applied pressure and socket geometry the contact springs on the removable carrier socket will undergo elastic deformation to realize a contact between the metal I/O pad on the DUT and the fixed socket contact (Figure 2).

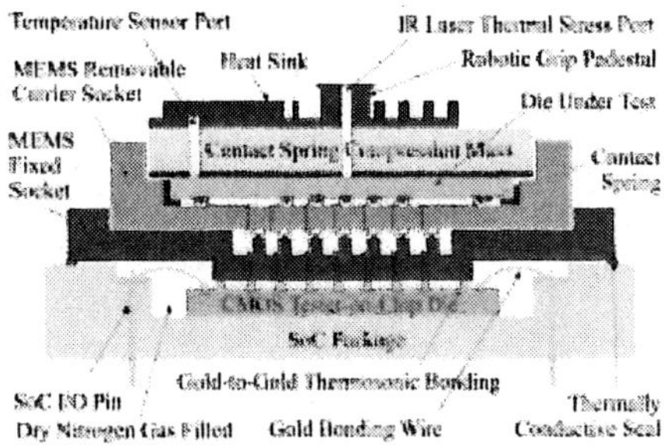

Figure 1. MEMS socket system conceptual configuration

A heat sink designed to meet the thermal management requirements is placed on the top of the load mass. The heat sink structure has vertical ports for the insertion of an IR laser thermal probe and also for thermal sensors to measure temperature. This stacking arrangement results in a low-space, low-contact resistance (as the contact path is short) and low-contact force interconnection environment. An alignment key is incorporated in the fixed and the removable socket structures to ensure proper alignment of the removable carrier and fixed sockets.

3. SPRING CONTACT DESIGN

The design objective of the cantilevered bridge-type spring contact mechanism was to develop a low-contact force, a low-contact resistance, a small area and a reliable pad pressure contact mechanism that would provide temporary connectivity between the I/O pad on the DUT and the contact

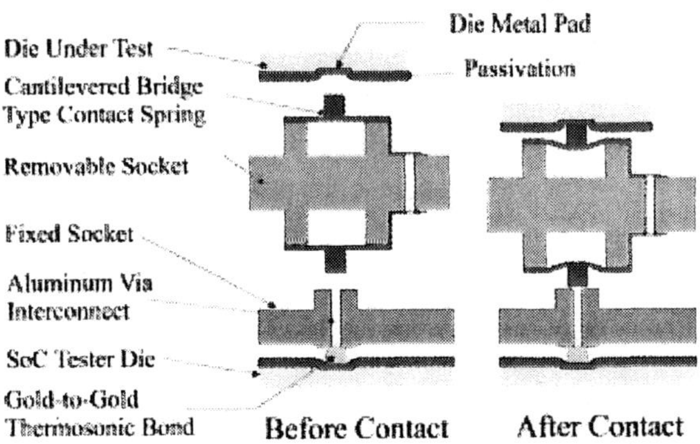

Figure 2. Schematic of the contact mechanism

electrode on the fixed socket. The pad's spring mechanism has to undergo elastic deformation while under pressure from the top load mass. Another important requirement is to maintain a contact surface that will remain reasonably flat even under possible torsional deformation to realize a higher contact area. The need for the second requirement arises from the fact that a contact pad on a cantilever beam type contact spring deforms due to an applied load where the top surface of the contact pad undergoes a shear that results in a curved surface (Figure 3). Thus, the whole pad area doesn't provide connectivity; instead a thin line contact is made resulting in a much higher contact resistance. Finite element analysis showed that a cantilevered bridge-type beam structure can provide a better contact, however, the bridge design must ensure that at the deformed state, the top surface of the contact pad on the spring does not deform into a concave surface where the contact area is limited only to the circumferential edge. Another consideration is to determine the Z-axis displacement range for a specified elastic deformation while achieving the required contact force. The stress at the deformed state must be less than the Von Mises stress [3] expressed as:

$$S_M = \left\{\frac{1}{2}\left[(\sigma_{p1} - \sigma_{p2})^2 + (\sigma_{p2} - \sigma_{p3})^2 + (\sigma_{p3} - \sigma_{p1})^2\right]\right\}^{1/2} \quad (1)$$

so as to ensure an elastic deformation. In (1) σ_{p1}, σ_{p2} and σ_{p3} represent the principle stresses. Polysilicon, a common MEMS structural material, due to it's lower ultimate tensile strength (σ_{UT}) of 1.2 GPa, is unsuitable for this application. Also, the use of polysilicon requires a copper or aluminum conductive trace over it, thus adding additional fabrication steps and concerns about thermal shear stress due to different coefficients of thermal expansion. Copper has a higher ultimate tensile strength (σ_{UT}) and very

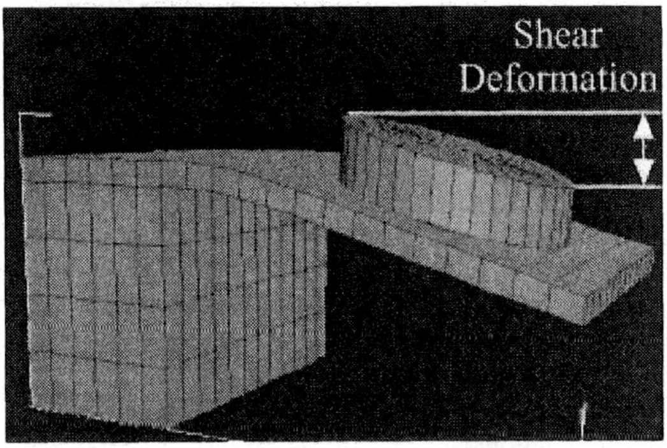

Figure 3. Shear deformation of a cylindrical contact pad on a cantilever beam

good electro-migration and conductivity properties and appears to be the best choice. Oxidation of the contacting surface of the pad can be avoided by applying a protective deposition of platinum. Based on these considerations a new contact spring geometry where a square thin copper plate with a contact pad on the top is connected to support posts by four cantilevered support beams has been designed as shown in Figure 4.

For analysis purposes, the structure can be viewed as a thin isotropic square plate of side length a, laterally loaded with a uniform load on a circular area at the center of the contact pad, that is elastically supported by 4 beams of width b at the edges parallel to the X-axis. For simplicity, the solid cylindrical contact pad on the plate is assumed to be a load mass and the total load on the plate is then the combination of the load due to the pad and the laterally applied uniform pressure on the top surface of the contact pad.. From classical structural analysis [4], the deflection of a thin plate due to a laterally applied uniform load is expressed as:

$$\frac{\partial^4 w}{\partial x^4} + 2\frac{\partial^4 w}{\partial x^2 \partial y^2} + \frac{\partial^4 w}{\partial y^4} = \frac{q}{D} \qquad (2)$$

where w is the deflection of the plate due to an applied load q and D is the flexural rigidity of the plate expressed as:

$$D = \frac{Eh^3}{12(1-\nu^2)} \qquad (3)$$

where E is the modulus of elasticity, ν is Poisson's ratio and h is plate

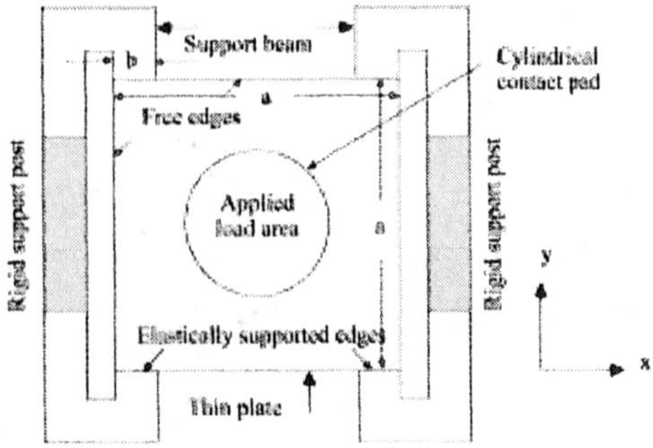

Figure 4. Cantilevered bridge-type spring contact geometry

thickness. The developed bending and torsional moments can be given as:

$$M_x = -D\left(\frac{\partial^2 w}{\partial x^2} + v\frac{\partial^2 w}{dy^2}\right) \tag{4}$$

$$M_y = -D\left(\frac{\partial^2 w}{\partial y^2} + v\frac{\partial^2 w}{dx^2}\right) \tag{5}$$

$$M_{xy} = -M_{yz} = D(1-v)\frac{\partial^2 w}{\partial x \partial y} \tag{6}$$

Shearing forces are expressed as:

$$Q_x = -D\frac{\partial}{\partial x}\left(\frac{\partial^2 w}{\partial x^2} + \frac{\partial^2 w}{\partial y^2}\right) \tag{7}$$

$$Q_y = -D\frac{\partial}{\partial y}\left(\frac{\partial^2 w}{\partial x^2} + \frac{\partial^2 w}{\partial y^2}\right) \tag{8}$$

To solve equation (2), appropriate boundary conditions are to be applied. For the contact spring geometry, the boundary conditions can be evaluated by considering the plate is elastically supported by 4 cantilever beams along the edges parallel to the X-axis by a distance b for each beam. The remaining edge sections along the edges parallel to the X-axis, viz. $x=a-2b$ and the edges parallel to the Y-axis are treated as free boundaries (Figure 4). Equation (2) can now be solved, first by assuming that all four edges of the plate are free and then applying equivalent bending moments due to support beams at the edge sections where the plate is elastically supported. For the elastically supported section, the deflection due to an applied load is not zero and is equal to the deflection of the beam. Also, the rotation of the edge interface of the support beam and the thin plate is equal to the twisting of the beam. If B and C are the flexural and the torsional rigidity of the beam, respectively, the pressure in the Z-axis direction transmitted from the plate to the supporting beam can be expressed as [4]:

$$-V_x = D\frac{\partial}{\partial x}\left[\frac{\partial^2 w}{\partial x^2} + (2-\nu)\frac{\partial^2 w}{\partial y^2}\right] \tag{9}$$

The differential equation for the deflection curve of the beam is [4]:

$$B\left(\frac{\partial^4 w}{\partial y^4}\right) = D\frac{\partial}{\partial x}\left[\frac{\partial^2 w}{\partial x^2} + (2-\nu)\frac{\partial^2 w}{\partial y^2}\right] \tag{10}$$

From a consideration of the rotational equilibrium of an element of the beam, it follows that [4]:

$$-C\frac{\partial}{\partial y}\left(\frac{\partial^2 w}{\partial x \partial y}\right) = D\left(\frac{\partial^2 w}{\partial x^2} + \nu\frac{\partial^2 w}{\partial y^2}\right) \tag{11}$$

Equations (10) and (11) provide the necessary boundary conditions for the edge sections of the p late where the plate is elastically supported by the beams. For the free edges, the appropriate boundary conditions are:

$$\left(\frac{\partial w^2}{\partial y^2} + \nu\frac{\partial^2 w}{\partial x^2}\right) = 0 \tag{12}$$

and:

$$\frac{\partial}{\partial x}\left[\frac{\partial^2 w}{\partial y^2} + (2-\nu)\frac{\partial^2 w}{\partial x \partial y}\right] = 0 \tag{13}$$

Applying the above boundary conditions, the deflection of the thin plate can be evaluated. Equation (9) provides the vertical shear force acting on the connecting edge of the U shaped cantilever beam and equation (10) provides the beam deflection.

For an externally applied load of 1.35 mN, the IntelliSuite FEA simulation result shows that the maximum Z-axis displacement of the contact pad is 5 μm (Figure 5). The thin square plate area is 60 x 60 μm^2 and the contact pad height is 5 μm and is 20 μm in radius. Each support arm is 70 μm long and 10 μm wide. At maximum displacement, the Von Mises stress is 1.47 GPa or little more than half of copper's yield strength (2.80 GPa). It is evident that the deformation of the spring under the specified force is within the elastic limit of copper and allows for a reasonable safety margin. The connectivity between the spring pad and the die pad is based on an area as opposed to a line contact.

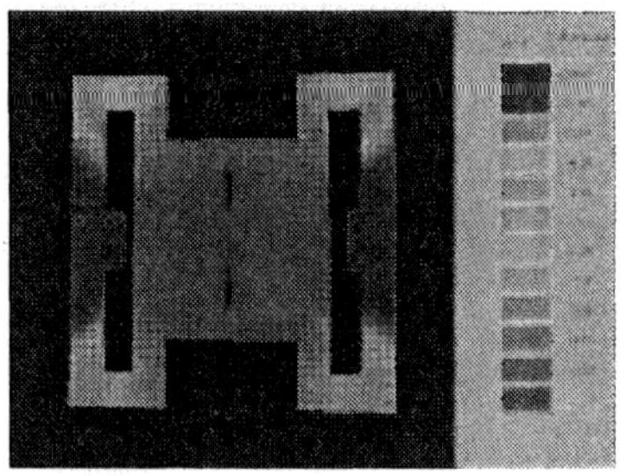

Figure 5. FEA simulation showing Z-axis displacement

4. CONTACT RESISTANCE CALCULATION

For a given set of material properties, surface topography and normal temperature the electrical resistance across the contact interface between two conductors decreases as the applied load increases. This occurs because real surfaces are not perfectly smooth and flat and may be considered as composed of a large number of small, but discrete contacting asperities. In a ductile material, as a normal force is applied across the contacting interface, the number of the surface asperities that support the applied load gradually increase due to their successive deformation. As a result the true contact area increases and in the limit, approaches the apparent contact area. The elastic or the elastic-plastic behavior of thin film contacting materials under pressure depends partially on the surface roughness of the contacting materials. The surface roughness depends on the particular deposition method employed, crystallographic properties of the thin film material, film thickness and deposition parameters such as, pressure, temperature, deposition time, etc. As surface roughness is a random phenomenon, it is not possible to formulate a deterministic analytical expression for asperity density, heights and asperity radii to evaluate an actual area of the contacting surfaces. An empirical method to estimate the asperity density based on statistical analysis has been found to be a suitable approach [5]. A modified version of Greenwood's formula [6] that incorporates an empirical model of asperity density for contact resistance evaluation is available in [5]. As it is stated in [5], if n is the number of contacting asperities and A_E is the apparent area of contact, the asperity density η can be expressed as:

$$\eta = (n / A_E) \tag{14}$$

If a is the average radius of contacting asperities, true contact area, A_C can be expressed as:

$$A_C = n(\pi a^2) = \frac{F}{\sigma_{YS}} \tag{15}$$

where F is the applied force and σ_{YS} is the yield strength of the softer contacting material. From equations (14) and (15), and relating F to pressure, P, $(F = PA_E)$, the average radius, a ,of contacting asperities can be expressed in terms of η as:

$$a = \sqrt{\frac{P}{\eta\pi\sigma_{YS}}} \tag{16}$$

In [5], it is also shown that average distance l between contacting asperities can also be expressed in terms of η as:

$$l = \left(\frac{1}{4n / A_E}\right)^{1/2} = \left(\frac{1}{4\eta}\right)^{1/2} \tag{17}$$

Using equations (14), (15), (16) and (17), Greenwood's basic formula of contact resistance [6] can now be expressed in terms of asperity density as [5]:

$$R_{CA} = R_C A_E = \left[\frac{\rho_1 + \rho_2}{\sqrt{\eta}}\left(\sqrt{\frac{\pi\sigma_{YS}}{2P}} + \frac{3k}{32}\right)\right] \tag{18}$$

where, R_C is the contact resistance, ρ_1 and ρ_2 are the bulk resistivities of the two contacting conductors, k is a shape factor, and R_{CA} is the unit area contact resistance expressed in ohm.m^2. The asperity density η of a thin film can be estimated from a statistical analysis of roughness and waviness profiles (motifs) of the SEM micrograph image of the thin film [7]. By substituting η and other parameters in equation (18) the contact resistance can be evaluated.

Detailed design specifications and MEMS fabrication procedures for the socket system can be found in [8].

5. SUMMARY

The design of a MEMS socket system to provide temporary or semi-permanent interconnection of MEMS, CMOS, photonics, etc. devices to a SoC test environment has been developed. A key-enabling element of the design is the low-contact resistance cantilevered bridge type spring contact structure. Contact spring failure due to fatigue is minimized, by ensuring that contact spring deformation is always limited to the elastic deformation range, so as to provide a longer lifecycle. Connectivity at high frequencies has been enhanced by developing a contact spring geometry with low inductive reactance so as to minimize impedance related problems.

ACKNOWLEDGEMENTS

The research has been made possible by the interest and support provided by the Gennum Corporation of Burlington, Ontario. The authors greatly acknowledge the generous support of the following partners: Canadian Microelectronics Corporation (CMC), MICRONET, NSERC and the IntelliSense Corporation, Wilmington, MA, USA.

REFERENCES

[1] Semiconductor Industry Association. *International Technology Roadmap for Semiconductors: 1999 Edition.* Austin, TX: International SEM ATECH, 1999, pp. 60, 231.

[2] K. Cheah, Y. M. Tan, J. Wei and C. K. Wong, "Gold to Gold Thermosonic Flip-Chip Bonding", in Proc. 2001 HD International Conference on High-Density Interconnect and Systems Packaging, Santa Clara, CA, 2001, pp. 165-170.

[3] Richard G. Budynas, *Advanced Strength and Applied Stress Analysis*, 2nd ed., New York: WCB McGraw-Hill, 1999, pp. 509-516.

[4] S. Timoshenko, S. Wonionowsky-Krieger, *Theory of Plates and Shells*, New York: McGraw-Hill Book Company Inc., 1959, pp. 79-224.

[5] S. S. Babu, M. L. Santella, Z. Feng, B. W. Riemer, and J. W. Cohron, "An Empirical model of the Effects of Pressure and Temperature on the Electrical Contact Resistance of Metals", Metals and Ceramics Division, Oak Ridge National Laboratory, Oak Ridge, Tech. Report, TN 37831-6096, 2001, pp. 7.

[6] J. A. Greenwood, "Constriction Resistance and the Real Area of Contact", British Journal of Applied Physics, Vol. 17, 1966, pp. 1621-1632.

Section 10.3

On the Application of Finite Element to Investigate the Reliability of Electrostatic Comb-Drive Actuators Utilized in Microfluidic and Space systems

Hesham Ahmed , Walied Moussa, Wael Badawy
University of Alberta and University of Calgary, Alberta, Canada

Abstract This study focuses on investigating the design parameters of lateral Electrostatic Comb -Drive actuators, where the effect of these parameters on the actuation performance is explored using Finite Element Analysis (FEA). This level of analysis is essent ial to the design process of system-on-a-chip MEMS applications, where a Comb -Drive can represent the main source of actuation within the chip system. Of particular interest to this study, is the application of the electrostatic Comb -Drive motor as a fluidic pump in on-a-chip systems used for drug delivery or in cooling of microprocessors used space vehicles. The commercial FE package ANSYS is utilized to construct a robust model and solve this multi-physics interaction problem using the Direct Coupled -Field Analysis. In this model, the thickness, gab and overlap between the Comb fingers are varied. The design of this model is also modulated to account for the change in the number of Comb fingers and the applied driving voltage. The calculated Comb displacement and generated electrostatic force are shown to be directly proportional to the number of Comb fingers. Moreover, the generated electrostatic force is found to be inversely proportional to the gap between the Comb fingers. As the thickness of these fingers increases the displacement is found to increase for a given value of the driving voltage. The electrostatic force is also shown to be proportional with the offset (i.e. Overlap) value between the fingers.

1. INTRODUCTION

Microactuator structures such as Comb -Drive are used as transducing elements in a wide variety of MEMS structures. Microactuators can be excited in several ways, including piezoelectric films, thermal expansion, shape memory alloy principle, magnetostatic forces and electrostatic forces.

Combined with capacitive (electrostatic) detection, the electrostatic excitation is an attractive approach for silicon microstructures because of the simplicity and compatibility with micromachining technology such as surface micromachining, which is the most common manufacturing technique for this type of micro devices. Accurate calculation of these electrostatic forces is an essential part of the design process of any microactuator. In Comb -Drives, the electrostatic force and the resulting Comb displacement can be controlled via changing the driving voltage. Recently, several studies have been carried out to obtain an efficient Comb -Drive design with high displacement and low driving voltage requirements [1].

Voltage controlled Comb-Drive actuators exert lateral electrostatic forces making them attractive for micropositioning applications, such as, xy -microstages [2], Micro mirrors used in the optical switches [3] and fluidic pump in on-a-chip drug delivery systems, [4]. The Comb -Drive can be fabricated via a surface-micromachining fabrication technology such as the Multi-User MEMS Process service (MUMPs). In this process, the Comb components are made entirely from a homogeneous, conducting polysilicon film. Fig. 1, shows a typical layout of the Comb -Drive structure where system components are illustrated.

2. THEORY

The driving principles of electrostatic Comb -Drive actuator is based on electrostatic characteristics [5]. As shown in Fig. 1, when voltage is applied to the fixed electrode (or fingers) while the movable electrodes (or fingers) are potential grounded, a potential difference results across the electrodes and they become electrically charged. This action will induce a certain capacitance in the charged electrodes. Thus, an electrostatic force is generated causing a displacement in the x direction. The driving force F created in the Comb can be expressed as:

$$F = \frac{\partial U}{\partial g} = \frac{1}{2} \frac{nt \epsilon_0 \epsilon_r V^2}{g} \qquad (1)$$

where U is the energy associated with the applied electric potential V, ϵ_r is the relative permittivity of the dielectric material between the two electrodes, ϵ_0 is the permittivity in the free space and equal to 8.85 pF/m, n is the number of the pairs of electrodes, t is the thickness of the electrodes, A is the area of the finger and g is the gap between two fingers, as shown in Fig. 1.

If the movable electrode is displaced following Hooke's law, the reaction force F_S induced in the suspension beam holding the movable electrodes, which represents a spring suspension system, will be related to the stiffness of the beam and the displacement and can be calculated from the following formula:

$$F_S = K_X x \qquad (2)$$

where, K_X is the spring stiffness in the x-direction (i.e. actuation direction) and x is the displacement. The suspension beams must be flexible enough in the direction of actuation, increasing the beam stiffness will require larger electrostatic force to cause the deflection and consequently requires a higher driving voltage. Formulas used to drive the spring stiffness are shown in Jaecklin et al. [1] and Rob et al. [6]. Furthermore, a suspension that is compliant in the direction of displacement and stiff in the orthogonal directions is required to achieve high displacement with lower driving voltage and can avoid many unfavourable operation problems. Therefore, different types of spring design have been applied in Comb -Drive actuators. Some of these designs include clamped-clamped beams, crab-leg flexure, and folded-beam flexure. Those designs are explained in details by Rob et al. [6]. In equilibrium position, the forces F_S and F have to be equal, where the displacement x can then be expressed as a function of the driving voltage (V), the gab (g), the thickness (t), the number of electrodes (n) and the stiffness (K_X) as:

$$x = \frac{1}{2}\frac{n t \epsilon_0 \epsilon_r V^2}{g K_X} \qquad (3)$$

Equation (3) shows that the relation between the actuator displacement and the applied driving voltage is not quite linear and as the applied voltage increases the displacement increases, normally this relation is known as the voltage-stroke relation of the Comb -Drive. The achievement of a smart design of the Comb -Drive, i.e. large displacement at low driving voltage, is an essential requirement that need to be fulfilled. Equations (1 to 3) summarize the design parameters of the Comb -Drive. These parameters are (i) the spring design and stiffness K_X and K_Y, (ii) the driving voltage V, (iii) the electrodes gap g, (iv) initial overlap L, (v) number (vi) thickness and (vii) width of the Comb fingers.

3. NUMERICAL APPROACH

To calculate the driving forces and displacement of different Comb -Drive designs, a robust finite element model has been constructed and solved within the multiphysics environment in the commercial finite element package ANSYS [7 and 8]. The construction of this model is automated to account for the variation of all the design parameters discussed in the previous section. To solve this multiphysics interaction problem, a direct finite element coupled-field analysis is applied.

Full dynamic FE simulation of coupled electrostatic structural problems is prohibitively expensive. Therefore, it is necessary to simplify the multiphysics modelling and provide accurate, high fidelity harmonic and time -domain solutions in a fast and efficient manner. This simplification is applied in the current work using the Reduced Order Modelling technique (ROM), which consist of using lumped finite elements (springs, masses, dampers, transducers, capacitors, inductors, resistors, etc.) and substructuring of large linear systems [9 and 10].

To start with, a linear Comb-Drive resonator shown in Fig. 1, was modeled several times with different geometries. In all analysed cases, the Comb -Drive is replaced with the TRANS126 transducer, [11]. This single transducer is then used for the drive and pick-up Combs. This analysis stage applied was focused on solving a 3D FE model of the Comb structure over the stroke-range of the device to compute the capacitance as a function of stroke using the electrostatic analysis method, mentioned earlier, which utilizes a built -in macro program in ANSYS named (CMATRIX), [9]. This stage produces the input parameters for the transducer element. Only uniaxial motion is considered at the resonator to calculate the change in the capacitance against the stroke. The second analysis stage performed was a static analysis associated with the substructure generation solver. In this analysis, a master node is defined at the centre of the model with a DOF in the direction of excitation and a static load applied in the direction of excitation. The stiffness matrix of the system is computed and the effective stiffness of the suspension system is then obtained from these results.

In the third analysis stage, a Modal Analysis is performed to extract the effective mass of the movable Comb-Drive. For most MEMS devices, there is a dominant response mode of vibration. This mode can be identified from the modal analysis by examining the largest value of the participation factor in the direction of excitation. This factor is a measure of how much a mode is contributing to the deflections in th e direction of excitation; and, the mass associated with it represents the effective mass at that mode in the excitation direction, which can be obtained easily from the analysis results. The damping factor is preset to correspond to a Q-factor of 100, which is used to calculate the damping ratio of the system; this damping ratio is an input to the COMB14 damper element in ANSYS, which is used in circuit simulations, [11].

The Fourth type of analysis was the Reduced Order Model analysis (ROM) where a lumped finite element model is built using the circuit simulation building capabilities of ANSYS. In this stage, a lumped model that can capture the characteristics behaviour of the global 3-D finite element model is used. In this circuit the input to the transducer element will be the excitation voltage and the output will be the structural DOF of the element end in the actuation direction. Voltage can be applied directly to the node of the transducer element or through the use of the general circuit element (CIRCU124), which is one of the lumped elements used to simulate the electrical systems. Element CIRCU124 includes sub-models of resistors, capacitors, inductors, current and voltage sources. The displacement in the movable Comb fingers was measured from the structural DOF of the element end in the actuation direction as the electrostatic energy is converted into mechanical energy via the transducer element.

4. RESULTS AND DISCUSSION

In the current study, the Direct FE Coupled-Field Analysis is applied to the robust electrostatic Comb-Drive model discussed earlier, where the effects of this Comb design parameters on the actuation performance is explored. Figure (2) and Fig (8) shows the variation of the Comb displacement and generated electrostatic force with the number of fingers at a driving voltage of 40 volts. In these figures, the calculated Comb displacement and generated electrostatic force are shown to be linearly proportional to the number of Comb fingers. For constant spring stiffness, Eq (2) shows tha t the electrostatic force should also vary linearly with the number of fingers in the studied Comb-Drive. In addition, when the driving voltage is increased, the comb displacement is shown in Fig (3) to nonlinearly increase for a comb design with 20 fingers. In this figure, the displacement rate of variation is shown to increase with the driving voltage. This is also valid for Fig (9) where the variation of the electrostatic force is shown.

The results in Fig (4) indicate that for a driving voltage of 45 volts, as the gap (g) between the Comb finger decreases, the induced capacitance nonlinearly increases and so is the induced Comb displacement. This effect of the Comb gab is also expected based on Eq (1). In this case, the displacement variation rate is shown to decrease with the increase of the gab (g). This effect is also indicated by Fig (10) for the electrostatic force. On the other hand, an increase in value of the overlap distance (L) is shown to increase the Comb displacement. This non-linear proportional relation between the Comb displacement and the offset or overlap distance (L) is shown in Fig (5) for a driving voltage of 40 volts and a comb gab (g) of 2 μm as it is shown also for the electrostatic force as indicated in Fig (11). In addition, the effect of the Comb -Drive finger thickness on its displacement value is shown in Fig (6). In this case, the finger thickness (t) is varied between 1, 2 or 3 μm. The driving voltage is then increased and the displacement is calculated. Figure (6) shows that the increase in the Comb displacement is nonlinearly proportional with the increase in the finger thickness, where the rate of displacement variation decreases for higher finger thickness. This effect on the electrostatic force is also seen at Fig (12). Finally, Fig (7) shows the displacement variation with the driving voltage at different spring designs. As the Comb-Drive folded-flexure length increases the spring stiffness in the actuation direction decreases, which increase the displacement of the Comb. Therefore, increasing the length of the folded-flexure beams is desirable to increase the Comb displacement.

5. CONCLUSION

In this work, the effect of the design parameters of lateral electrostatic Comb-Drive actuators on the actuation performance is explored using the Direct FE Coupled-Field Analysis. The induced Comb displacement and electrostatic force are shown to be directly proportional to the number of Comb fingers. Also, as the driving voltage increase the displacement is shown to increase but up to a certain limit where side instability or formally known as "Sticking" occurs, this phenomenon is discussed in detail by Rob et al. [6]. Furthermore, the Comb displacement is found to be inversely proportional to the gap between the Comb fingers. This relation is also limited by the sticking phenomena. Moreover, as the thickness of these fingers increases the displacement is found to increase for a given value of the driving voltage. The Comb displacement is also shown to be nonlinearly proportional with the offset (i.e. Overlap) value between the fingers. For the spring design analysis, the spring stiffness in the actuation direction is shown to decrease with the increase of the Comb-Drive folded-flexure length. Since this results in the increase of the Comb displacement, increasing the length of the folded-flexure beams was concluded to be desirable.

REFERENCES

[1] Jaecklin V P, Linder C, de Rooij N F, Moret J M "Micromechanical comb actuators with low driving voltage" Journal of Micromechanics and Microengineering. Vol. 2, (1992) p 250-255

[2] Ty Harness and Richard R A Syms “Characteristic modes of Electrostatic Comb-Drive X-Y Microactuators”, J. Micromech. Microeng. 10 (2000), pp. 7-14
[3] Meng-Hsiung Kiang, Olav Solgaard, Kam Y. Lau and Richard S. Muller “Electrostatic Comb-Drive-Actuated Micromirros for laser-Beam Scanning and Positioning”, J. Microelectromech. Syst. Vol.7 n.1, pp. 27-37, March 1998
[4] Sandia National Laboratories “Sandia Lab News”, Vol.53 n.13, June 29 ,2001
[5] Wiliam C. Tang, Tu-Cuong H. Nguyen, and Roger T. Howe "Laterally Driven Polysilicon Resonant Microstructures", Proc. 1989 IEEE Micro Electro Mechanical Systems, pp. 53-59, February 1989
[6] Rob Legtenberg, A W Groeneveld and M Elwenspoek “Comb-Drive actuators for large displacements”, J. Micromech. Microeng. 6 (1996), pp. 320-329.
[7] S . Groothuis, "Analyzing Microminiature Devices", ANSYS Solutions, Vol. 1, No. 1, Spring 1999.
[8] D . Ostergaard, "Tooling-Up for MicroElectroMechanical Systems (MEMS)", ANSYS Solutions, Vol. 1, No. 3, Winter 1999.
[9] ANSYS Release 5.7 User Manuals, 2000.
[10] Dale Ostergaad, Miklos Gyimesi "Finite Element Based Reduced Order Modeling of Micro Electro Mechanical Systems MEMS", ANSYS Inc. Southpointe, 275 Technology Dr., Canonsburg, PA 15317, USA
[11] M. Gyimesi, D. Ostergaard, "Electro-Mechanical Transducer for MEMS in ANSYS", Proc. 1999 Int. Conf. on Modeling and Simulation of Microsystems, April 19- 21, 1999.

Section 10.4

AN HDL MODEL FOR A VACUUM-SEALED MICROMACHINED PRESSURE SENSOR

Rafik S. Guindi
Electronics & Communication Engineering, Cairo University, Egypt

Abstract: The development of behavioural models for micro-electromechanical systems (MEMS) is of great importance as it allows system-level simulation for MEMS components. This paper describes a behavioural model for a vacuum-sealed micro machined pressure sensor that can be integrated with electrical components within an electrical simulator using an analog hardware description language. Detailed description of the model as well as simulation results are presented.

1. INTRODUCTION

New developments in Micro-electromechanical Systems (MEMS) span across such diverse applications as inertial navigation systems [1], digital mirror displays [2], DNA analysis systems [3], and RF distributed sensor networks [4]. These systems incorporate truly mixed technologies, integrating digital and analog electronics, mechanical structures, electromechanical actuators, and fluidic chambers. There is a growing demand for CAD tools to support rapid design of systems involving physical interactions between the various domains, which would enable MEMS engineers to build larger systems and allow them to concentrate on higher-level design issues [5].

No rapid design process is available today for MEMS. Only one or two CAD iterations involving simple functional simulations are usually attempted during prototype design. As a result, fabrication replaces simulation in the iterative loop. This is very expensive since fabricated prototypes often do not meet performance specifications, and sometimes are not even functional. Full verification of designs requires months of effort, and design optimization is not realistic in all but the simplest of cases. These problems inhibit growth in the field, particularly in the use of MEMS for low-cost, low-volume application specific sensors, integrated on the same chip as attendant electronic information processing and communication capabilities [5].

From here arises the importance of developing behavioural models for MEMS is which both electrical and mechanical parts can be simulated within the same simulator. The ability to simulate mechanical systems on an electrical simulator makes the complete simulation of mixed mechanical/electrical systems such as sensors and interface circuits possible by making use of behavioural mathematical modelling.

In this paper, a behavioural model for a vacuum-sealed silicon micro machined pressure sensor [6] is presented. The model was implemented using the HDL-A™ hardware description language [7], and simulated within the ELDO™ simulator, which is a SPICE-like electrical simulator. First we describe the structure of the pressure sensor, and then the model is presented along with modelling details and simulation results.

2. SENSOR STRUCTURE AND OPERATING PRINCIPLE

Figure 1 shows the structure of the servo capacitive vacuum sensor [6]. It consists of a silicon wafer sandwiched between two glass substrates forming a vacuum-sealed reference cavity. A sensing electrode on the lower glass plate and the silicon form a capacitor, with a capacitance *CS* that varies according to the applied pressure. The electrode on the upper glass and the silicon form another capacitor *CA*, the servo capacitor.

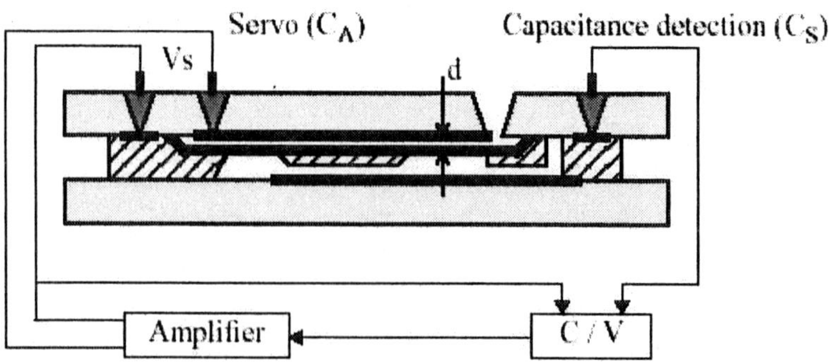

Figure 1: The Pressure Sensor in a Closed Loop.

When the device is operated in a closed-loop mode, a C / V converter circuit generates a voltage proportional to the sensed capacitance *CS*. This voltage is amplified to *Vs* which is fed to the servo electrode, producing an electrostatic force *Fe* that opposes the deflection of the silicon plate due to the pressure *P*.

Under balanced conditions, *Fe* = *P*, and the movable silicon electrode is maintained in its initial position, which gives the system a very wide dynamic range. The electrostatic force per unit area is given by:

$$Fe = \frac{(\varepsilon_o Vs^2)}{2d^2} = (P) \qquad (1)$$

from which we obtain the relation between the servo voltage *Vs* and the pressure *P*:

$$Vs = \sqrt{(2d^2 P)/\varepsilon_o} \qquad (2)$$

3. MODELING THE PRESSURE SENSOR

Entity and architecture

The pressure sensor is modeled as an entity with 4 external pins (see Figure 2):

- an input pin for the external pressure, designated (P)
- an input pin for the servo voltage, designated (Vs)
- an output pin for the sensing capacitance, designated (Cs)
- a common electrical node (Gnd).

The (P) input is declared as a 'mechanical' pin, meaning that the Kirchhoff across' and 'through' quantities in the electrical simulator are analogous to the 'translational displacement' and 'force' respectively, as applied at that pin. All other pins are declared 'electrical'.

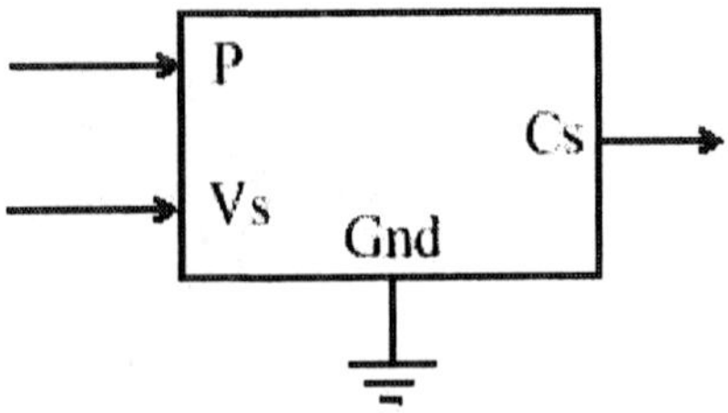

Figure 2: Pressure Sensor Model Entity.

The architecture of the sensor model processes the inputs in several stages to produce the value of the sensing capacitor *CS*. First, the input voltage *Vs* is used to compute the value of the electrostatic force *Fe* according to equation (1). Since this force acts in opposition to the pressure *P*, it is the difference between them that produces a deflection in the movable membrane. This deflection is calculated, and hence the change in the output (sensing) capacitance is computed. If the sensor is to be used in an open-loop mode, there will be no input voltage applied, and the deflection in the movable membrane will be a function of the applied pressure only. Figure 3 shows the architecture of the model in relation with the external circuitry.

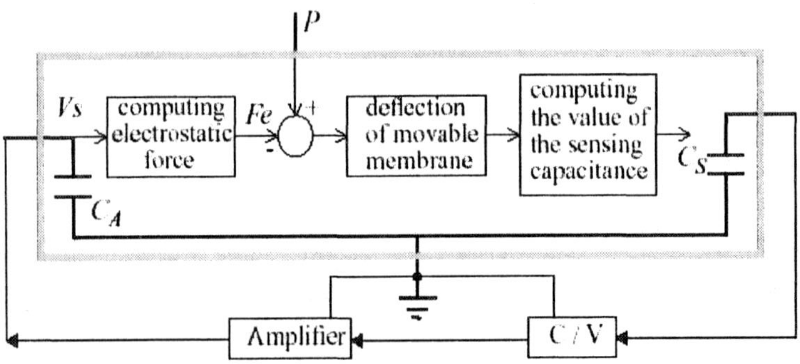

Figure 3: Sensor Model Architecture Placed in a Servo Loop.

Modelling the plate deflection

The main challenge is to properly model the deflection of the silicon plate under the effect of the external pressure. Since the plate is fixed at all four sides, the applied pressure deforms the plate in such a way that the central part is pushed towards the bottom plate, hence altering the capacitance, while the edges of the plate remain fixed. To model this deflection, we consider first the deflection of a beam of span L with two fixed supports at its ends (see Figure 4). Under the effect of a uniformly distributed load W, the deflection of the beam is governed by the equation:

$$\frac{d^4z}{dx^4} = \frac{W}{EI} \quad (3)$$

where E is the beam's Young modulus of elasticity, and I its moment of inertia. The four boundary conditions needed to solve equation (3) are given by:

$$\begin{array}{lll} z = 0 & \text{for } x = 0 \text{ and } x = L & \\ dz/dx = 0 & \text{for } x = 0 \text{ and } x = L & \end{array} \quad (4)$$

By successive integration of the differential equation (3), and using the boundary conditions (4), a solution for the deflection z as a function of x is:

$$z(x) = kW[x^2(L-x)^2] \quad (5)$$

where k is a constant.

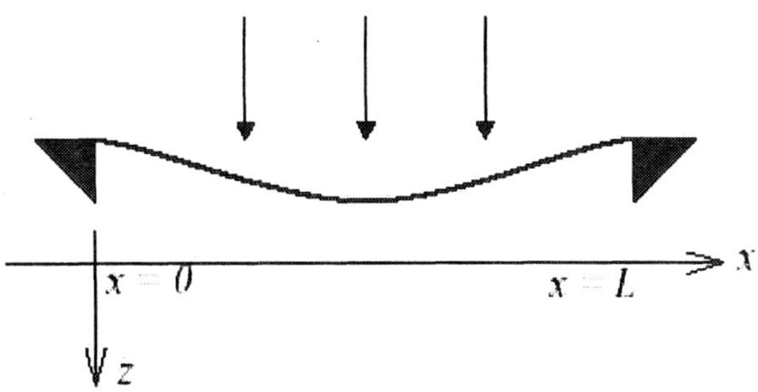

Figure 4: Deflection of a Beam Due to a Distributed Load .

The above result can be extended into the y-dimension, yielding the relation:

$$z(x, y) = KW[x^2(L_x - x)^2][y^2(L_y - y)^2] \qquad (6)$$

where K is a constant, and Lx and Ly are the dimensions of the plate along the x and y directions respectively, as shown in Figure 5.

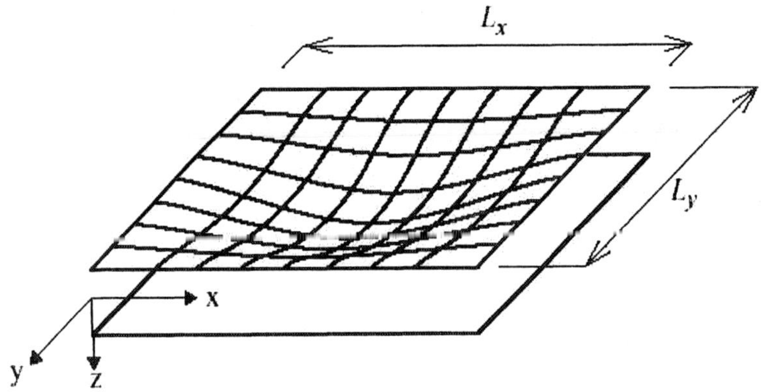

Figure 5: Deflection of the Sensor Plate when Subjected to Pressure.

Computing the value of the sensing capacitance

To simplify the computation, and to avoid having to integrate over the surface of the deflected plate, we divide the plate into n by m equal rectangles. We calculate the capacitance of each rectangle by assuming that it experiences a uniform displacement in the z-direction equal to the average deflection at its center, as given by equation (6). This is illustrated in one dimension in Figure 6. Each capacitance has a value of

$$C_{ij} = \varepsilon_o \frac{L_x}{n} \frac{L_y}{m} \frac{1}{(d_o - z_{ij})} \qquad (7)$$

where do is the distance between the middle and bottom plates of the sensor when no external pressure is applied, and zij is the displacement computed from equation (6), with zij given by:

$$z_{ij} = z\left(\frac{L_x(2i-1)}{2n}, \frac{L_y(2j-1)}{2m}\right) \quad (8)$$

The total value of the sensing capacitance *Cs* is the sum of the (*n m*) individual capacitances, given by:

$$C_S = \frac{\varepsilon_o L_x L_y}{nm} \sum_{i=1}^{n} \sum_{j=1}^{m} \frac{1}{(d_o - z_{ij})} \quad (9)$$

In equations (7-9), an upper limit is imposed on the value of *zij*, such that it will always be less than *do*. This limit is due to the physical structure of the pressure sensor, and leads to a maximum value at which *CS* saturates.

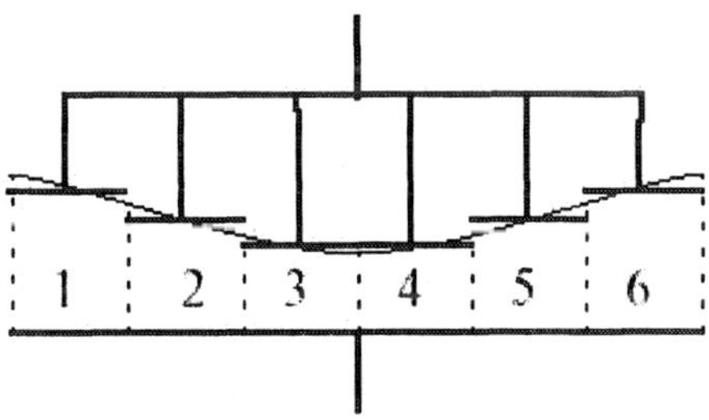

Figure 6: The Total Sensing Capacitance Modeled as a Number of Parallel Capacitors.

4. SIMULATION

Model parameters

Various model parameters are summarized in Table 1. Some parameters (d, do, Lx, Ly) are required, others (n, m) have default values that can be overridden by the user. The number of segments (n and m) used to compute the sensing capacitance determine the accuracy of the model, and can be used for calibration according to the mode of operation of the sensor in the circuit:

- 1. Open-loop mode:

In this mode the *Vs* input to the sensor is not used. The pressure is the only input to the model, and the computed value of the sensing capacitance is the intended output. In this mode, with no electrostatic force to counteract the input pressure, the silicon plate experiences the most deformation, and a large number of segments is needed to accurately compute the value of *CS*..

Parameter	Description	Default
d	Distance between middle and top plates	Required
d_o	Distance between middle and bottom plates	Required
L_x	Length of middle (movable) plate	Required
L_y	Width of middle plate	Required
n	Number of segments along plate length	5
m	Number of segments along plate width	5

Table 1: Model Parameters.

- 2. Open-loop with variable *Vs*:

In this mode, the value of *Vs* is changed until the generated electrostatic force cancels out the effect of the input pressure. Under this condition, the value of *CS* is equal to its value under zero pressure condition, and the middle plate experiences no deformation. The sensor circuit is effectively operated as a null bridge, and the output of the circuit is given by the value of *Vs* that returns *CS* to its original value. In this mode, the number of plate segments n and m is irrelevant, and we can take $n = m = 1$

- 3. Servo mode:

In servo mode, the plate experiences deformation due to the difference between the input pressure and the electrostatic force generated by *Vs*. This deformation is less than in the open-loop case, and therefore a smaller number of plate segments can be used with reasonable accuracy. The value of *Vs* is taken as the output of the sensor circuit.

Results

The following graphs are obtained with $Lx = Ly = 4$ mm, and $d = do = 3.6$ μm. Figure 7(a) illustrates the effect of increasing the number of plate segments on the computed value of the sensing capacitance *CS* with $Vs = 0$. The effect of increasing the value of *Vs* is shown in Figure 7(b), which illustrates the increase in the dynamic range of the sensor with larger values of *Vs*. The pressure response characteristic in closed-loop operation is shown in Figure 8.

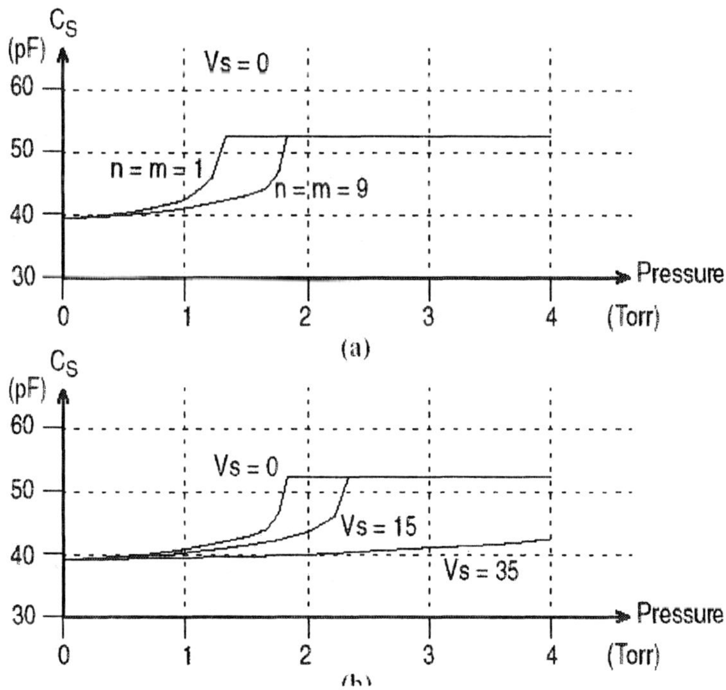

Figure 7: (a) Capacitance Change in Open-Loop Mode. (b) Effect of Changing *Vs*.

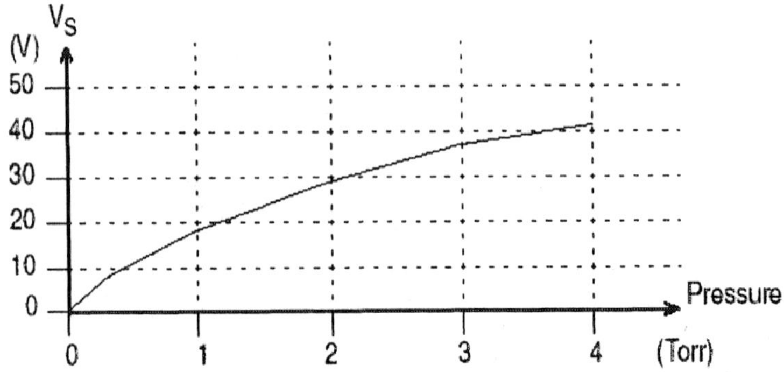

Figure 8: Output Voltage in Servo-Mode.

5. CONCLUSION

In this paper we have presented a behavioural model for a microleletromechanical pressure sensor. The mechanical movable parts have been accurately modelled, and the electromechanical interaction successfully captured in a simple HDL architecture. The model is accurate yet flexible enough to represent the sensor along with the external circuitry within an electrical simulator to perform a full system simulation in a variety of operational modes. The model can be used within top-down methodologies, in system design, system level simulation, and optimization.

ACKNOWLEDGEMENT

I would like to acknowledge the contributions of my students M.E. Gheith A.H. Bashir, T.A. Moselhy, and M.E. Salem. Your enthusiasm made this project a memorable experience.

REFERENCES

1. N.Yazadi, F. Ayazi, and K. Najafi, "Micromachined inertial sensors", *Proc. of the IEEE Special issue on integrated sensors, microactuators, and microsystems (MEMS)*, Vol. 86, pp. 1640-1659, August 1998.

2. P.F. Van Kessel, L.J. Hornbeck, R.E. Meier, and M.R. Douglass, "A MEMS-based projection display", *Proc. of the IEEE Special issue on integrated sensors microactuators, and microsystems (MEMS)*, Vol. 86, pp. 1640-1659, August 1998.
3. C.H. Mastrangelo, M.A. Burns, and D.T. Burke, "Microfabricated devices for genetic diagnostics", *Proc. of the IEEE Special issue on integrated sensors, microactuators, and microsystems (MEMS)*, Vol. 86, pp. 1640-1659, August 1998.
4. C.T.-C. Nguyen, L.P.B. Katehi, and G.M. Rebeiz, "Micromachined devices for wireless communication", *Proc. of the IEEE Special issue on integrated sensors, microactuators, and microsystems (MEMS)*, Vol. 86,pp. 1640 -1659, August 1998.
5. T. Mukherjee and G.K. Fedder, "Structured design of microelectromechanical systems", *Proc. of the IEEE Special issue on integrated sensors ,microactuators, and microsystems (MEMS)*, Vol. 86, pp. 1640 -1659,August 1998.
6. M. Esahi, S. Sugiyama, K. Ikeda, Y. Wang, and H. Miyashita, "Vacuumsealed micromachined pressure sensors" *Proc. of the IEEE Special issue on integrated sensors, microactuators, and microsystems (MEMS)*, Vol. 86, pp. 1640-1659, August 1998.
7."HDL-ATM Reference Manual", v1.4_1, Mentor Graphics, March 1995.

Section 10.5

Performance Analysis of MEMS-based Inertial Sensors for Positioning Applications

Walid Abdel-Hamid, Aboelmagd Noureldin, Dr. Naser El-Sheimy and Dr. Gerard Lachapellee
University of Calgary, Department of Geomatics Engineering, Calgary, Alberta, CANADA.

Abstract: The last two decades have shown an increasing trend in the use of positioning and navigation technologies in land vehicle applications. Cost and space constraints are currently driving manufacturers of vehicles to investigate the applicability of MEMS-based inertial sensors for the development of low cost and small size navigation and guidance systems. The aim of this paper is to investigate the performance of MEMS-based inertial sensors and to analyse their impact on the short-term positioning accuracy. In addition, this article suggests some real-time digital signal processing methods based on the wavelet techniques to improve the performance of these MEMS-based sensors.

Key words: MEMS-bases inertial sensors, inertial navigation, noise modelling, and wavelets multi-resolution analysis.

1. INTRODUCTION

Micro-electro-mechanical sensors (MEMS) are currently utilized in many applications including inertial measurement units (IMU) for navigation applications. The small size and the low cost of these units are highly beneficial for special navigation applications, such as vehicle and personnel location. MEMS-based inertial sensors can, in general, be classified according to the ways of detecting mass position, the mode of operation and the fabrication processes [Schwarz & El-Sheimy, 1999]. All MEMS-based gyros make use of the Coriolis principle through either a single rotating accelerometer or a pair of high frequency vibrating accelerometers and thus are based on Accelerometer technology [Hulsing & MacGugan, 1993].

Classification of MEMS -based Inertial Sensors

The MEMS sensors are classified in terms of the mass position detection criterion as Piezoresistive signal pick-off, Capacitive and Piezoelectric. Table (1) summarizes the motion sensing technique as well as the advantages and the disadvantages of each category. Meanwhile, table (2) shows the main two methods of the fabrication process of MEMS-based inertial sensors [e.g. Varadan at al, 2001, Smits & Balleto, 1994, Zhou, 1998].

Table 1. Classification of MEMS sensors according to mass position detection

Category	Motion sensing Technique	Advantages	Disadvantages
Piezoresistive signal pick-off.	deflection of the piezoresistive material	cheap, simple interface and wide range	low precision and high sensitivity to temperature
Capacitive Sensors	the change in capacitance of two electrodes	More accurate, excellent linearity & low temp. sensitivity	great complexity, EM interference & small dynamic range
Piezoelectric Sensors	Produces an electric charge due to change in motion	relatively wide dynamic range	low sensitivity to small changes in motion

Table 2. Classification of MEMS sensors according to fabrication processes

Category	Sensing Element	Advantages	Disadvantages
Polysilicon Micromachining	A thin silicon structure located on the surface of a die	Suites well for integration with other sensors	Relatively low accuracy and small bandwidth
Bulk Micromachining:	a single crystal inside a block sandwiched between two layers	Better accuracy than polysilicon micromachining	Larger size, more expensive & not good for integration

Current Accuracies of MEMS -Based Sensors

The performance characteristics of MEMS-based inertial sensors are generally defined in terms of the bias stability, scale factor (SF) stability and output noise. The present characteristics of some of the commercially available MEMS-based accelerometers and gyroscopes are respectively shown on Tables (3) and (4) [Schwarz and El -Sheimy, 1999].

2. MEMS-BASED INERTIAL SENSORS ERRORS

As mentioned earlier bias and scale factor errors as well as output noise and axes misalignment determine the performance of inertial sensors. The bias error consists basically of two parts. The first part is a deterministic part that can be evaluated by utilizing some special calibration procedures. The second part is the residual random component at the sensor output, which remains after removing the deterministic part estimated by calibration. In addition to the non-orthogonalties between the sensor axes, the misalignment between the vehicle axes and the sensitive axes of the inertial sensors are usually not a major source of errors in case of six degrees of freedom MEMS systems (IMU) like the bias error and the output noise.

Table 3. Accuracy of some currently produced MEMS accelerometers

Manufacturer or Institute	Product	Performance Characteristics		
		Bias Stability	SF Stability	Output noise
Summit	15203A	---	0.01%/°C	0.225 mg/√Hz
Cross-bow	LF series	±30 mg	---	0.10 mg/√Hz
Charles Stark Draper Laboratory	Capacitive bulk wafer	±100 μg	500 ppm (RMS)	---
Litton Guidance and Control	Capacitive force rebalance	±35 μg	100 ppm (RMS)	---
Allied Signal Inc.	Resonant vibrating beam	±200 μg	60 ppm (RMS)	---

Table 4. Accuracy of some currently produced MEMS gyroscopes

Manufacturer or Institute	Product	Performance Characteristics		
		Bias Stability	SF Stability	Output noise
BEI Systron Donner	QRS11	7.2 deg/ hr	0.03%/°C	0.01 °/sec/√Hz
British Aerospace	FG314 Dry Tuned	5 deg/ hr	-410 ppm/°C	0.01 °/sec/√Hz
Inertial Science	RRS75	72 deg/ hr	50 ppm/°C	0.01 °/sec/√Hz

Allied Signal Inc.	2accelerometers on a single chip	10 deg/ hr	100 ppm (RMS)	---
Charles Stark Draper Lab.	Tuning Fork & (VWOG) gyro	100-300 deg/hr	500 ppm (RMS)	---

Gyros Measurement Errors

Gyroscope is an angular rate sensor providing either angular rate in case of rate sensing type or attitude in case of rate integrating type. The following model represents most of the errors contained in a single gyroscope measurement

$$I_{\omega} = \omega + b_{\omega} + S\omega + N\omega + \varepsilon(\omega) \qquad (1)$$

where I_{ω} is the vector of measurements in (deg/hr), ω is the vector of true angular velocities (deg/hr), b_{ω} is the vector of gyroscope biases (deg/hr), S is the gyro scale factor in matrix form, N is the matrix representing the non-orthogonality of the gyro triad and $\varepsilon(\omega)$ is the vector of the gyro sensor noise (deg/hr)

Accelerometer Measurement Errors

The accelerometer error sources are quite similar to those of gyroscopes except for the scale factor, which contains two components (linear and non-linear). The following specific force model represents the errors contained in a single accelerometer measurement [Schwarz and Wei, 2000]

$$I_f = f + b_f + S_1 f + S_2 f + Nf + \ g + \varepsilon(f) \qquad (2)$$

where I_f is the vector of measurements in (m/s^2), f is the vector of true specific forces (m/s^2), b_f is the vector of accelerometer biases (deg/hr), S_1 is the linear scale factor in matrix form, S_2 is the non-linear scale factor in matrix form, N is the matrix representing the non-orthogonality, δg is the deviation from the theoretical gravity value (m/s^2) and $\varepsilon(\omega)$ is the vector of the accelerometer noise (m/s^2)

3. STOCHASTIC MODELLING OF NOISE SEQUENCE

The accurate estimation of the vehicle position and attitude necessitates modelling the noise components associated with the output of MEMS-based inertial sensor using an appropriate stochastic model. Most of the present positioning techniques model this noise component as 1st order Gauss-Markov process [e.g. Ohlmeyer et al, 1998]. However, it is essential to study the stochastic properties of such noise component in order to determine its correlation over time. In general, stochastic analysis classifies any random process as either white (uncorrelated) or correlated (colour) noise component based on studying the corresponding autocorrelation sequence [Gelb 1974]. Gauss-Markov (of first or higher orders) processes are usually utilized to model correlated noise components.

White Noise

White Noise, expressed sometimes as zero order Gauss-Markov, has a particularly simple (constant) power spectral density which means that the power is uniformly distributed over all frequency components (ω) in the full infinite range [Brown, 1985]. The power spectral density and the autocorrelation function of white noise can be expressed as

$$\Phi_{xx}(\omega) = \Phi_o = const. \qquad \text{.........................(3)}$$

$$_{xx}(\ ) = \Phi_o\ (\ ) \qquad \text{.........................(4)}$$

Where is the time shift and ( ) is the Dirac function equals zero at any value of except at (τ=0).

Correlated Noise

The most common way of modelling the coloured noise, or can be expressed as correlated noise, is the Gauss-Markov Processes. There are an infinite number of random processes orders called the Gauss-Markov processes family. The power spectral density and the autocorrelation function of the first order Gauss-Markov has their own shape [Fig. 1] and can be expressed as

$$\Phi_{xx}(\omega) = \frac{2\ {}_1\ {}^2}{\omega^2 + {}^2_1} \qquad \text{.........................(5)}$$

$$_{xx}(\ ) = {}^2 e^{-\ _1|\ |} \qquad \text{.........................(6)}$$

Where β_1 is the inverse of the correlation time and σ^2 is the variance of the random process.

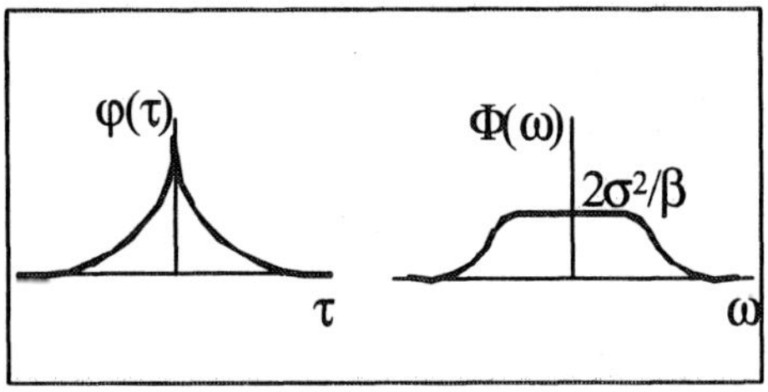

Figure 1. Power spectral density and autocorrelation function of first order Gauss Markov Random Process

The second order Gauss-Markov process has a different shape power spectral density and autocorrelation functions [Fig. 2] and they can be expressed as

$$\Phi_{xx}(\omega) = \frac{4 \; {}^{3}_{2} \; {}^{2}}{(\omega^{2} + \; {}^{2}_{2})^{2}} \qquad \ldots\ldots\ldots\ldots\ldots\ldots\ldots(7)$$

$$_{xx}(\;) = \; {}^{2}e^{- \; {}^{2}|\;|}(1 + \; {}_{2}|\;|) \qquad \ldots\ldots\ldots\ldots(8)$$

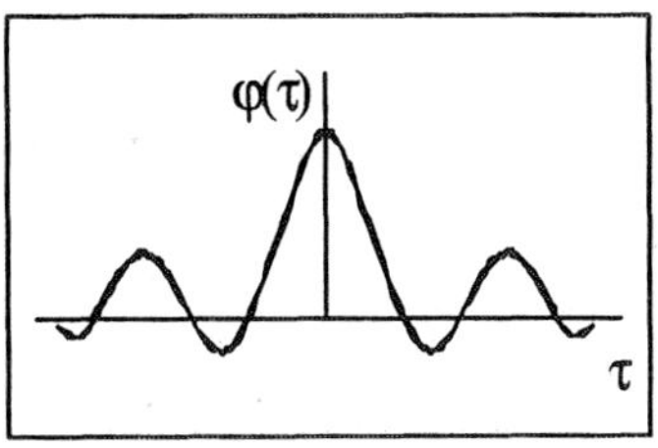

Figure 2. Autocorrelation function of second order Gauss-Markov Random Process

4. PERFORMANCE IMPROVEMENT UTILIZING WAVELET PRE-FILTERING METHOD

This section describes the use of wavelet de-noising methods for pre-filtering the MEMS-based inertial sensor measurements in order to reduce their output noise. Generally, the wavelet transformation of a time-domain signal is defined in terms of the projections of this signal onto a family of functions that are all normalized dilations and translations of a wavelet function. The wavelet transform is, therefore, defined as follows [Strang and Nguyen, 1997]:

$$X_V = \frac{1}{\sqrt{}} \int_{-\infty}^{\infty} x(t) \left(\frac{t-V}{} \right) \qquad \text{.................(9)}$$

Where, X_V is the wavelet transform of $x(t)$ and and V are the continuous dilation and translation parameters. Since the wavelet theory was developed for orthonormal wavelet bases, and V are given as 2^{-m} and $n2^{-m}$, respectively, where m and n are the dilation and translation indices. In discrete domain, the implementation of the wavelet transform is based on a bank of discrete time filters that have essentially half-band lowpass and highpass characteristics.

The above wavelet transformation is applied to the output of MEMS-based accelerometers and gyroscopes to remove the high frequency noise components at their outputs and to detect the overall trend of the signal. This trend represents the low frequency part of the signal. In terms of wavelet analysis, this corresponds to the greatest scale (dilation) indices. As the value of the dilation index increases, the resolution decreases, thus giving a better estimate of the unknown trend. In other terms, if we processed the inertial sensor signal with several level of decomposition utilizing these lowpass and highpass filters, the higher frequency components can be removed and the trend of the sensor signal remains.

5. RESULTS

In order to study the stochastic properties of MEMS-based inertial sensors, static experiments for about one hour long were performed. At each experiment, the measurements of the AHRS300CA IMU [Crossbow, San Jose, CA] were collected at different data rates (1, 10 and 20 Hz). This IMU contains low cost MEMS-based inertial sensors including three-axis accelerometers and three-axis gyroscopes mounted in three mutually orthogonal directions. These measurements have been processed to investigate the appropriate noise-modelling criterion for these sensors and to determine the effectiveness of the suggested wavelet de-nosing technique in limiting their output uncertainties. Prior to collecting the above measurements, the six sensors of the desired IMU were calibrated using 10 six-position static calibration tests [Salychev, 1998]. The mean value and repeatability of each accelerometer bias and scale factor as well as each gyro drift were computed. Table (5) presents such biases and repeatability values for Z-accelerometer and Z-gyro only as they will be used in the following sub-sections.

Table 5. Bias and scale factor errors and their repeatability values for Z-axis sensors

Sensor	Z-Accelerometer		Z-Gyroscope
Error	Bias (mg)	SF (ppm)	Drift (°/sec)
Mean Value	5.4567	1136.20	0.1877
Repeatability	0.2313	132.257	0.0455

Noise Modelling

In the current study, the noise characteristics of the AHRS300CA IMU have been studied to evaluate the Markov random process order of each gyro and accelerometer noise component. The autocorrelation sequence of each noise component corresponding to each used data rate was computed and plotted. In addition, the obtained autocorrelation results were averaged using intervals of ten seconds and plots were regenerated for the seek of easier interpretation. Obtained autocorrelation sequences, after averaging, of the Z-axis accelerometer and gyroscope outputs, are shown in figures (3) and (4).

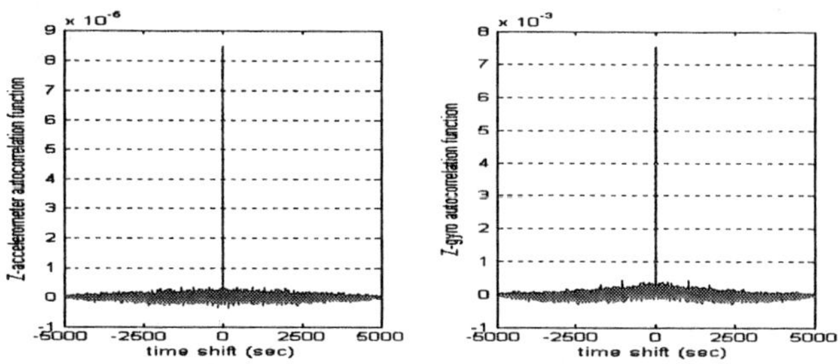

Figure 3. Autocorrelation sequence of Z-axis accelerometer and Gyro at 1 Hz data rate

Studying autocorrelation sequences in the above two figures and based on the formerly given brief theoretical background we can have some remarks. The observed noise behaviour may depend on the output data rate of MEMS inertial sensor. The second order behaviour of both Z-gyroscope and Z-accelerometer appears clearly when looking at the 10-Hz data rate output rather than for the 1-Hz output. The reason for that is the short-term rapid changes happening in between observation intervals that may not be observable when using low output data rates. This may consequently lead to a false indication of the Gauss-Markov order of such noise components when being modelled in any filtering algorithm.

Pre-Filtering Results

The raw measurements of the three gyroscopes and the three accelerometer of the AHRS300CA were averaged out at 1 second-interval after removing deterministic parts of bias and scale factor errors. Figure 5.a shows a sample of such compensated output data for the Z-accelerometer while the other sample [Figure 6.a] is related to Z gyro measurements. The raw measurements were then processed by up to 20 levels of decomposition utilizing the "*db8*" wavelet function.

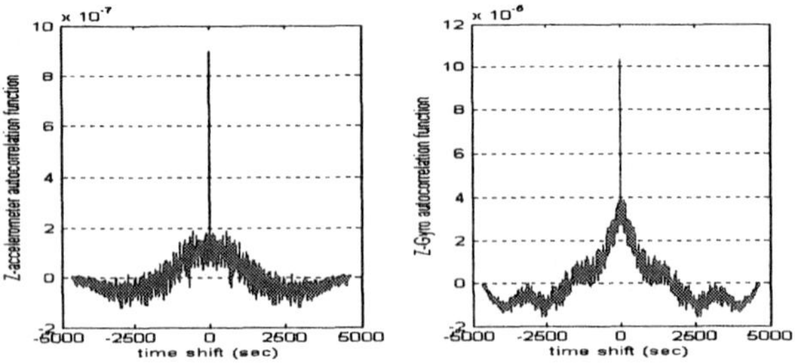

Figure 4. Autocorrelation sequence of Z-axis accelerometer and Gyro at 10 Hz data rate

The Z-accelerometer measurement were analysed by this wavelet function and the results are shown on Figures 5.b to 5.d after the 4th, 8th and 13th levels of decomposition. Obviously, increasing the level of decomposition significantly reduced the noise level. However, it can be depicted that the signal reconstructed after 13 levels of decomposition became no longer capable of representing the original signal due to the excessive decompositions, which filtered out part of the useful signal in addition to the noise.

In fact, the effectiveness of the wavelet de-noising criterion is more evident when applied to the gyroscopes because of their relatively low signal to noise ratio (SNR). Figures 6.b to 6.d shows the gyroscope measurements (wz) after the 7th, 12th and 18th levels of decomposition. Apparently, the 12th level of decomposition has achieved significant noise reduction with detection of the long-term evolution of the gyroscope measurement. In addition, it can be determined that the 18th level of decomposition destroyed the gyroscope original signal due to excessive low-pass filtering.

6. CONCLUSIONS

This article presented a stochastic quantitative analysis of MEMS-based inertial sensors. The analysis of the autocorrelation sequences of the noise components associated with the output of MEMS-based inertial sensors indicated that the conventional way of modelling these components as first order Gauss-Markov processes may lead to an appropriate representation of their behavior. The results have shown that they could better be modeled as second order Gauss-Markov processes. Although the suggested second-order modeling scheme may increase the modeling complexity and delay the convergence of the estimation algorithm, significant improvement in the estimation accuracy can be achieved. In addition, this article has suggested the utilization of wavelet de noising techniques to reduce the output uncertainty of the MEMS-based inertial sensors. This digital signal processing method can be integrated with the MEMS-based inertial sensors and the GPS chip to provide efficient real-time pre-filtering criterion

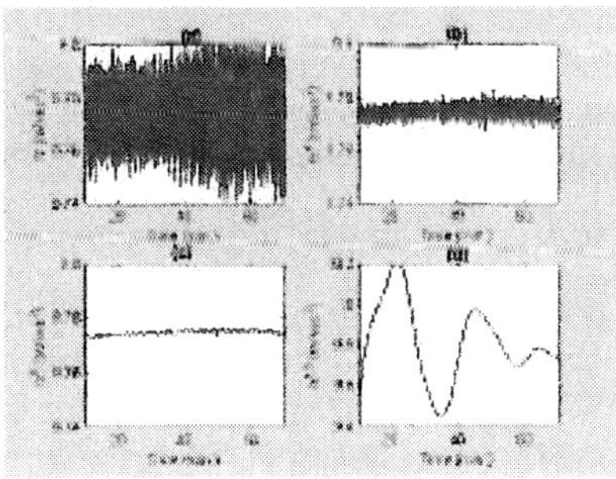

Figure 5. The Accelerometer *fz* raw measurement and after de-noising at different levels of decomposition.

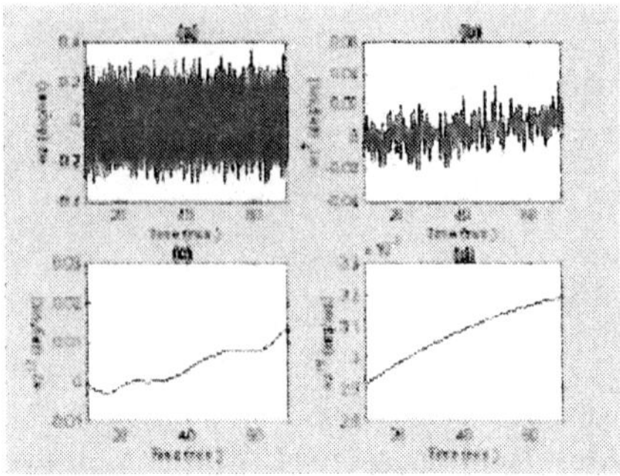

Figure 6. The gyroscope ω_z raw measurement and after de-noising at different levels of decomposition.

REFERENCES

[1]Brown, R. G., and P. C. Hwang, **"Introduction to random signals and applied Kalman filtering"**, Second edition, John Wiley & Sons Inc., 1985.

[2]Gelb, A., **"Applied optimal estimation"**, Text Book, The Analytic Science Corporation, 1974.

Hulsing, R., and D. MacGugan, **"Miniature IMU based on micro-machined Coriolis sensor"**, Proceedings of ION NTM, pp (353-360), 1993.

[3]Ohlmeyer, E. J., T. R. Pepitone, and B. L. Miller, **"Assessment of integrated GPS/INS for the EX-171 extended range guided munitions"**, a paper declared as a work for the US Government, 1998

[4]Salychev, O., "Inertial systems in navigation and geophysics", 1998.

[5]Schwarz, K. P., and M. Wei, **"INS/GPS integration for geodetic applications"**, Lecture Notes ENGO 623, Dept of Geomatics Engineering, U of C, 2000.

[6]Schwarz, K. P., and N. El-Sheimy, **"Future Positioning and Navigation Technologies"**, A study performed under the scientific service agreement, Dept of Geomatics Engineering, U of C, March 1999.

[7]Smith, J. G., and A. Balleto, **"Dynamic Admittance Matrix of piezoelectric cantilever bimorphs"**, Journal of Micro-electro-mechanical Systems, Vol. 3, No. 3, Sep. 1994

[8]Strang G. and Nguyen T., **"Wavelets and Filter Banks"**, Wellesley-Cambridge Press, 1997.

[9]Varadan, V. K., V. V. Varadan, H. Subramanian, **"Fabrication, characterization and testing of wireless MEMS-IDT based micro-accelerometer"**, Sensors and Actuators, A: physical, Vol. 90, pp (7-19), May 2001

[10]Zhou, Y., **"Layout Synthesis of Accelerometers"**, M.Sc. Project Report, Department of Electrical and Computer Engineering, Carnegie Mellon University, August 1998.

Index

A

adder 112, 324, 355, 365
address 120, 147, 166, 271-278, 282, 289-290, 296, 395, 397, 399, 427
algorithm 12, 75, 78, 79, 85, 90-92, 101, 117, 137, 148, 150-151, 162, 186, 188, 209, 210, 212, 216, 239, 243, 250, 251, 289, 299, 302, 303, 313, 322, 330, 331, 345, 374, 375, 377, 381, 386, 387, 388, 389, 390, 391, 392, 415, 419, 421, 423, 427, 454, 507, 509
ALU 365
AMS 407
analysis 14, 27, 32, 35, 37, 38, 46, 47, 51, 62, 63, 64, 69, 86, 88, 89, 90, 91, 92, 96, 117, 120, 136, 137, 154, 155, 161, 183, 294, 311, 314, 317, 324, 355, 356, 360, 364, 365, 410, 412, 414, 422, 427, 430, 454-455, 458-459, 464, 469, 470, 474, 475, 477, 480, 481, 483, 485, 498, 502, 505, 508
analyzing 105, 114
application-specific36, 72, 208, 224, 232, 233, 235, 236, 242, 388
architecture 7, 15, 17, 25, 52, 93, 101, 113, 119, 146, 156, 162, 171, 205, 224, 255-256, 270-271, 281-282, 284-285, 287, 295-299, 309, 322-323, 325, 335, 339, 368, 394-396, 401, 406, 489
as soon as possible 313
ASAP 256, 310-311, 313, 316
ASIC. 1, 3, 6, 15, 25-26, 39, 48, 195, 428
associativity 88, 93
ASSP 334
asynchronous circuit 11
atomic 265

B

basic ..31, 40, 42, 45, 46, 52, 54, 55, 87, 173, 187, 235, 245, 274, 324, 456, 459, 464, 475
behavioral 49, 53, 56-58, 107, 111-113, 125, 127, 161, 172, 424
BIST 16, 19, 147, 148, 149, 150, 155, 156, 369, 375, 379, 381
bit-level 12
bit-serial 327
block 18, 108, 340
block diagram .. 108, 234, 235, 271, 285, 287, 289, 293, 303-307, 324, 327, 338, 340, 353
block size......88, 311, 314, 315, 320, 321
blocks....................................... 203, 400
boolean function 123, 128
bus .. 15, 19, 178, 282, 297, 393-396, 426

C

capacitances ... 300, 379, 415, 417, 424, 492
changing 6, 8, 50, 54, 410, 425, 461, 478
channel................................ 27, 28, 427
chip. 2-3, 9, 11-19, 30, 31, 35, 39, 75, 85, 86, 93, 118, 145-148, 150, 155, 158-159, 193, 210, 221, 259, 263, 266, 267, 270, 273, 296, 298, 300, 324, 335-336, 340, 342, 355, 375, 383, 385, 403-407, 410, 422, 426, 428, 430, 444, 457, 463, 477, 478, 486, 501, 509
chrominance 279, 280
circular... 470
CLB 185, 199, 202, 203
clock distribution 431, 432
clock period 168
clock skew 169, 431
clocked..................... 395, 397, 398, 422
clustered.. 234
CMOS ... 5, 10, 12, 17-18, 105, 110-112, 255-256, 259-268, 298-300, 307, 309, 323-324, 332-336, 340-342, 355-357, 362-366, 369, 402, 430, 433, 436, 442-443, 454-456, 466, 467, 475
coarse-grain 198, 199
combining..................... 24, 57, 260, 461
communications................ 18, 27, 29, 82
component 13, 77, 145, 272, 278, 289, 410
computation .. 10, 12, 25, 31-36, 49, 120, 127, 199, 207, 276, 285-287, 296, 299, 326, 332, 355, 390, 395, 413, 445, 454, 491
correlation................................ 502, 503
co-verification.................... 172-179, 183
CPL.. 356
CPU 3, 15, 78, 86, 178-179, 294
creating 32, 229, 261-263, 313
critical path 11, 47, 136, 238, 362, 416
current 1, 3, 10-12, 21-22, 42, 45, 67, 75, 85, 101-105, 119, 128, 150-151, 155-157, 172, 200, 210, 232, 247, 252, 262, 265-266, 273, 275, 289, 300, 323, 367, 399, 403, 407-412, 421, 444, 448, 449, 451, 456, 459-463, 480, 481-482, 506

D

data 52, 114-115, 144-147, 156, 162, 164, 178-181, 216, 228, 265, 269, 271-278, 296-297, 311-312, 327, 354, 395, 397
DCT.................. 239-240, 285, 295, 326
debugging... 16, 131, 134, 168, 183, 185
decoder 25, 48, 313, 397
delay 10-14, 31, 51, 86, 88, 96, 151, 195, 201, 300, 314-315, 324, 355-356, 360, 362-364, 373, 407-416, 421- 425, 431, 444, 446, 455, 509
delay-insensitive............................... 425
delivering................................... 15, 156
design challenges...................... 6, 10, 14
design flow .. 92, 172, 196, 209-215, 235, 237, 403-405, 421, 425-426
design methodologies 4, 6
design process 6, 9, 45-46, 101, 116, 131, 157, 181, 430, 457, 464, 477, 478, 485
design rules.................................. 40, 45
design styles.............................. 133, 245
DFT ... 100, 145, 146, 148, 150, 151, 155, 326, 369, 370
digital filter 31, 402
discrete cosine transform .. 232, 239, 243, 285, 326
discrete fourier transform.................. 326
distributed arithmetic .. 323-326, 333, 334
DSP synthesis 9
dual-port 76
dynamic power dissipation 369
dynamic programming...................... 415

E

embedded software 6, 157, 172, 221, 405

embedding 153, 260
equivalent 21, 24, 113, 295, 327, 333, 347, 350, 435, 437, 454, 463, 472
euclidean distance 89
executable specifications 7
exhaustive 87, 99, 104

F
fabrication process 24, 32, 499
FFT 12, 29-32
FIFOs 51
filter 31, 40, 315-321, 390, 395, 508
filter bank 390
finite impulse response 326
FIR filter 18, 40, 368, 394-397, 400, 402
flexibility 156, 254
FPGA 24-25, 36, 48, 72-82, 158, 184-186, 190-198, 200- 206, 244, 406
full-search 11
functional level 209, 212, 217

G
gate ... 36, 53-54, 108, 195-196, 278, 356, 465
generating ... 229, 262, 323-325, 384, 427

H
hardware ... 7, 15-19, 36, 48, 52, 54, 60, 61, 70, 76, 78, 82, 96, 100, 113-115, 132, 137, 165, 170-171, 182-183, 194, 219, 221, 254, 324, 345, 401, 427
HDTV 22, 48
hierarchical 32, 57-58, 120, 147, 158, 184, 185, 188, 190-199, 200-203, 274

I
IIR 29
integer linear programming 423
interconnect 165, 367, 385-388, 392, 403-419, 421-441, 444-467
Interface 19, 96, 395, 397, 456, 466
Inverter 109, 111, 439, 442
IP 1, 4-8, 13-14, 19, 23, 26-28, 32, 35, 38, 39, 47, 85-86, 97, 100, 145, 159, 286, 323, 395-401, 406-407, 467

J
junction 265, 333

K
kernel. 6-7, 16, 31, 51, 157, 172-179, 214

L
latency .. 6, 9, 10-14, 118, 201-203, 255, 270, 274, 277, 281, 286, 300, 368, 383
layout ... 198, 323-326, 330-332, 410-419, 421-423, 478
leakage 6, 10, 300
libraries ... 7, 28, 36, 49-58, 99, 108, 113, 185, 222, 256, 309
linear 17, 99, 106, 116, 137, 148, 286, 319, 335-339, 345, 375, 387, 412, 423, 427, 455, 480, 482, 502
list scheduling 385, 386-389
liveness 104, 121
LMS 18
logic. 1, 3-4, 13-17, 30-31, 52, 61, 66-68, 70, 73, 82, 99, 101-106, 110, 117, 121, 132, 136, 148, 151, 154, 184-186, 198, 200-208, 240, 244-250, 259, 262, 267, 278, 287, 300, 323-331, 356, 358, 365, 367, 369-380, 401, 406-407, 410, 419, 422, 439
loop 136, 261-263, 410, 412, 485, 487, 489, 493, 494
low-power10-13, 266, 365-367, 384, 401, 419, 425-427
luminance 279

M
mapping. 11, 23, 137, 292, 302, 305, 379, 384-385, 388, 392, 426
master-slave 148

memory.4-6, 9, 13, 16, 23, 30-31, 66, 74-78, 80-81, 86, 88, 93-94, 105, 118, 120, 126-127, 136-137, 142, 147-148, 158-159, 165-84, 190-193, 200, 204, 214-215, 224, 226, 233, 255-256, 259-260, 261-265, 267-268, 270-281, 284-300, 304-308, 310-321, 355, 396, 399, 405, 407, 424, 426, 462, 478
memory architecture.. 256, 270-271, 281, 284-286, 298-299, 305, 308
MEMS..... 405-406, 456-469, 475-478, 481, 484-486, 496-502, 505, 507-508, 510
methodology ...5, 7, 8, 35-36, 38, 46-47, 49, 53, 58, 61-66, 70, 102, 128, 155, 157, 161, 172, 178, 181, 209-211, 214-215, 244, 401, 403, 405, 422
minimization.............................. 12, 254
mixed-signal...................................... 406
model 5-7, 15, 29, 35-38, 44, 49, 51, 55, 72-79, 93-99, 106, 116-120, 123-133, 137, 139, 142, 147, 157-162, 172-174, 178-187, 189-190, 195, 201, 203, 207, 210, 214, 221-229, 231, 234, 256, 274, 310-314, 321, 345-351, 375, 385, 412- 413, 415, 422-423, 431-435, 444, 453-457, 474, 476, 477, 480-482, 485-489, 492-496, 501-502
model parameters.............................. 492
motion estimation........ 11, 285, 295, 305
MPEG 3, 25-28, 48, 256, 298-300, 305, 308-310, 316-322
multiplexer-based............................. 146
multiplication..... 31, 290, 299, 303, 327, 402
multiplier... 31, 105, 198, 200, 203, 240, 326, 365, 395, 402

N

NAND gate.............................. 371, 372
Networks 19, 28, 156, 194, 254, 283, 442
Noise............409-411, 428, 502-503, 506
NOR gate.. 372
normalized........................ 239, 447, 505

O

OBDD..330-332
operation 12, 14, 31, 77-79, 81, 134, 138, 175, 233-234, 236, 260, 262, 264, 271-277, 280, 287, 289, 304, 323-326, 329, 332-333, 339, 355, 378, 396-399, 408, 417, 479, 493-494, 498
optimizing.......................... 162, 222, 426
orthogonality.......................... 77, 80, 502
oxide.................................... 5, 6, 14, 265

P

parallel processing...... 81, 161, 168, 208, 245, 252
partitioning 11, 36, 40, 45, 60-65, 70, 74, 125-127, 147, 207, 209-214, 219, 248-250
performance.... 1, 2, 6, 8, 9-11, 13, 15-16, 20-28, 37, 78, 85-86, 96, 100, 102, 127, 130-133, 136-143, 151, 154, 155, 157, 179, 181, 197, 200, 205, 214, 219, 233, 235, 239, 256, 260, 263, 270, 272, 279, 280-286, 299-300, 308, 313, 317, 324, 349, 355-356, 360-368, 383, 389, 394-397, 400, 403-408, 415, 419, 421-422, 425, 428, 457-458, 460, 464, 477, 482- 483, 486, 498, 500
performance characteristics 141, 263, 500
performance requirements 1, 285
performance simulation 130, 138
peripherals.................... 14, 178-179, 180
pipelining............. 11, 270, 276-277, 282
platform-based design................. 32, 406
polysilicon 469, 483, 500
power analysis.................................. 399
power consumption. 2, 10, 12, 21, 28, 30, 94, 200-203, 235, 260, 298-300, 308, 324, 335, 342, 355-356, 360-370, 379- 380, 383, 385, 392, 396, 400, 424
power dissipation... 10-11, 14, 21, 23, 25, 299, 300, 340, 367-368, 370, 404, 430-435, 438-441, 462
power reduction........... 30, 300, 404, 441

power requirements 13
programmable DSP.............. 1, 20-22, 27
propagation delay 414, 453

R
RAM .. 9, 73, 76, 118, 123-124, 127, 158, 184-185, 189-193, 198, 202-203, 259, 268-269, 282
real-time processing........... 157-159, 168
recursive 404, 444-445
reducing... 12, 14, 36, 60, 126, 153, 160, 234, 299-300, 404, 408, 414, 426, 431, 441
refresh 273
register allocation 224
regular................................ 23, 52, 55, 239
reliability 425, 456-459, 463, 477
representations 117
resistance 476
resource allocation 9
retiming 278
Reuse 5, 35-36, 38-39, 41, 43, 48-50, 52, 53, 60, 156, 219
RISC 20-22, 24, 26, 29, 229, 242
RLS.. 18
routing 31, 73, 147, 154, 198-201, 368, 383, 387, 389, 405, 413, 415-417, 421-424, 428
RTL ... 7, 15, 36, 49-50, 53, 58, 100, 128, 136, 159, 161, 172-173, 279, 281, 398

S
safety 104, 106, 121, 473
sampling rate 353
scaling..... 5, 32, 266, 324, 329, 346-347, 349, 353, 367, 383, 408-409
scattered 223, 228
schedule... 311, 313-316, 320-321, 368, 384, 387-388, 390, 392
scheduling..9, 12, 16, 137, 151, 154, 155, 209, 242-243, 368, 384-392
selecting 40, 378
short channel.. 5
short-circuit... 10, 404, 430-434, 438, 441
short-circuit power dissipation... 432, 438
SIMD 296
Simulations 95, 400
SNR 508
SOC design.... 36, 99, 100, 159-160, 167
spatial correlation 373
SPICE 332, 436, 438-439, 445, 486
static4, 8-10, 54, 165, 222, 259, 465, 481, 505
switching activity. 93, 367, 395, 397, 421
synthesis 9, 12-13, 40, 45, 50, 54, 82, 122, 129, 136-137, 158, 170, 184-185, 190, 193, 197, 208, 214, 244-245, 250-253, 278, 287, 294-295, 331, 368, 383-384, 388-391, 400, 408, 424, 428, 430
systolic array 106

T
technology 3-5, 10-13, 20, 21, 23, 25, 40, 73, 85, 96, 196, 215, 221, 255-256, 259- 265, 267, 270-273, 286, 298, 307, 309, 321, 324, 325, 335-336, 340, 342, 356, 361-362, 364, 367, 379, 385, 392, 406-408, 420, 424, 428, 456-457, 462-463, 466-478, 499
testbenches 48, 70
testing ... 16, 99, 103, 145-149, 154, 185, 216, 247, 370, 378, 406, 456, 466, 510
thickness .. 265, 434, 457, 471, 474, 477, 479-480, 482-483
throughput...... 2, 6, 10, 30, 32, 130, 277, 281, 300, 355, 385, 461
top-down...................... 14, 407, 457, 496
trace 168, 243, 469
transaction 114, 128, 194, 253, 321-322, 334, 365, 402, 427, 442, 454-455, 465
transformation..... 31, 101, 299, 301-304, 367, 369, 371, 374, 377, 380, 423, 424
two-dimensional 239, 243, 319
two-phase.. 388

V

VC 5
Verification 15, 19, 36, 40, 48, 60-63, 66, 70, 99, 100, 103, 113-116, 128, 130, 144, 247
verifying 36, 99, 102, 117, 124, 126
Virtual Socket Interface Alliance...........5
Viterbi algorithm 239, 240
Viterbi decoder 27
VLIW...................223, 233-234, 242-243
VLSI . 10-11, 17-19, 33, 82, 87, 103, 113-115, 156, 194, 242, 266, 269, 298, 309, 334-335, 367, 381, 383, 385, 392-393, 401, 425-428, 442-444, 454
voltage ... 5-6, 10-12, 255, 259, 261-262, 266, 299, 300, 324, 329, 332, 335, 337- 338, 340-342, 361, 366, 408, 410, 433, 436, 437, 446-453, 457, 462, 477, 478, 479, 480, 481, 482, 483, 487, 488

W

wavelet transform 310, 505